Optimal Design of
Queueing
Systems

Shaler Stidham, Jr.

University of North Carolina
Chapel Hill, North Carolina, U. S. A.

CRC Press
Taylor & Francis Group
Boca Raton   London   New York

CRC Press is an imprint of the
Taylor & Francis Group, an **informa** business
A CHAPMAN & HALL BOOK

Chapman & Hall/CRC
Taylor & Francis Group
6000 Broken Sound Parkway NW, Suite 300
Boca Raton, FL 33487-2742

Library of Congress Cataloging-in-Publication Data

Stidham, Shaler.
    Optimal design for queueing systems / Shaler Stidham Jr.
       p. cm.
    "A CRC title."
    Includes bibliographical references and index.
    ISBN 978-1-58488-076-9 (alk. paper)
       1. Queueing theory. 2. Combinatorial optimization. I. Title.

T57.9.S75 2009
519.8'2--dc22
                                                       2009003648

Visit the Taylor & Francis Web site at
http://www.taylorandfrancis.com

and the CRC Press Web site at
http://www.crcpress.com

Contents

List of Figures

Preface

What began a long time ago as a comprehensive book on optimization of queueing systems has evolved into two books: this one on optimal design and a subsequent book (still in the works) on optimal control of queueing systems.

In this setting, "design" refers to setting the parameters of a queueing system (such as arrival rates and service rates) before putting it into operation. By contrast, in "control" problems the parameters are control variables in the sense that they can be varied dynamically in response to changes in the state of the system.

The distinction between design and control, admittedly, can be somewhat artificial. But the available material had outgrown the confines of a single book and I decided that this was as good a way as any of making a division.

Why look at design models? In principle, of course, one can always do better by allowing the values of the decision variables to depend on the state of the system, but in practice this is frequently an unattainable goal. For example, in modern communication networks, real-time information about the buffer contents at the various nodes (routers/switches) of the network would, in principle, help us to make good real-time decisions about the routing of messages or packets. But such information is rarely available to a centralized controller in time to make decisions that are useful for the network as a whole. Even if it were available, the combinatorial complexity of the decision problem makes it impossible to solve even approximately in the time available. (The essential difficulty with such systems is that the time scale on which the system state is evolving is comparable to, or shorter than, the time scale on which information can be obtained and calculations of optimal policies can be made.) For these and other reasons, those in the business of analyzing, designing, and operating communication networks have turned their attention more and more to *flow control*, in which quantities such as arrival (e.g., packet-generation) rates and service (e.g., transmission) rates are computed as time averages over periods during which they may be reasonably expected to be constant (e.g., peak and off-peak hours) and models are used to suggest how these rates can be controlled to achieve certain objectives. Since this sort of decision process involves making decisions about rates (time averages) and not the behavior of individual messages/packets, it falls under the category of what I call a design problem. Indeed, many of the models, techniques, and results discussed in this book were inspired by research on flow and routing control that has been reported in the literature on communication networks.

Of course, flow control is still control in the sense that decision variables can change their values in response to changes in the state of the system, but the states in question are typically at a higher level, involving congestion averages taken over time scales that are much longer than the time scale on which such congestion measures as queue lengths and waiting times are evolving at individual service facilities. For this reason, I believe that flow control belongs under the broad heading of design of queueing systems.

I have chosen to frame the issues in the general setting of a queueing system, rather than specific applications such as communication networks, vehicular traffic flow, supply chains, etc. I believe strongly that this is the most appropriate and effective way to produce applicable research. It is a belief that is consistent with the philosophy of the founders of operations research, who had the foresight to see that it is the underlying structure of a system, not the physical manifestation of that structure, that is important when it comes to building and applying mathematical models.

Unfortunately, recent trends have run counter to this philosophy, as more and more research is done within a particular application discipline and is published in the journals of that discipline, using the jargon of that discipline. The result has been compartmentalization of useful research. Important results are sometimes rediscovered in, say, the communication and computer science communities, which have been well known for decades in, say, the traffic-flow community.

I blame the research funding agencies, in part, for this trend. With all the best intentions of directing funding toward "applications" rather than "theory," they have conditioned researchers to write grant proposals and papers which purport to deal with specific applications. These proposals and papers may begin with a detailed description of a particular application in which congestion occurs, in order to establish the credibility of the authors within the appropriate research community. When the mathematical model is introduced, however, it often turns out to be the $M/M/1$ queue or some other old, familiar queueing model, disguised by the use of a notation and terminology specific to the discipline in which the application occurs.

Another of my basic philosophies has been to present the various models in a unified notation and terminology and, as much as possible, in a unified analytical framework. In keeping with my belief (expressed above) that queueing theory, rather than any one or several of its applications, provides the appropriate modeling basis for this field, it is natural that I should have adopted the notation and terminology of queueing theory. Providing a unified analytical framework was a more difficult task. In the literature optimal design problems for queueing systems have been solved by a wide variety of analytical techniques, including classical calculus, nonlinear programming, discrete optimization, and sample-path analysis. My desire for unity, together with space constraints, led me to restrict my attention to problems that can be solved for the most part by classical calculus, with some ventures into elementary nonlinear programming to deal with constraints on the design variables. A side benefit of this self-imposed limitation has been that, although the book is mathematically rigorous (I have not shied away from stating results as theorems and giving complete proofs), it should be accessible to anyone with a good undergraduate education in mathematics who is also familiar with elementary queueing theory. The downside is that I have had to omit several interesting areas of queueing design, such as those involving discrete decision variables (e.g., the number of servers) and several interesting and powerful analytical techniques, such as sample-path analysis. (I plan to include many of these topics in my queueing control book, however, since they are relevant also in that context.)

The emphasis in the book is primarily on qualitative rather than quantitative insights. A recurring theme is the comparison between optimal designs resulting from different objectives. An example is the (by-now-classical) result that the individually optimal arrival rate is typically larger than the socially optimal arrival rate.* This is a result of the fact that individual customers, acting in self-interest, neglect to consider the *external effect* of their decision to enter a service facility: the cost of increased congestion which their decision imposes on other users (see, e.g., Section 1.2.4 of Chapter 1). As a general principle, this concept is well known in welfare economics. Indeed, a major theme of the research on queueing design has been to bring into the language of queueing theory some of the important issues and qualitative results from economics and game theory (the Nash equilibrium being another example). As a consequence this book may seem to many readers more like an economics treatise than an operations research text. This is intentional. I have always felt that students and practitioners would benefit from an infusion of basic economic theory in their education in operations research, especially in queueing theory.

Much of the research reported in this book originated in vehicular traffic-flow theory and some of it pre-dates the introduction of optimization into queueing theory in the 1960s. Modeling of traffic flow in road networks has been done mainly in the context of what someone in operations research might call a "minimum-cost multi-commodity flow problem on a network with non-linear costs". As such, it may be construed as a subtopic in nonlinear programming. An emphasis in this branch of traffic-flow theory has been on computational techniques and results. Chapters 7 and 8 of this book, which deal with networks of queues, draw heavily on the research on traffic-flow networks (using the language and specific models from queueing theory for the behavior of individual links/facilities) but with an emphasis on qualitative properties of optimal solutions, rather than quantitative computational methods.

Although models for optimal design of queueing systems (using my broad definition) have proliferated in the four decades since the field began, I was surprised at how often I found myself developing new results because I could not find what I wanted in the literature. Perhaps I did not look hard enough. If I missed and/or unintentionally duplicated any relevant research, I ask for-

---

* But see Section 7.4.4 of Chapter 7 for a counterexample.

bearance on the part of those who created it. The proliferation of research on queueing design, together with the explosion of different application areas each with its own research community, professional societies, meetings, and journals, have made it very difficult to keep abreast of all the important research. I have tried but I may not have completely succeeded.

A word about the organization of the book: I have tried to minimize the use of references in the text, with the exception of references for "classical" results in queueing theory and optimization. References for the models and results on optimal design of queues are usually given in an endnote (the final section of the chapter), along with pointers to material not covered in the book.

Acknowledgements

I would like to thank my editors at Chapman Hall and CRC Press in London for their support and patience over the years that it took me to write this book. I particularly want to thank Fred Hillier for introducing me to the field of optimization of queueing systems a little over forty years ago. I am grateful to my colleagues at the following institutions where I taught courses or gave seminars covering the material in this book: Cornell University (especially Uma Prabhu), Aarhus University (especially Niels Knudsen and Søren Glud Johansen), N.C. State University (especially Salah Elmaghraby), Technical University of Denmark, University of Cambridge (especially Peter Whittle, Frank Kelly, and Richard Weber), and INRIA Sophia Antipolis (especially François Baccelli and Eitan Altman). My colleagues in the Department of Statistics and Operations Research at UNC-CH (especially Vidyadhar Kulkarni and George Fishman) have provided helpful input, for which I am grateful. I owe a particular debt of gratitude to the graduate students with whom I have collaborated on optimal design of queueing systems (especially Tuell Green and Christopher Rump) and to Yoram Gilboa, who helped teach me how to use MATLAB® to create the figures in the book. Finally, my wife Carolyn deserves special thanks for finding just the right combination of encouragement, patience, and (at appropriate moments) prodding to help me bring this project to a conclusion.

CHAPTER 1

Introduction to Design Models

Like the descriptive models in "classical" queueing theory, optimal design models may be classified according to such parameters as the arrival rate(s), the service rate(s), the interarrival-time and service-time distributions, and the queue discipline(s). In addition, the queueing system under study may be a network with several facilities and/or classes of customers, in which case the nature of the flows of the classes among the various facilities must also be specified.

What distinguishes an optimal design model from a traditional descriptive model is the fact that some of the parameters are subject to decision and that this decision is made with explicit attention to economic considerations, with the preferences of the decision maker(s) as a guiding principle. The basic distinctive components of a design model are thus:

1. the decision variables,

2. benefits and costs, and

3. the objective.

Decision variables may include, for example, the arrival rates, the service rates, and the queue disciplines at the various service facilities. Typical benefits and costs include rewards to the customers from being served, waiting costs incurred by the customers while waiting for service, and costs to the facilities for providing the service. These benefits and costs may be brought together in an objective function, which quantifies the implicit trade-offs. For example, increasing the service rate will result in less time spent by the customers waiting (and thus a lower waiting cost), but a higher service cost. The nature of the objective function also depends on the horizon (finite or infinite), the presence or absence of discounting, and the identity of the decision maker (e.g., the facility operator, the individual customer, or the collective of all customers).

Our goal in this chapter is to provide a quick introduction to these basic components of a design model. We shall illustrate the effects of different reward and cost structures, the trade-offs captured by different objective functions, and the effects of combining different decision variables in one model. To keep the focus squarely on these issues, we use only the simplest of descriptive queueing models – primarily the classical $M/M/1$ model. By further restricting attention to infinite-horizon problems with no discounting, we shall be able to use the well-known steady-state results for these models to derive closed-form expressions (in most cases) for the objective function in terms of the decision variables. This will allow us to do the optimization with the simple and familiar tools of differential calculus. Later chapters will elaborate on each of the models introduced in this chapter, relaxing distributional assumptions and considering more general cost and reward structures and objective functions. These more general models will require more sophisticated analytical tools, including linear and nonlinear programming and game theory.

We begin this chapter (Sections 1.1 and 1.2) with two simple examples of optimal design of queueing systems. Both examples are in the context of an isolated $M/M/1$ queue with a linear cost/reward structure, in which the objective is to minimize the expected total cost or maximize the expected net benefit per unit time in steady state. In the first example the decision variable is the service rate and in the second, the arrival rate. The simple probabilistic and cost structure makes it possible to use classical calculus to derive analytical expressions for the optimal values of the design variables.

The next three sections consider problems in which more than one design parameter is a decision variable. In Section 1.3, we consider the case where both the arrival rate and service rate are decision variables. Here a simple analysis based on calculus breaks down, since the objective function is not jointly concave and therefore the first-order optimality conditions do not identify the optimal solution. (This will be a recurring theme in our study of optimal design models, and we shall explore it at length in later chapters.) Section 1.4 revisits the problem of Section 1.2 – finding optimal arrival rates – but now in the context of a system with two classes of customers, each with its own reward and waiting cost and arrival rate (decision variable). Again the objective function is not jointly concave and the first-order optimality conditions do not identify the optimal arrival rates. Indeed, the only interior solution to the first-order conditions is a saddle-point of the objective function and is strictly dominated by *both* boundary solutions, in which only one class has a positive arrival rate. Finally, in Section 1.5, we consider the simplest of networks – a system of parallel queues in which each arriving customer must be routed to one of several independent facilities, each with its own queue.

A final word before we start. In a design problem, the values of the decision variables, once chosen, cannot vary with time nor in response to changes in the *state* of the system (e.g., the number of customers present). Design problems have also been called *static control* problems, in contrast to *dynamic control* problems in which the decision variables can assume different values at different times, depending on the observed state of the system. In the literature a static control problem is sometimes called an *open-loop control* problem, whereas a dynamic control problem is called a *closed-loop control* problem. We shall simply use the term *design* for the former and *control* for the latter type of problem.

1.1 Optimal Service Rate

Consider an $M/M/1$ queue with arrival rate λ and service rate μ. That is, customers arrive according to a Poisson process with parameter λ. There is a single server, who serves customers one at a time according to a FIFO (First-In-First-Out) queue discipline. Service times are independent of the arrival process and i.i.d. with an exponential distribution with mean μ^{-1}. Suppose that λ is fixed, but μ is a decision variable.

Examples

1. A machine center in a factory: how fast a machine should we install?

2. A communication system: what should the transmission rate in a communication channel be (e.g., in bits/sec.)?

Performance Measures and Trade-offs.

Typical performance measures are the number of customers in the system (or in the queue) and the waiting time of a customer in the system (or in the queue). If the system operates for a long time, then we might be interested in the long-run average or the expected steady-state number in the system, waiting time, and so forth. All these are measures of the level of *congestion*. As μ increases, the congestion (as measured by any of these quantities) decreases. (Of course this property is not unique to $M/M/1$ systems.) Therefore, to minimize congestion, we should choose as large a value of μ as possible (e.g., $\mu = \infty$, if there is no finite upper bound on μ). But, in all real systems, increasing the service rate costs something. Thus there is a trade-off between decreasing the congestion and increasing the cost of providing service, as μ increases. One way to capture this trade-off is to consider a simple model with linear costs.

1.1.1 A Simple Model with Linear Service and Waiting Costs

Suppose there are two types of cost:

(i) a service-cost rate, c (cost per unit time per unit of service rate); and

(ii) a waiting-cost rate h (cost per unit time per customer in system).

In other words, (i) if we choose service rate μ, then we pay a service cost $c \cdot \mu$ per unit time; (ii) a customer who spends t time units in the system accounts for $h \cdot t$ monetary units of waiting cost, or equivalently, the system incurs $h \cdot i$ monetary units of waiting cost per unit time while i customers are present. Suppose our objective is to minimize the long-run average cost per unit time. Now it follows from standard results in descriptive queueing theory (or the general theory of continuous-time Markov chains) that the long-run average cost equals the expected steady-state cost, if steady state exists (which is true if and only if $\mu > \lambda$). Otherwise the long-run average cost equals ∞. Therefore, without loss of generality let us assume $\mu > \lambda$.

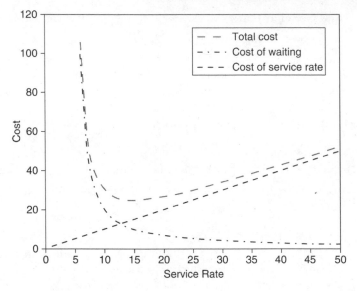

Figure 1.1 *Total Cost as a Function of Service Rate*

Let $C(\mu)$ denote the expected steady-state total cost per unit time, when service rate μ is chosen. Then

$$C(\mu) = c \cdot \mu + h \cdot L(\mu) \,,$$

where $L(\mu)$ is the expected steady-state number in system. For a *FIFO M/M/1* queue, it is well known (see, e.g., Gross and Harris [79]) that

$$L(\mu) = \lambda W(\mu) = \frac{\lambda}{\mu - \lambda} \,, \qquad (1.1)$$

where $W(\mu)$ is the expected steady-state waiting time in system.* Thus our optimization problem takes the form:

$$\min_{\{\mu : \mu > \lambda\}} C(\mu) = c \cdot \mu + h \cdot \left(\frac{\lambda}{\mu - \lambda} \right) \,. \qquad (1.2)$$

Note that

$$C''(\mu) = \frac{2h\lambda}{(\mu - \lambda)^3} > 0 \,, \text{ for all } \mu > \lambda \,,$$

so that $C(\mu)$ is convex in $\mu \in (\lambda, \infty)$. Moreover, $C(\mu) \to \infty$ as $\mu \downarrow \lambda$ and as $\mu \uparrow \infty$. (See Figure 1.1.) Hence we can solve this problem by differentiating $C(\mu)$ and setting the derivative equal to zero:

$$C'(\mu) = c - \frac{h\lambda}{(\mu - \lambda)^2} = 0 \,. \qquad (1.3)$$

---

* The expression (1.1) holds more generally for any work-conserving queue discipline that does not use information about customer service times. See, e.g., El-Taha and Stidham [60].

This yields the following expression for the unique optimal value of the service rate, denoted by μ^*:

$$\mu^* = \lambda + \sqrt{\frac{\lambda h}{c}} \ . \tag{1.4}$$

The optimal value of the objective function is thus given by

$$C(\mu^*) = c\left(\lambda + \sqrt{\lambda h/c}\right) + \lambda h/\sqrt{\lambda h/c} = c\lambda + \sqrt{\lambda h c} + \sqrt{\lambda h c} \ .$$

This expression has the following interpretation. The term $c \cdot \lambda$ represents the *fixed* cost of providing the minimum possible level of service, namely, $\mu = \lambda$. The next two terms – both equal to $\sqrt{\lambda h c}$ – represent, respectively, the service cost and the waiting cost associated with the optimal "surplus" service level, $\mu^* - \lambda$. Note that an optimal solution divides the *variable* cost equally between service cost and waiting cost.

More explicitly, if one reformulates the problem in equivalent form with the *surplus* service rate, $\tilde{\mu} := \mu - \lambda$, as the decision variable and removes the fixed-cost term, $c\lambda$, from the objective function, then the new objective function, denoted by $\tilde{C}(\tilde{\mu})$, takes the form

$$\tilde{C}(\tilde{\mu}) = c\tilde{\mu} + h\lambda/\tilde{\mu} \ . \tag{1.5}$$

The optimal value of $\tilde{\mu}$ is given by

$$\tilde{\mu}^* = \sqrt{\frac{\lambda h}{c}} \ ,$$

and the optimal value of the objective function by

$$\tilde{C}(\tilde{\mu}^*) = c\sqrt{\lambda h/c} + \lambda h/\sqrt{\lambda h/c} = \sqrt{\lambda h c} + \sqrt{\lambda h c} \ .$$

It is the particular structure of the objective function (1.5) – the sum of a term proportional to the decision variable and a term proportional to its reciprocal – that leads to the property that an optimal solution equates the two terms, a property that of course does not hold in general when one is minimizing the sum of two cost terms. The general condition for optimality (cf. equation (1.3)) is that the *marginal increase* in the first term should equal the *marginal decrease* in the second term, not that the terms themselves should be equal. It just happens in this case that the latter property holds when the former does.

Readers familiar with inventory theory will note the structural equivalence of the objective function (1.5) to the objective function in the classical economic-lot-size problem and the resulting similarity between the formula for $\tilde{\mu}^*$ and the economic-lot-size formula.

1.1.2 Extensions and Exercises

1. *Constraints on the Service Rate.* Suppose the service rate is constrained to lie in an interval, $\mu \in [\underline{\mu}, \bar{\mu}]$. Characterize the optimal service rate, μ^*,

in this case. Do the same for the case where the feasible values of μ are discrete: $\mu \in \{\mu_1, \mu_2, \ldots, \mu_m\}$.

2. *Nonlinear Waiting Costs.* Suppose in the above model that the customer's waiting cost is a nonlinear function of the time spent by that customer in the system: $h \cdot t^a$, if the time in system equals t, where $a > 0$. (Note that for $a < 1$ the waiting cost $h \cdot t^a$ is concave in t, whereas for $a > 1$ it is convex in t.) Set up and solve the problem of choosing μ to minimize the expected steady-state total cost per unit time, $C(\mu)$. For what values of a is $C(\mu)$ convex in μ?

3. *General Service-Time Distribution.* Consider an *M/GI/1* model, in which the generic service time S has mean $\mathrm{E}[S] = 1/\mu$ and second moment $\mathrm{E}[S^2] = 2\beta/\mu^2$, where $\beta \geq 1/2$ is a given constant and μ is the decision variable. (Thus the coefficient of variation of service time is given by $\sqrt{var(S)}/\mathrm{E}[S] = \sqrt{2\beta - 1}$, which is fixed.) In this case the Pollaczek-Khintchine formula yields

$$W(\mu) = \frac{1}{\mu} + \frac{\lambda\beta}{\mu(\mu - \lambda)} \ .$$

Set up the problem of determining the optimal service rate μ^*, with linear waiting cost rates. For what values of β is $C(\mu)$ convex? If possible, find a closed-form expression for μ^* in terms of the parameters, λ, c, h, and β. (The easy cases are when $\beta = 1$ (e.g., exponentially distributed service time) and $\beta = 1/2$ (constant service time, $S \equiv 1/\mu$).)

1.2 Optimal Arrival Rate

Now consider a *FIFO M/M/1* queue in which the service rate μ is fixed and the arrival rate λ is a decision variable.

Examples

1. A machine center: at what rate λ should incoming parts (or subassemblies) be admitted into the work-in-process buffer?

2. A communication system: at what rate λ should messages (or packets) be admitted into the buffer before a communication channel?

Performance Measures and Trade-offs

As λ increases, the throughput (number of jobs served per unit time) increases. (For $\lambda < \mu$, the throughput equals λ; for $\lambda \geq \mu$, the throughput equals μ.) This is clearly a "good thing." On the other hand, the congestion also increases as λ increases, and this is just as clearly a "bad thing." Again a simple linear model offers one way of capturing the trade-off between the two performance measures.

1.2.1 A Simple Model with Deterministic Reward and Linear Waiting Costs

Suppose there is a deterministic reward r per entering customer and (as in the previous model) a waiting cost per customer which is linear at rate h per unit time in the system. Let $B(\lambda)$ denote the expected steady-state net benefit per unit time. Then

$$B(\lambda) = \lambda \cdot r - h \cdot L(\lambda) , \qquad (1.6)$$

where $L(\lambda)$ is the steady-state expected number of customers in the system, expressed as a function of the arrival rate λ. As in the previous section, we have $L(\lambda) = \lambda W(\lambda)$, where $W(\lambda)$ is the steady-state expected waiting time in the system, and (assuming a first-in, first-out (*FIFO*) queue discipline) $W(\lambda)$ is given by

$$W(\lambda) = \frac{1}{\mu - \lambda} , \quad 0 \le \lambda < \mu ,$$

with $W(\lambda) = \infty$ for $\lambda \ge \mu$. Again it follows from standard results in descriptive queueing theory that the long-run average cost equals the expected steady-state cost, if steady state exists (which is true if and only if $\lambda < \mu$). Otherwise the long-run average cost equals ∞. Therefore, without loss of generality we assume $\lambda < \mu$.

For the *M/M/1* model, the problem thus takes the form:

$$\max_{\{\lambda \in [0,\mu)\}} r \cdot \lambda - h \cdot \left(\frac{\lambda}{\mu - \lambda} \right) . \qquad (1.7)$$

The presence of the constraint, $\lambda \ge 0$, makes this problem more complicated than the example of the previous section. Since $B(\lambda) \to -\infty$ as $\lambda \uparrow \mu$, we do not need to concern ourselves about the upper limit of the feasible region. But we must take into account the possibility that the maximum occurs at the lower limit, $\lambda = 0$.

Let λ^* denote the optimal arrival rate. Note that

$$B''(\lambda) = \frac{-2h\mu}{(\mu - \lambda)^3} < 0 , \text{ for all } \mu > \lambda ,$$

so that $B(\lambda)$ is strictly concave and differentiable in $0 \le \lambda < \mu$. Therefore its maximum occurs either at $\lambda = 0$ (if $B'(0) \le 0$) or at the unique value of $\lambda > 0$ at which $B'(\lambda) = 0$ (if $B'(0) > 0$).

It then follows from (1.6) that λ^* is the unique solution in $[0, \mu)$ to the following conditions:

(Case 1) $\qquad \lambda = 0 \qquad$, if $r \le hL'(0)$; $\qquad\qquad (1.8)$
(Case 2) $\quad r = hL'(\lambda) \quad$, if $r > hL'(0)$. $\qquad\qquad (1.9)$

Now for the *M/M/1* queue,

$$L'(\lambda) = \frac{\mu}{(\mu - \lambda)^2} ,$$

so that $B'(0) \le 0$ if $r \le h/\mu$ and $B'(0) > 0$ if $r > h/\mu$. Therefore

(Case 1) $\qquad \lambda^* = 0 , \qquad\qquad$ if $r \le h/\mu$;

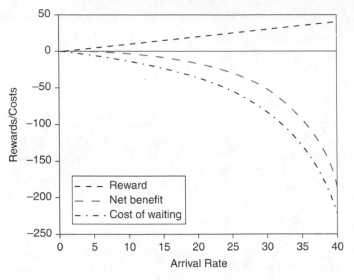

Figure 1.2 *Optimal Arrival Rate, Case 1: $r \leq h/\mu$*

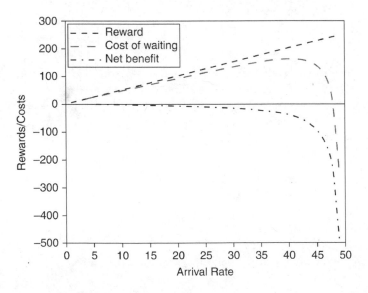

Figure 1.3 *Optimal Arrival Rate, Case 2: $r > h/\mu$*

$$\text{(Case 2)} \quad \lambda^* = \mu - \sqrt{\mu h/r} \,, \quad \text{if } r > h/\mu \,;$$

The two cases are illustrated in Figures 1.2 and 1.3, respectively.

Since $\mu - \sqrt{\mu h/r} > 0$ if and only if $r > h/\mu$, we can combine Cases 1 and 2 as follows:

$$\lambda^* = \left(\mu - \sqrt{\mu h/r} \right)^+ ,$$

where $x^+ := \max\{x, 0\}$. Note that in Case 1 we have $h/\mu \geq r$; that is, the expected waiting cost is at least as great as the reward even for a customer who enters service immediately. Hence it is intuitively clear that $\lambda^* = 0$: there is no economic incentive to admit any customer. If $r > h/\mu$, then it is optimal to allocate λ so that the surplus capacity, $\mu - \lambda$, equals the square root of $\mu h/r$.

1.2.2 Extensions and Exercises

1. *Constraints on the Arrival Rate.* Suppose the feasible set of values for λ is the interval, $[\underline{\lambda}, \bar{\lambda}]$, where $0 \leq \underline{\lambda} < \bar{\lambda} \leq \infty$. The problem now takes the form:

$$\max_{\{\lambda \in [\underline{\lambda}, \bar{\lambda}]\}} \{\lambda \cdot r - hL(\lambda)\} . \qquad (1.10)$$

Since $B(\lambda) = -\infty$ for $\lambda \geq \mu$, we can rewrite the problem in equivalent form as

$$\max_{\{\lambda \in [\underline{\lambda}, \min\{\bar{\lambda}, \mu\}]\}} \left\{\lambda \cdot r - h\left(\frac{\lambda}{\mu - \lambda}\right)\right\} . \qquad (1.11)$$

(Note that the feasible region reduces to $[\underline{\lambda}, \mu)$ when $\bar{\lambda} \geq \mu$.) Characterize the optimal arrival rate, λ^*, for this problem.

2. *General Service-Time Distribution.* Consider an *M/GI/1* model, in which the generic service time S has mean $\mathrm{E}[S] = 1/\mu$ and second moment $\mathrm{E}[S^2] = 2\beta/\mu^2$, where $\beta \geq 1/2$ is given. The Pollaczek-Khintchine formula yields

$$W(\lambda) = \frac{1}{\mu} + \frac{\lambda\beta}{\mu(\mu - \lambda)} .$$

Set up the problem of determining the optimal arrival rate, λ^*, with deterministic reward and linear waiting cost. Show that λ^* is again characterized by (1.8) and (1.9), and use this result to derive an explicit expression for λ^*, in terms of the parameters, μ, β, r, and h.

1.2.3 An Upper Bound on the Optimal Arrival Rate

Note that

$$B(\lambda) = \lambda r - h\lambda W(\lambda) = \lambda(r - hW(\lambda)) , \qquad (1.12)$$

so that $B(\lambda) > 0$ for positive values of λ such that $r > hW(\lambda)$ and $B(\lambda) \leq 0$ for values of λ such that $r \leq hW(\lambda)$. If $r \leq hW(0)$ then $r \leq hW(\lambda)$ for all $\lambda \in [0, \mu)$, since $W(\cdot)$ is an increasing function. In this case $\lambda^* = 0$. Otherwise, we can restrict attention, without loss of optimality, to values of λ such that $r > hW(\lambda)$. In the *M/M/1* case, $W(\lambda) = 1/(\mu - \lambda)$, so that $r \leq hW(0)$ if and only if $r \leq h/\mu$. Moreover, $r = hW(\lambda)$ if and only if $\lambda = \mu - h/r$. These observations motivate the following definition.

Define $\bar{\lambda}$ by:

$$\text{(Case 1)} \qquad \bar{\lambda} = 0 \;, \qquad \text{if } r \leq h/\mu \;; \qquad\qquad (1.13)$$

$$\text{(Case 2)} \quad \bar{\lambda} = \mu - h/r \;, \quad \text{if } r > h/\mu \;; \qquad\qquad (1.14)$$

Since $B(\lambda) \geq 0$ for $0 \leq \lambda \leq \bar{\lambda}$, and $B(\lambda) \leq 0$ for $\bar{\lambda} < \lambda < \mu$, it follows that $\bar{\lambda}$ is an upper bound on λ^*. Moreover, in some contexts $\bar{\lambda}$ can be interpreted as the *individually optimal* (or *equilibrium*) arrival rate, as we shall see presently.

1.2.4 Social vs. Individual Optimization

In our discussion of performance measures and trade-offs, we have been implicitly assuming that the decision maker is the operator of the queueing facility, who is concerned both with maximizing throughput and minimizing congestion. Our reward/cost model assumes that each entering customer generates a benefit r to the facility and that it costs the facility h per unit time per customer in the system. In this section we offer alternative possibilities for who the decision maker(s) might be. But first we must resolve another issue.

We have also been implicitly assuming that the decision maker (whoever it is) can freely choose the arrival rate λ from the interval $[0, \mu)$. How might such a choice be implemented? Here is one possibility.

Suppose that potential customers arrive according to a Poisson process with mean rate Λ ($\Lambda \geq \mu$). A potential customer joins (or is accepted) with probability a and balks (or is rejected) with probability $1 - a$. The accept/reject decisions for successive customers are mutually independent, as well as independent of the number of customers in the system. That is, it is not possible to observe the contents of the queue before the accept/reject decision is made. As a result, customers enter the system according to a Poisson arrival process with mean rate $\lambda = a\Lambda$.[†] Moreover, a customer who enters with probability a when the arrival rate equals λ receives an expected net benefit equal to

$$a(r - hW(\lambda)) + (1 - a)0 = a(r - hW(\lambda)) \;.$$

Now let us consider the possibility that the decision makers are the customers themselves, rather than the facility operator. We discuss this possibility in the next two subsections.

1.2.4.1 Socially Optimal Arrival Rate

Suppose now that benefits and costs accrue to individual customers and the decision maker represents the collective of all customers. In this case, a reasonable objective for the decision maker is to maximize the expected net benefit received per unit time by the collective of all customers: $B(\lambda) = \lambda(r - hW(\lambda))$. This is precisely the objective function that we have been considering. In this

---

[†] Note that the assumption that $\Lambda \geq \mu$ ensures that the feasible region for λ is the interval $[0, \mu)$, as in our original formulation.

context, our probabilistic interpretation of the choice of λ still makes sense. That is, the decision maker, acting on behalf of the collective of all customers, admits each potential arrival with probability $a = \lambda/\Lambda$.

The optimal arrival rate λ^* can now be interpreted as *socially optimal*, since it maximizes *social welfare*, that is, the expected net benefit received per unit time by the collective of all customers, namely $B(\lambda)$. To emphasize this interpretation, we shall henceforth write "λ^s" instead of "λ^*". In the *M/M/1* case, then, the socially optimal arrival rate is given by

$$\lambda^s = (\mu - \sqrt{\mu h/r})^+ . \tag{1.15}$$

The system controller can implement λ^s by admitting each potential arrival with probability $a^s := \lambda^s/\Lambda$ and rejecting with probability $1 - a^s$.

1.2.4.2 Comparison with Individually Optimal Arrival Rate

This interpretation of λ^s as the socially optimal arrival rate suggests the following question: how does the socially optimal arrival rate compare to the *individually optimal* arrival rate that results if each individual potential arrival, acting in its own interest, decides whether or not to join?

Suppose (as above) that potential customers arrive according to a Poisson process with arrival rate Λ ($\Lambda \geq \mu$) and each joins the system with probability a and balks with probability $1-a$. Each customer who enters the system when the arrival rate is λ receives a net benefit $r - hW(\lambda)$. A customer who balks receives nothing. As is always the case with design (static control) models, we assume that the decision ($a = 0, 1$) must be made without knowledge of the actual state of the system, e.g., the number of customers present.

Now, however, the criterion for choice of a is purely selfish: each customer is concerned only with maximizing its own expected net benefit. Since a single individual's action has a negligible effect on the system arrival rate λ, each potential customer can take λ as given. For a given λ, the individually optimizing customer seeks to maximize its expected net benefit,

$$a(r - hW(\lambda)) + (1 - a) \cdot 0 ,$$

by an appropriate choice of a, $0 \leq a \leq 1$. Thus, the customer will join with probability $a = 1$, if $r > hW(\lambda)$; join with probability $a = 0$, if $r < hW(\lambda)$; and be indifferent among all a, $0 \leq a \leq 1$, if $r = hW(\lambda)$.

Motivated by the concept of a *Nash equilibrium*, we define an *individually optimal* (or *equilibrium*) arrival rate, λ^e (and associated joining probability $a^e = \lambda^e/\Lambda$), by the property that no individual customer trying to maximize its own expected net benefit has any incentive to deviate unilaterally from λ^e (a^e). From the above observations, it follows that $\lambda^e = 0$ ($a^e = 0$) if $r \leq hW(0)$ (Case 1), whereas if $r > hW(0)$ (Case 2) then $\lambda^e = a^e\Lambda$ is the (unique) value of $\lambda \in (0, \mu)$ such that

$$r = hW(\lambda) . \tag{1.16}$$

To see this, first note that in Case 1 the expected net benefit from choosing a positive joining probability, $a > 0$, is $a(r - hW(0))$, which is less than or equal to zero, the expected net benefit from the joining probability $a^e = \lambda^e/\Lambda = 0$. Hence, in Case 1 there is no incentive for a customer to deviate unilaterally from $a^e = 0$. In Case 2, since $r - hW(\lambda^e) = 0$, the expected net benefit is

$$a(r - hW(\lambda^e)) + (1 - a) \cdot 0 = 0 \ ,$$

and hence does not depend on the joining probability a. Thus, customers are indifferent among all joining probabilities, $0 \le a \le 1$, so that once again there is no incentive to deviate from $a^e = \lambda^e/\Lambda$.

Since $W(\lambda) = 1/(\mu - \lambda)$ in the $M/M/1$ case, we see that the individually optimal arrival rate λ^e coincides with $\bar{\lambda}$ as defined by (1.13) and (1.14). But we have shown that $\lambda^* = \lambda^s \le \bar{\lambda} = \lambda^e$. In other words, the *socially optimal* arrival rate, λ^s, is less than or equal to the individually optimal arrival rate, λ^e.

The following theorem summarizes these results:

Theorem 1.1 *The socially optimal arrival rate is no larger than the* individually optimal *arrival rate:* $\lambda^s \le \lambda^e$. *Moreover,* $\lambda^s = \lambda^e = 0$, *if* $r \le h/\mu$, *and* $0 < \lambda^s < \lambda^e$, *if* $r > h/\mu$.

A review of our arguments above will show that this property is not restricted to $M/M/1$ systems and is in fact quite general. In fact, this theorem is valid for *any* system (for example, a $GI/GI/1$ queue) in which the following conditions hold:

1. $W(\lambda)$ is strictly increasing in $0 \le \lambda < \mu$;

2. $W(\lambda) \uparrow \infty$ as $\lambda \uparrow \mu$;

3. $W(0) = 1/\mu$.

1.2.5 Internal and External Effects

Suppose $r > h/\mu$. It follows from (1.12) that

$$B'(\lambda) = r - [h \cdot W(\lambda) + h \cdot \lambda W'(\lambda)] \ ,$$

and that λ^s is found by equating $h \cdot W(\lambda) + h \cdot \lambda W'(\lambda)$ to r, whereas (cf. (1.16)) λ^e is found by equating $h \cdot W(\lambda)$ to r. We can interpret $h \cdot W(\lambda)$ as the *internal effect* and $h \cdot \lambda W'(\lambda)$ as the *external effect* of a marginal increase in the arrival rate. The quantity $h \cdot W(\lambda)$ is the waiting cost of the marginal customer who joins when the arrival rate is λ. It is "internal" in that it is a cost borne only by the customer itself. On the other hand, the quantity $h \cdot \lambda W'(\lambda)$ is the marginal increase in waiting cost incurred by all the customers as a result of a marginal increase in the arrival rate. It is "external" to the marginal joining customer, since it is a cost which that customer does not incur. The fact that $\lambda^s \le \lambda^e$ (that is, customers acting in their own interest join the system more frequently than is socially optimal) is due to an individually optimizing customer's failure to take into account the external effect of its decision to enter. The formula for λ^e only takes into account the internal effect of the decision to enter, that is the customer's own waiting cost, $hW(\lambda)$. By contrast, the formula for λ^s takes into account both the internal effect, $hW(\lambda)$, and the external effect, $h\lambda W'(\lambda)$.

It follows that individually optimizing customers can be induced to behave in a socially optimal way by charging each entering customer a fee or *congestion toll* equal to the external effect, $h\lambda W'(\lambda)$. In this way arrival control can be *decentralized*, in the sense that each individual customer can be left to make its own decision. (Again, note that these results hold for any system in which $W(\lambda)$ is a well defined function satisfying conditions (1)–(3). See Chapter 2 for further analysis and generalizations.)

1.3 Optimal Arrival Rate and Service Rate

Now let us consider an *M/M/1* queue in which both the arrival rate λ and the service rate μ are decision variables. We shall use a reward/cost model that combines the features of the models of the last two sections. There is a reward r per entering customer, a waiting cost h per unit time per customer in the system, and a service cost c per unit time per unit of service rate. The objective function (to be maximized) is the steady-state expected net benefit per unit time, $B(\lambda, \mu)$, that is,

$$B(\lambda, \mu) = \lambda \cdot r - h \cdot L(\lambda, \mu) - c \cdot \mu , \ 0 \le \lambda < \mu ,$$

with $B(0,0) = 0$. (Note that $B(\lambda, \mu)$ has a discontinuity at $(0,0)$.) If $c \ge r$, then obviously the optimal solution is $\lambda^* = \mu^* = 0$, with net benefit $B(0,0) = 0$, since for all $0 \le \lambda < \mu$ we have $B(\lambda, \mu) < 0$. Henceforth we shall assume that $c < r$, in which case we can exclude the point $(0,0)$ and restrict attention to the region $\{(\lambda, \mu) : 0 \le \lambda < \mu\}$, since it contains pairs (λ, μ) for which $B(\lambda, \mu) > 0$. Note that $B(\lambda, \mu)$ is continuously differentiable over this region.

Following the program of the previous two sections, let us use the first-order optimality conditions to try to identify the optimal pair, (λ^*, μ^*). Differentiating $B(\lambda, \mu)$ with respect to λ and μ and setting the derivatives equal to zero leads to the equations,

$$\frac{\partial}{\partial \lambda} B(\lambda, \mu) = r - h \cdot \frac{\partial}{\partial \lambda} L(\lambda, \mu) = 0 ,$$

$$\frac{\partial}{\partial \mu} B(\lambda, \mu) = -h \cdot \frac{\partial}{\partial \mu} L(\lambda, \mu) - c = 0 .$$

Since $L(\lambda, \mu) = \lambda/(\mu - \lambda)$, for $0 \le \lambda < \mu$, we have

$$\frac{\partial}{\partial \lambda} L(\lambda, \mu) = \frac{\mu}{(\mu - \lambda)^2} , \ \frac{\partial}{\partial \mu} L(\lambda, \mu) = \frac{-\lambda}{(\mu - \lambda)^2} ,$$

from which we obtain the following two simultaneous equations for λ and μ,

$$\frac{h \cdot \mu}{(\mu - \lambda)^2} = r ,$$

$$\frac{h \cdot \lambda}{(\mu - \lambda)^2} = c,$$

the unique solution to which is

$$\lambda = \frac{h \cdot c}{(r - c)^2}, \ \mu = \frac{h \cdot r}{(r - c)^2}. \tag{1.17}$$

Note that this solution is feasible (that is, $\lambda < \mu$) since $c < r$.

To recapitulate, under the assumption that $c < r$, we have identified a unique interior point of the feasible region $(0 < \lambda < \mu)$ that satisfies the first-order optimality conditions. Surely this must be the optimal solution. After all, we have simply brought together the two models and analyses of the previous sections, in which μ and λ, respectively, were decision variables and in the course of which we verified that our objective function, $B(\lambda, \mu)$, is both concave in λ and concave in μ. What we have not verified, however, is joint concavity in (λ, μ). Without joint concavity, we cannot be sure that a solution to the first-order optimality conditions is a local (let alone a global) maximum.

In fact $B(\lambda, \mu)$ is *not* jointly concave in (λ, μ), because $L(\lambda, \mu) = \lambda/(\mu - \lambda)$ is not jointly convex . To check for joint convexity, we must evaluate

$$\Delta := \left(\frac{\partial^2 L}{\partial \lambda^2}\right)\left(\frac{\partial^2 L}{\partial \mu^2}\right) - \left(\frac{\partial^2 L}{\partial \lambda \partial \mu}\right)^2$$

and check whether Δ is nonnegative. Since

$$\frac{\partial^2 L}{\partial \lambda^2} = \frac{2\mu}{(\mu - \lambda)^3},$$

$$\frac{\partial^2 L}{\partial \mu^2} = \frac{2\lambda}{(\mu - \lambda)^3},$$

$$\frac{\partial^2 L}{\partial \lambda \mu} = \frac{-(\lambda + \mu)}{(\mu - \lambda)^3},$$

we have

$$\begin{aligned}
\Delta &= \left(\frac{2\mu}{(\mu - \lambda)^3}\right)\left(\frac{2\lambda}{(\mu - \lambda)^3}\right) - \left(\frac{-(\lambda + \mu)}{(\mu - \lambda)^3}\right)^2 \\
&= \frac{1}{(\mu - \lambda)^6}\left[4\lambda\mu - (\lambda^2 + 2\lambda\mu + \mu^2)\right] \\
&= \frac{1}{(\mu - \lambda)^6}\left[-(\lambda^2 - 2\lambda\mu + \mu^2)\right] \\
&= \frac{1}{(\mu - \lambda)^6}\left[-(\mu - \lambda)^2\right] \\
&= \frac{-1}{(\mu - \lambda)^4} < 0
\end{aligned}$$

Thus $L(\lambda, \mu)$ is not jointly convex and therefore $B(\lambda, \mu)$ is not jointly concave in (λ, μ).

It follows that the stationary point (1.17) identified by the first-order conditions does not necessarily yield the global maximum net benefit. To gain further insight, let us evaluate $B(\lambda, \mu)$ at this stationary point. Substituting the expressions from (1.17) into the formula for $B(\lambda, \mu)$ and simplifying, we obtain (after simplifying)

$$B(\lambda, \mu) = -\frac{h \cdot c}{r - c} < 0 = B(0, 0) \ .$$

So the proposed solution in fact yields a negative net benefit! It is therefore dominated by the point $(0, 0)$ (do nothing) and we know that we can do even better than that when $c < r$.

To see how much better, let us examine the problem from a slightly different perspective. Define the traffic intensity ρ (as usual) by $\rho := \lambda/\mu$ and rewrite the net benefit as a function of λ and ρ:

$$\tilde{B}(\lambda, \rho) := r \cdot \lambda - \frac{h \cdot \rho}{1 - \rho} - \frac{c \cdot \lambda}{\rho} \ .$$

Now fix a value of ρ such that

$$\frac{c}{r} < \rho < 1 \ .$$

Then we have

$$\tilde{B}(\lambda, \rho) = \lambda \cdot (r - \frac{c}{\rho}) - \frac{h \cdot \rho}{1 - \rho} \ .$$

The second term is constant and the first term is positive and can be made arbitrarily large by choosing λ sufficiently large. Thus $B(\lambda, \rho) \to \infty$ as $\lambda \to \infty$ and hence there is no finite optimal solution to the problem. Rather, one can obtain arbitrarily large net benefit by judiciously selecting large values of both λ and μ.

Of course these observations raise serious questions about the realism of our model. We shall address these questions later (in Chapter 5). In the meantime, we need to understand what went wrong with our approach based on finding a solution to the first-order optimality conditions.

As we saw, the net-benefit function in this model fails to be jointly concave because it contains a congestion-cost term that is proportional to $L(\lambda, \mu)$, the expected steady-state number of customers in the system, which fails to be jointly convex. This congestion-cost term can be written as

$$h \cdot L(\lambda, \mu) = \lambda(h \cdot W(\lambda, \mu)) \ ,$$

where $W(\lambda, \mu)$ is the expected steady-state waiting of a customer in the system. In other words, we have a congestion cost per unit time that takes the form

(no. customers arriving per unit time) × (congestion cost per customer) .

While the congestion cost per customer (in this case, $h/(\mu - \lambda)$) is jointly convex, the result of multiplying by λ is to destroy this joint convexity.

As we shall see in later chapters, this type of congestion cost and its associated non-joint-convexity are not an anomaly but in fact are typical in queueing optimization models. As a result one must be very careful when applying classical economic analysis based on first-order optimality equations. It is not enough to simply assume that the values of the parameters are such that there exists a finite optimal solution in the interior of the feasible region, which then must satisfy the first-order conditions (because they are necessary for an interior maximum). We have seen in the present example that there may be no such interior optimal solution, no matter what the parameter values are. Moreover, there may be an easily identified solution to the first-order conditions which one is tempted to identify as optimal but which may in fact be far from optimal.

The literature contains a surprising number of examples in which these kinds of mistakes have been made.

1.4 Optimal Arrival Rates for a Two-Class System

Now suppose we have an *M/M/1* queue in which there are two classes of customers. The service rate μ is fixed but the arrival rates of the two classes (denoted λ_1 and λ_2) are decision variables. Customers are served in order of arrival, regardless of class, so that the expected steady-state waiting time in the system is the same for both classes and is a function, $W(\lambda)$, of the total arrival rate, $\lambda := \lambda_1 + \lambda_2$. Recall that in the *M/M/1* case $W(\lambda)$ is given by

$$W(\lambda) = \frac{1}{\mu - \lambda} \ , \ \lambda < \mu \ ; \ W(\lambda) = \infty \ , \ \lambda \geq \mu \ . \tag{1.18}$$

We shall assume a reward/cost model like that of Section 1.2, but with class-dependent rewards and waiting cost rates. Specifically, there is a reward r_i per entering customer of class i, and a waiting cost h_i per unit time per customer of class i in the system. The objective is to maximize the steady-state expected net benefit per unit time:

$$\max_{\{\lambda, \lambda_1, \lambda_2\}} \quad B(\lambda_1, \lambda_2) = r_1\lambda_1 + r_2\lambda_2 - (\lambda_1 h_1 + \lambda_2 h_2)W(\lambda)$$

$$\text{s.t.} \quad \lambda_1 + \lambda_2 = \lambda$$

$$\lambda_1 \geq 0 \ , \ \lambda_2 \geq 0$$

As in the single-class model considered in Section 1.2, if all rewards and costs accrue to the customers, a solution $(\lambda_1^s, \lambda_2^s)$ to this optimization problem will be *socially optimal*, in the sense of maximizing the aggregate net benefit accruing to the collective of all customers. Moreover, if potential customers of class i arrive according to a Poisson process with mean rate $\Lambda_i \geq \mu$, then a socially optimal allocation can be implemented by admitting each class-i arrival with probability $a_i^s = \lambda_i^s/\Lambda_i$.

The following Karush-Kuhn-Tucker (*KKT*) first-order conditions are *necessary* for $(\lambda_1, \lambda_2, \lambda)$ to be optimal for this problem (see, e.g., Bazaraa et al. [16]):

$$r_i \;=\; h_i W(\lambda) + \delta \text{ and } \lambda_i > 0 \tag{1.19}$$

$$\text{or } r_i \;\leq\; h_i W(\lambda) + \delta \text{ and } \lambda_i = 0 \tag{1.20}$$

for $i = 1, 2$, and

$$\lambda \;=\; \lambda_1 + \lambda_2 \,, \tag{1.21}$$

$$\delta \;=\; (\lambda_1 h_1 + \lambda_2 h_2) W'(\lambda) \,. \tag{1.22}$$

Now consider this system from the perspective of individual optimization. Suppose a fixed, arbitrary toll, δ, is charged to each entering customer. Each customer of class i takes $W(\lambda)$ as given and chooses the probability a_i of joining to maximize

$$a_i \cdot (r_i - h_i W(\lambda) - \delta) + (1 - a_i) \cdot 0 \,, \ a_i \in [0, 1] \,.$$

In other words, a class-i customer who joins receives the net benefit, $r_i - h_i W(\lambda)$, minus the toll, δ, paid for the use of the facility. A customer who balks receives (pays) nothing. Then it is easy to see that arrival rates, $\lambda_i = a_i \cdot \Lambda_i$, that satisfy equations (1.19) and (1.20) will be individually optimal for the customers of both classes. Moreover, for the given toll δ, a solution to (1.19), (1.20), and (1.21) is a Nash equilibrium.

As expected, equation (1.22) reveals that the socially optimal toll is just the *external effect*, defined (as usual) as the marginal increase in the total delay cost incurred as a result of a marginal increase in the flow, λ. By charging this socially optimal toll, the system operator can induce individually optimizing customers to behave in a socially optimal way, thereby making the Nash-equilibrium allocation coincide with the socially optimal allocation $(\lambda_1^s, \lambda_2^s, \lambda^s)$ (cf. Section 1.2).

1.4.1 Solutions to the Optimality Conditions: the M/M/1 Case

Let us now examine the properties of the solution(s) to the *KKT* conditions, using the explicit expression (1.18) for $W(\lambda)$ for an *M/M/1* system. The problem of finding a socially optimal allocation of flows takes the form

$$\max_{\{\lambda_1, \lambda_2\}} \quad r_1 \lambda_1 - \frac{h_1 \lambda_1}{\mu - \lambda_1 - \lambda_2} + r_2 \lambda_2 - \frac{h_2 \lambda_2}{\mu - \lambda_1 - \lambda_2}$$

$$\text{s.t.} \quad \lambda_1 + \lambda_2 < \mu$$

$$\lambda_1 \geq 0 \,, \ \lambda_2 \geq 0$$

Without loss of generality, we may assume that $\mu = 1$. (Equivalently, measure flows in units of fraction of the service rate μ.) Let $a := r_1 / h_1$, $b := r_2 / h_2$, $c := h_1 / h_2$. Then an equivalent form for the above problem is

$$\max_{\{\lambda_1, \lambda_2\}} \quad c \left(a \lambda_1 - \frac{\lambda_1}{1 - \lambda_1 - \lambda_2} \right) + b \lambda_2 - \frac{\lambda_2}{1 - \lambda_1 - \lambda_2} \tag{1.23}$$

$$\text{s.t.} \quad \lambda_1 + \lambda_2 < 1$$

$$\lambda_1 \geq 0 \; , \; \lambda_2 \geq 0$$

For an interior optimal solution, equation (1.19) must be satisfied for $i = 1, 2$. The unique solution to these equations is given by

$$\tilde{\lambda}_1 = \frac{b(c-1)}{(ca-b)^2} - \frac{1}{c-1}$$

$$\tilde{\lambda}_2 = \frac{c}{c-1} - \frac{ca(c-1)}{(ca-b)^2}$$

It can be shown that this pair $(\tilde{\lambda}_1, \tilde{\lambda}_2)$ is an interior point $(\tilde{\lambda}_1 > 0, \tilde{\lambda}_2 > 0, \tilde{\lambda}_1 + \tilde{\lambda}_2 < 1)$ if the parameters satisfy the following conditions:

$$b > a > 1 \; ;$$

$$c > \frac{b-1}{a-1} \; ;$$

$$a < \frac{(ca-b)^2}{(c-1)^2} < b \; .$$

So, for an $M/M/1$ system in which the parameters satisfy these conditions, we have established that the first-order optimality conditions have a unique interior-point solution. This result tempts us to conclude that this solution is indeed optimal. But the model of Section 1.3, in which the unique interior-point solution to the optimality conditions turned out to be nonoptimal, should serve as a warning to proceed more cautiously. The question remains whether there are other, non-interior-point solutions to the KKT conditions and whether one of these could yield a higher value of the objective function. Put another way: are the KKT conditions sufficient as well as necessary for an optimal solution to our problem?

1.4.2 Are the KKT Conditions Sufficient?

To answer this question, let us return to the problem in its original form. The objective function takes the following form (after substituting for λ from the equality constraint),

$$B(\lambda_1, \lambda_2) = r_1 \lambda_1 + r_2 \lambda_2 - f(\lambda_1, \lambda_2) \; ,$$

where $f(\lambda_1, \lambda_2) := (\lambda_1 h_1 + \lambda_2 h_2) W(\lambda_1 + \lambda_2)$. That is, $f(\lambda_1, \lambda_2)$ is the total delay cost per unit time expressed as a function of λ_1 and λ_2. The KKT conditions will be sufficient for social optimality if $B(\lambda_1, \lambda_2)$ is jointly concave in (λ_1, λ_2), which is true if and only if $f(\lambda_1, \lambda_2)$ is jointly convex in (λ_1, λ_2). It is easily verified that $f(\lambda_1, \lambda_2)$ is convex in λ_1 and convex in λ_2. To check for joint convexity, we evaluate

$$\Delta := \left(\frac{\partial^2 f}{\partial \lambda_1^2} \right) \left(\frac{\partial^2 f}{\partial \lambda_2^2} \right) - \left(\frac{\partial^2 f}{\partial \lambda_1 \partial \lambda_2} \right)^2$$

and find that $\Delta = -((h_1 - h_2)W'(\lambda_1 + \lambda_2))^2$, which is strictly negative unless $h_1 = h_2$, that is, unless the customer classes are homogeneous with respect to their sensitivity to delay. Thus $f(\lambda_1, \lambda_2)$ is *not* in general a jointly convex function of λ_1 and λ_2. Indeed, the conditions for joint convexity fail at *every* point in the feasible region if the customer classes are heterogeneous, that is, if $h_1 \neq h_2$. It follows that $B(\lambda_1, \lambda_2)$ fails to be jointly concave unless $h_1 = h_2$.

Remark 1 Note that we did not use the specific functional form (1.18) of $W(\lambda)$ in our demonstration of the nonconvexity of $f(\lambda_1, \lambda_2)$. The only properties that we used were that the delay $W(\lambda)$ for each customer is an increasing, convex, and differentiable function of the sum of the flows, and that the delay cost per unit time for each class i is the product of the flow, λ_i, and the delay cost per customer, $h_i W(\lambda)$. All these properties are weak and hold for many queueing models, not just for the $M/M/1$ case. As we shall see in Chapters 4 and 5, nonconvexity is a widely encountered phenomenon in models for the design of queues with more than one decision variable.

The nonconcavity of the objective function, $B(\lambda_1, \lambda_2)$, leads one to suspect that the first-order *KKT* conditions, (1.19)–(1.22), may not be sufficient for an optimal allocation. In particular, an interior-point solution to these conditions – such as the one found in the previous subsection – might not be optimal. Let us now examine that question. First observe that such a solution must lie on the line $\lambda_1 + \lambda_2 = \lambda$, where λ satisfies

$$r_1 - h_1 W(\lambda) = r_2 - h_2 W(\lambda) . \tag{1.24}$$

Along this line both the total flow λ and the net benefit, $B(\lambda_1, \lambda_2)$, are constant: $B(\lambda_1, \lambda_2) = B$, say. In particular, the two extreme points on this line, namely, $(\lambda, 0)$, and $(0, \lambda)$, share this net benefit; that is,

$$B(\lambda, 0) = B(0, \lambda) = B .$$

But

$$B(\lambda, 0) \leq B(\lambda_1^*, 0) ,$$
$$B(0, \lambda) \leq B(0, \lambda_2^*) ,$$

where λ_i^* is the optimal flow allocation to class i when only that class receives positive flow $(i = 1, 2)$.

Thus we see that any interior solution to the first-order *KKT* conditions is dominated by *both* the optimal single-class allocations. In other words, the system achieves at least as great a net benefit by allocating all flow to a single class, *regardless of which class*, than by using an interior allocation satisfying the first-order conditions!

Our next observation has to do with external effects, congestion tolls, and equilibrium properties. First note that charging each user a toll δ (per unit of flow) equal to the external effect, that is,

$$\delta = (\lambda_1 h_1 + \lambda_2 h_2)W'(\lambda_1 + \lambda_2) ,$$

makes $(\tilde{\lambda}_1, \tilde{\lambda}_2)$ a Nash equilibrium for individually optimizing customers: no customer of either class has an incentive to deviate from this allocation, assuming that all other customers make no change. Thus, we see that, even by charging the "correct" toll (namely, a toll equal to the external effect), we cannot be certain that the customers will be directed to a socially optimal flow allocation. Rather, the resulting allocation, even though it is a Nash equilibrium, may be dominated by both of the optimal single-class allocations.

Thus we have a dramatic example of the pitfalls of marginal-cost pricing (that is, pricing based on first-order optimality conditions) when the customer classes are heterogeneous in their sensitivities to congestion.

As an example, let us return to the $M/M/1$ example of Section 1.4.1. Let $a = 4$, $b = 9$, and $c = 4$. In this case, the solution to the first-order conditions is

$$\tilde{\lambda}_1 = 0.218 \; ; \; \tilde{\lambda}_2 = 0.354 \; .$$

The optimal single-user flow allocations are $\lambda_1^s = 0.500$ and $\lambda_2^s = 0.667$. The objective function values of these three flow allocations are:

$$
\begin{aligned}
B(\tilde{\lambda}_1, \tilde{\lambda}_2) &= 3.81 \\
B(\lambda_1^s, 0) &= 4.00 \\
B(0, \lambda_2^s) &= 4.00
\end{aligned}
$$

Thus we have an illustration of the general result derived above: the interior-point equilibrium flow allocation is dominated by both optimal single-user allocations.

For this example, Figure 1.4 and Figure 1.5 show, respectively, a contour plot and graph of the response surface of the net benefit function, $B(\lambda_1, \lambda_2)$.

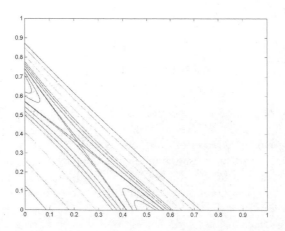

Figure 1.4 *Net Benefit: Contour Plot*

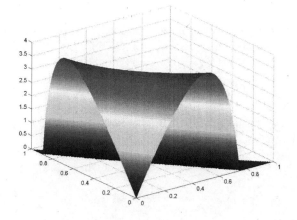

Figure 1.5 *Net Benefit: Response Surface*

1.5 Optimal Arrival Rates for Parallel Queues

Now let us consider n independent *M/M/1* queues, with service rates μ_j and arrival rates λ_j, $j = 1, \ldots, n$. Suppose that the μ_j are fixed and that the λ_j are design variables. Our objective is to minimize the steady-state expected number of customers in the system, subject to a constraint that the total arrival rate should equal a fixed value, λ. Thus the problem takes the form

$$\text{min} \quad \sum_{j=1}^{n} \frac{\lambda_j}{\mu_j - \lambda_j}$$

$$\text{s.t.} \quad \sum_{j=1}^{n} \lambda_j = \lambda \tag{1.25}$$

$$0 \le \lambda_j < \mu_j \, , \, j = 1, \ldots, n \, .$$

We can interpret this problem as follows. Suppose customers arrive to the system according to a Poisson process with mean arrival rate λ. We must decide how to split this arrival process among n parallel exponential servers, each with its own queue. The splitting is to be done probabilistically, independently of the state and past history of the system. That is, each arriving customer is sent to queue j with probability $a_j = \lambda_j/\lambda$, so that the arrival process to queue j is Poisson with mean arrival rate λ_j.

We shall use a Lagrange multiplier to eliminate the constraint on the total arrival rate. The Lagrangean problem is:

$$\text{min} \quad \sum_{j=1}^{n} \frac{\lambda_j}{\mu_j - \lambda_j} - \alpha \sum_{j=1}^{n} \lambda_j \tag{1.26}$$

$$\text{s.t.} \quad 0 \le \lambda_j < \mu_j \, , \, j = 1, \ldots, n \, .$$

The solution is parameterized by α, which can be interpreted as the imputed reward per unit time per unit of arrival rate. Problem (1.26) is separable, so we can minimize the objective function separately for each facility. For facility j, the problem takes the form of the single-facility arrival-rate-optimization problem of Section 1.2, with $r = \alpha$, $h = 1$. The solution is:

$$\lambda_j = \lambda_j^s(\alpha) := (\mu_j - \sqrt{\mu_j/\alpha})^+ , \ j = 1, \ldots, n . \tag{1.27}$$

This solution will be optimal for the original problem if α is chosen so that $\sum_{j=1}^n \lambda_j^s(\alpha) = \lambda$.

Thus an optimal allocation satisfies the following conditions ($j = 1, \ldots, n$):

$$L_j'(\lambda_j) \ = \ \frac{\mu_j}{(\mu_j - \lambda_j)^2} = \alpha , \text{ if } \lambda_j > 0 , \tag{1.28}$$

$$L_j'(\lambda_j) \ = \ \frac{1}{\mu_j} \geq \alpha , \text{ if } \lambda_j = 0 , \tag{1.29}$$

for some α such that $\sum_{j=1}^n \lambda_j = \lambda$.

These results can be used to solve the original problem (1.25) graphically. First, plot each $\lambda_j^s(\alpha)$ as a function of α, as shown in Figure 1.6. Define

$$\lambda^s(\alpha) := \sum_{j=1}^n \lambda_j^s(\alpha) ,$$

so that $\lambda^s(\alpha)$ is the total arrival rate in an optimal solution of problem (1.26) corresponding to Lagrange multiplier α. We can now find the optimal solution to the original problem for a particular value of λ by drawing a horizontal line from the vertical axis at level λ and finding its intersection with the graph of $\lambda^s(\alpha)$, then drawing a vertical line to the α axis. Where this line intersects the graph of $\lambda_j^s(\alpha)$, we obtain $\lambda_j^s = \lambda_j^s(\lambda)$, the optimal value of λ_j for the original problem with total arrival rate λ.

We can derive an explicit solution for the λ_j^s in terms of the parameter λ (denoted $\lambda_j^s(\lambda)$, $j = 1, \ldots, n$) in the following way. First, order the μ_j so that $\mu_1 \geq \mu_2 \geq \cdots \geq \mu_n$. From (1.27) it can be seen that $\lambda^s(\alpha)$ is a continuous, strictly increasing function of α, for $\alpha \geq \mu_1^{-1}$. In this range, therefore, $\lambda^s(\alpha)$ has an inverse, which we denote by $\alpha(\lambda)$. We solve for $\alpha(\lambda)$ separately over the intervals induced by $\mu_1^{-1} \leq \alpha \leq \mu_2^{-1}$, $\mu_2^{-1} \leq \alpha \leq \mu_3^{-1}, \ldots$. In particular, for $\mu_1^{-1} \leq \alpha \leq \mu_2^{-1}$,

$$\lambda_1^s(\alpha) \ = \ \mu_1 - \sqrt{\mu_1/\alpha} ,$$
$$\lambda_j^s(\alpha) \ = \ 0 , \ j = 2, \ldots, n .$$

Thus $\lambda_1^s(\alpha) = \lambda$ in this range, so that

$$\sqrt{\frac{1}{\alpha}} = \frac{\mu_1 - \lambda}{\sqrt{\mu_1}} , \tag{1.30}$$

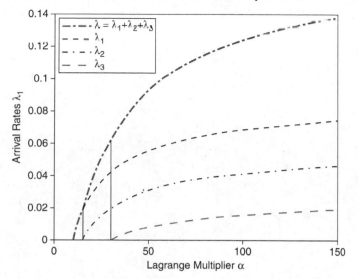

Figure 1.6 *Arrival Control to Parallel Queues: Parametric Socially Optimal Solution*

and hence

$$\lambda_1^s(\lambda) = \mu_1 - \frac{\sqrt{\mu_1}}{\sqrt{\mu_1}}(\mu_1 - \lambda) = \lambda \ .$$

But it follows from (1.30) that $\mu_1^{-1} \leq \alpha \leq \mu_2^{-1}$ if and only if $0 \leq \lambda \leq \mu_1 - \sqrt{\mu_1\mu_2}$.

Summarizing, for $r_1 := 0 \leq \lambda \leq r_2 := \mu_1 - \sqrt{\mu_1\mu_2}$, we have

$$\lambda_1^s(\lambda) \ = \ \lambda \ ,$$
$$\lambda_j^s(\lambda) \ = \ 0 \ , j = 2, \ldots, n \ .$$

Continuing this argument, we can deduce the general form of the solution for $\lambda_j^s(\lambda)$, $j = 1, \ldots, n$. In general, define $r_k := \sum_{i=1}^{k}(\mu_i - \sqrt{\mu_i\mu_k})$, $k = 1, \ldots, n$, $r_{n+1} := \sum_{i=1}^{n} \mu_i$. Then, for $k = 1, \ldots, n$, if $r_k \leq \lambda \leq r_{k+1}$,

$$\lambda_j^s(\lambda) \ = \ \mu_j - \left(\frac{\sqrt{\mu_j}}{\sum_{i=1}^{k} \sqrt{\mu_i}} \right) \left(\sum_{i=1}^{k} \mu_i - \lambda \right) \ , j = 1, \ldots, k \ ,$$
$$= \ 0 \ , j = k+1, \ldots, n \ .$$

Note that each λ_j^s is piecewise linear in λ. Figure 1.7 gives a typical illustration. Note that, once $\lambda_j^s(\lambda)$ is positive, its rate of increase is nonincreasing in λ (thus $\lambda_j^s(\lambda)$ is concave in $\lambda \geq r_j$) and that the rates of increase of the $\lambda_j^s(\lambda)$ for fixed λ are nondecreasing in j.

Individually Optimal Allocation

The allocation described above assumes that the allocation of total "demand," λ, to the various facilities is made in accordance with the system-wide

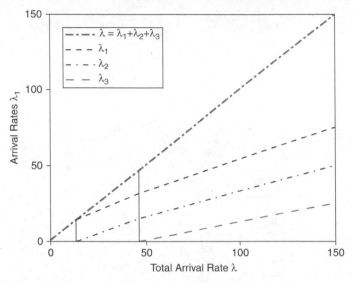

Figure 1.7 *Arrival Control to Parallel Queues: Explicit Socially Optimal Solution*

objective of minimizing the total rate of waiting per unit time: $\sum_{j=1}^{n} L_j(\lambda_j) = \sum_{j=1}^{n} \lambda_j/(\mu_j - \lambda_j)$. An equivalent way of viewing this problem is to visualize each arriving customer having a probability, $a_j = \lambda_j/\lambda$, of joining facility j, $j = 1, \ldots, n$, where the $a'_j s$ are to be chosen (by an omnipotent system designer) to minimize the steady-state expected waiting time of an arbitrary customer:

$$\sum_{j=1}^{n} \left(\frac{\lambda_j}{\lambda} \right) \left(\frac{1}{\mu_j - \lambda_j} \right) = \frac{1}{\lambda} \sum_{j=1}^{n} L_j(\lambda_j)$$

Now let us consider an allocation $(\lambda_1, \ldots, \lambda_n)$ (equivalently, a set of joining probabilities $(a_1, \ldots, a_n)$) from the point of view of an individual customer who wishes to minimize his expected waiting time. Under the allocation in question, an arriving customer chooses facility j with probability $a_j = \lambda_j/\lambda$; conditional on joining facility j, the expected waiting time is $(\mu_j - \lambda_j)^{-1}$. (As is always the case in design models, we assume that the fixed mean service rates μ_j and the arrival rates λ_j associated with the given allocation are known and the system is in steady state, but the exact number of customers at each facility cannot be observed.) The customer's unconditional expected waiting time is therefore $\sum_{j=1}^{n} a_j(\mu_j - \lambda_j)^{-1}$. As usual we call an allocation $(\lambda_1, \ldots, \lambda_n)$ (or a set of joining probabilities $(a_1, \ldots, a_n)$) *individually optimal* if no customer, acting in its own interest, has an incentive to deviate unilaterally from the allocation. This will be the case if and only if $(\mu_j - \lambda_j)^{-1} = (\mu_k - \lambda_k)^{-1}$ for all j, k such that $\lambda_j > 0$ and $\lambda_k > 0$, and $(\mu_j - \lambda_j)^{-1} \leq \mu_k^{-1}$, if $\lambda_j > 0$ and $\lambda_k = 0$. Otherwise, e.g., if $(\mu_j - \lambda_j)^{-1} > (\mu_k - \lambda_k)^{-1}$ for some j, k such that $\lambda_j > 0$, an arriving customer could strictly reduce its expected waiting time

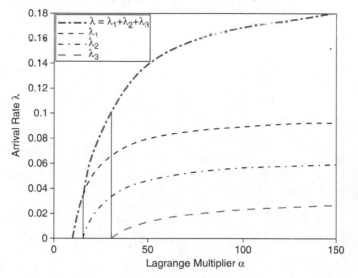

Figure 1.8 *Arrival Control to Parallel Queues: Parametric Individually Optimal Solution*

by joining facility j with probability $a'_j := 0$ and facility k with probability $a'_k := a_j + a_k$, rather than $a_j = \lambda_j/\lambda$ and $a_k = \lambda_k/\lambda$, respectively.

In other words, an individually optimal allocation satisfies the following conditions, for $j = 1, \dots, n$:

$$W_j(\lambda_j) \;=\; \frac{1}{\mu_j - \lambda_j} = \alpha \text{ , if } \lambda_j > 0 \; ; \tag{1.31}$$

$$W_j(\lambda_j) \;=\; \frac{1}{\mu_j} \geq \alpha \text{ , if } \lambda_j = 0 \; ; \tag{1.32}$$

for some $\alpha > 0$ such that $\sum_{j=1}^n \lambda_j = \lambda$.

We would like to compare such an allocation, denoted $\lambda_j^e(\alpha)$, or $\lambda_j^e(\lambda)$, to the *socially optimal* allocation, $\lambda_j^s(\alpha)$, or $\lambda_j^s(\lambda)$. First observe from (1.31) and (1.28) that an individually optimal allocation equates *average costs*, $1/(\mu_j - \lambda_j)$ (internal effects), whereas a socially optimal allocation equates *marginal costs*, $\mu_j/(\mu_j - \lambda_j)^2 = 1/(\mu_j - \lambda_j) + \lambda_j/(\mu_j - \lambda_j)^2$ (internal plus external effects), at all open facilities j.

In terms of α, the individually optimal allocation can be written as

$$\lambda_j^e(\alpha) = (\mu_j - 1/\alpha)^+ \; , \; j = 1, \dots, n \; .$$

Figure 1.8 illustrates the behavior of $\lambda_j^e(\alpha)$, assuming $\mu_1 \geq \mu_2 \geq \cdots \geq \mu_n$. Now α must be chosen so that $\sum_{j=1}^n \lambda_j^e(\alpha) = \lambda$, in order to find $\lambda_j^e(\lambda)$, $j = 1, \dots, n$. This can be done in the same way as for socially optimal allocations. (The details are left to the reader.) In general, define $s_k := \sum_{i=1}^k (\mu_i - \mu_k)$, $k = 1, \dots, n$, $s_{n+1} := \sum_{i=1}^n \mu_i$. Then the individually optimal allocation is as

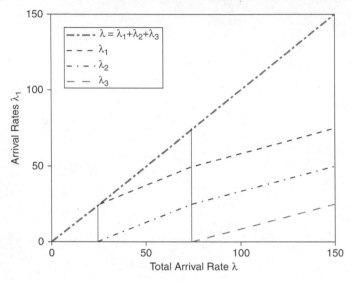

Figure 1.9 *Arrival Control to Parallel Queues: Explicit Individually Optimal Solution*

follows: for $k = 1, \ldots, n$, if $s_k \leq \lambda \leq s_{k+1}$, then

$$\lambda_j^e(\lambda) = \mu_j - [\sum_{i=1}^{k} \mu_i - \lambda]/k , \, j = 1, \ldots, k ,$$
$$= 0 , \, j = k + 1, \ldots, n .$$

Figure 1.9 illustrates the behavior of the individually optimal facility arrival rates as a function of the total arrival rate. Note that the positive $\lambda_j^e(\lambda)$ are piecewise linear in λ, with nonincreasing slope. The slopes of all positive $\lambda_j^e(\lambda)$ are equal in this case.

In Figure 1.10, the individually optimal allocation is superimposed on the socially optimal allocation, for purposes of comparison. As a general observation, we can say that the individually optimal allocation assigns more (fewer) customers to faster (slower) servers than the socially optimal allocation. More specifically, for the example in Figure 1.10, the individually optimal allocation always assigns more arrivals to facility 1, the fastest one, and fewer arrivals to facility 3, the slowest one, than the socially optimal allocation does. As λ increases, facility 2 first receives fewer, then more, arrivals in the individually optimal than in the socially optimal allocation. Thus, facility 2 plays the role of a "slower" server in light traffic and a "faster" server in heavy traffic.

1.6 Endnotes

Over the past forty years, there have been a number of survey papers and books that discuss optimal control of queues, including Sobel [181], Stid-

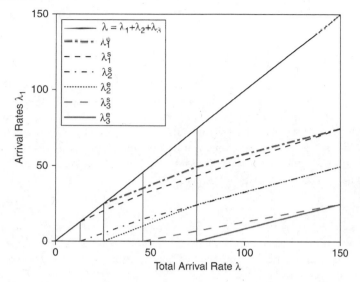

Figure 1.10 *Arrival Control to Parallel Queues: Comparison of Socially and Individually Optimal Solutions*

ham and Prabhu [191], Crabill, Gross, and Magazine [46], Serfozo [174], Stidham [184], [185], [186], Kitaev and Rykov [111], and Hassin and Haviv [86]. Optimal design is touched on in some of these references but, to the best of my knowledge, the present book is the first to provide a comprehensive treatment of optimal design of queues.

Section 1.1

The model and results in this section were introduced in a pioneering paper by Hillier [93]. Indeed, the emergence of optimization of queueing systems (both design and control) as a legitimate subject for research owes a great deal to Hillier and his PhD students in operations research at Stanford University, beginning in the mid 1960s.

Section 1.2

Edelson and Hildebrand [59] introduced the basic model of this section. They compared the socially optimal toll with the *facility optimal* toll, that is, the toll that maximizes revenue to the toll collector (e.g., the facility operator). They showed that the two are equal when all customers received the same reward, r, from joining and receiving service. When customers are heterogeneous – that is, the reward is a random variable, R – the facility optimal toll (arrival rate) is in general larger (smaller) than the socially optimal toll (arrival rate) (see Chapter 2).

Section 1.3

Surprisingly, I could find no published reference in which exactly this model is considered. The material in this section is taken largely from my class notes for a course on Optimization of Queueing Systems which I have taught, in various versions, since the early 1970s. Dewan and Mendelson [54] considered a model for combined choice of the arrival and service rate, but with heterogeneous rewards only. They examined only the solution to the necessary first-order optimality conditions, without considering the possibility that these conditions might not be sufficient. (In the examples they presented, the conditions were, fortuitously, always sufficient.) Stidham [187] considered essentially the same model as Dewan and Mendelson [54] and pointed out the possible failure of the objective function to be jointly concave and the resulting insufficiency of the first-order conditions.

Section 1.4

The model and results of this section come primarily from Stidham [189], which considered a more general model for a multiclass network of queues. Chapter 4 expands on the material in this section and Chapter 8 considers the extension to networks.

Section 1.5

This model was introduced in an unpublished paper (Stidham [182]) and then elaborated and extended in Bell and Stidham [18]. We return to the topic of parallel queues in Chapter 6.

CHAPTER 2

Optimal Arrival Rates in a Single-Class Queue

The model we study in this chapter is a generalization of the model introduced in Section 1.2 of the introductory chapter. We observed there that many of the salient features of the optimal arrival-rate model with deterministic reward and linear waiting cost do not depend on the system being an $M/M/1$ queue operating in steady state. For example, the individually optimal arrival rate λ^e is an upper bound on the socially optimal arrival rate λ^s for any queueing system satisfying the following conditions:

1. $W(\lambda)$ is strictly increasing in $0 \leq \lambda < \mu$;
2. $W(\lambda) \uparrow \infty$ as $\lambda \uparrow \mu$;
3. $W(0) = 1/\mu$.

Moreover, an individually optimizing customer who enters the system should be charged a toll equal to the external effect in order to render its behavior optimal for the system as a whole.

To what extent do properties like these continue to hold when one relaxes the assumptions that the system is operating in steady state and that all entering customers earn the same reward r and incur a waiting cost at the same constant rate h while in the system? We shall address these questions, and many others, in this chapter.

2.1 A Model with General Utility and Cost Functions

We consider a service facility operating over a finite or infinite time interval. At this stage, rather than specify a particular queueing model (we shall later consider specific examples), we prefer to describe the system in general terms, keeping structural and stochastic assumptions at a minimum.

The essential ingredients are:

- the arrival rate λ – the average number of customers entering the system per unit time during the period (the decision variable);

- the average (gross) utility per unit time, $U(\lambda)$, during the period;

- the average waiting cost per customer, $G(\lambda)$, during the period;

- the admission fee or toll, δ, paid by each entering customer.

The meaning of the word "average" depends on the specific model context. For example, it may mean a sample-path time average or (in the case of an infinite time period) the expectation of a steady-state random variable. These ingredients are now discussed in more detail.

The arrival rate λ measures the average number of customers arriving and joining the system per unit time during the period of interest. As indicated, λ is a decision variable. The set of feasible values for λ is denoted A. Our default assumption will be that $A = [0, \infty)$. Alterations to our model and results to allow for more general feasible sets are usually straightforward and will be left to the reader.

To capture the benefit of having a higher throughput, there is a utility function, $U(\lambda)$, which measures the average gross value received per unit time as a function of the arrival rate λ. For example, in the model considered in Section 1.2 with a deterministic reward, r, per entering customer, $U(\lambda) = r \cdot \lambda$, so that the utility function is linear. Our default assumption is that $U(\lambda)$ is nondecreasing, differentiable, and concave in $\lambda \geq 0$. We allow $U'(0) = \infty$. In Section 2.3 we show how a value function of this form can arise when there is a renewal process of potential arriving customers (with rate $\Lambda < \infty$) and a probabilistic joining rule is followed. We can accommodate a finite upper bound, Λ, on λ by defining $U(\lambda) = U(\Lambda)$ for $\lambda > \Lambda$. But this definition may not be compatible with the differentiability assumption, unless $U'(\Lambda) = 0$. Later (in Section 2.2) we shall examine the effects of relaxing some of the regularity conditions satisfied by $U(\lambda)$, including differentiability.

Balanced against the benefit of throughput is the cost to customers caused by the time they spend in the system. For a given λ, $G(\lambda)$ denotes the average waiting cost of a job, averaged over all customers who arrive during the period in question. Our default assumption is that $G(\lambda)$ takes values in $[0, \infty]$ and is strictly increasing and differentiable in $\lambda \geq 0$. We allow $G(\lambda)$ to equal ∞ in order to accommodate, for example, a single-server system with service rate μ, in which we typically have $G(\lambda) = \infty$ for $\lambda \geq \mu$. This convention makes it unnecessary to include the constraint $\lambda < \mu$ explicitly in our formulation.

For example, in the case of an infinite time period, it might be that

$$G(\lambda) = \mathrm{E}[h(\boldsymbol{W}(\lambda))] , \tag{2.1}$$

where $h(t)$ is the waiting cost incurred by a job that spends a length of time t in the system and, for each $\lambda \geq 0$, $\boldsymbol{W}(\lambda)$ is the steady-state random waiting time in the system for the queueing system induced by λ. In the case of a linear waiting cost, $h(t) = h \cdot t$,

$$G(\lambda) = h \cdot W(\lambda) ,$$

where $W(\lambda) := \mathrm{E}[\boldsymbol{W}(\lambda)]$. For the example of an $M/M/1$ queue operating in steady state considered in Section 1.2 of Chapter 1, we have

$$G(\lambda) = \frac{h}{\mu - \lambda} .$$

Let $H(\lambda) = \lambda G(\lambda)$. Then $H(\lambda)$ is a measure of the average waiting cost incurred by the system per unit time, inasmuch as it equals the product of the average number of customers arriving per unit time and the average waiting cost per customer.* We shall assume that $H(\lambda)$ is a convex function of $\lambda \geq 0$. (Note that the assumption that $G(\lambda)$ is strictly increasing and differentiable implies that $H(\lambda)$ is also strictly increasing and differentiable.)

All these properties are weak and common in the queueing literature. A sufficient condition for $H(\lambda)$ to be convex is that $G(\lambda)$ is convex, which is simply an assumption that each customer's marginal cost of waiting does not decrease as the arrival rate increases. As an illustration this property holds in our canonical example: an $M/M/1$ queue with linear waiting costs and *FIFO* queue discipline.

Remark 1 Customers might be sensitive to losses rather than (or in addition to) delays. (This situation can arise in a system with a finite buffer, in which an arriving customer who finds the buffer full is lost.) In this case, $G(\lambda)$ might measure the cost incurred if a customer is lost because of buffer overflow. In the special case in which $G(\lambda) = h \cdot P(\lambda)$, $P(\lambda)$ might measure the steady-state probability that a customer is lost (or the fraction of customers lost) and h the sensitivity of customers to such a loss. Although we shall continue to refer to "delay sensitivity" or "waiting costs" throughout the discussion of this model, the reader should keep in mind that the results also apply to other measures of congestion, such as losses.

In addition to incurring the waiting cost $G(\lambda)$, an entering customer may have to pay an admission fee (or toll) δ. In the present model, the sum of the toll and the waiting cost constitutes the *full price* of admission, which we denote in general by π, or $\pi(\lambda)$, when we want to emphasize its dependence on λ for a fixed δ. Thus we have

$$\pi(\lambda) = \delta + G(\lambda) .$$

Remark 2   The concept of the full price of admission is common to many models for arrival-rate selection, as we shall see. In more complicated systems, such as a set of parallel facilities (Chapter 6) or a network of queues (Chapters 7 and 8), the derivation of the full price is more complicated, as it may involve choices among alternate facilities or routes. But the analysis of the arrival-rate selection problem is basically the same as in the single-facility, single-class model considered in this chapter. Consequently we shall develop much of the theory for the present model in a general framework that will allow our results to be carried over to the more complicated models in subsequent chapters without unnecessary repetition. When we are operating in this general framework, we shall simply assume that $\pi = \pi(\lambda)$ is a given strictly increasing and differentiable function of λ.

As noted, the arrival rate λ is a decision variable. The solution to the

---

* In queueing terms, the relation – waiting cost per unit time = (arrival rate) × (waiting cost per customer) – is a special case of $H = \lambda G$, the generalization of $L = \lambda W$, which holds under weak assumptions. (See El-Taha and Stidham [60], Chapter 6. In the case of linear waiting cost, it just follows from $L = \lambda W$ itself.)

decision problem depends on who is making the decision. The decision may be made by the individual customers, each concerned only with its own net utility (*individual optimality*), or by a system operator, who might be interested in maximizing the aggregate net utility to all customers (*social optimality*) or in maximizing profit (*facility optimality*).

2.1.1 Individually Optimal (Equilibrium) Arrival Rate

We first consider the decision problem from the point of view of an arriving customer concerned only with its own net utility, which it wishes to maximize (individual optimality). Suppose we are given the full price of admission, $\pi(\lambda)$, as a function of $\lambda \geq 0$. We assume that $\pi(\cdot)$ is strictly increasing and differentiable.

For a particular value π of the full price of admission, an arriving customer concerned only with maximizing its own net utility will join if the value it receives from joining exceeds π, balk if it is lower, and be indifferent between joining and balking if it equals π. The marginal utility, $U'(\lambda)$, may be interpreted as the value received by the marginal user when the arrival rate is λ. (See below for more discussion and motivation of this interpretation.) At an individually optimal arrival rate, the marginal user will be indifferent between joining and balking, so that

$$U'(\lambda) = \pi , \tag{2.2}$$

if this equation has a solution in $A = [0, \infty)$. If $U'(0) < \pi$, then there is no solution to (2.2) in A; in this case no user has any incentive to join and we set $\lambda = 0$. If $U'(0) \geq \pi$, then since $U'(\lambda)$ is continuous and nonincreasing in λ, there is a solution to (2.2) in A. (We assume that $\lim_{\lambda \to \infty} U'(\lambda) < \pi$, in order to avoid trivialities.) Thus for a fixed price π an individually optimal arrival rate is characterized by the following equations:

$$U'(\lambda) \leq \pi , \text{ and } \lambda \geq 0 \tag{2.3}$$
$$U'(\lambda) = \pi , \text{ if } \lambda > 0 \tag{2.4}$$

Now if $\pi = \pi(\lambda)$ for a value of λ satisfying these conditions, then the system is in equilibrium: no individually optimizing customer acting unilaterally will have any incentive to deviate from its current action. In this case we have

$$U'(\lambda) \leq \pi(\lambda) , \text{ and } \lambda \geq 0 \tag{2.5}$$
$$U'(\lambda) = \pi(\lambda) , \text{ if } \lambda > 0 \tag{2.6}$$

These are the equilibrium conditions which uniquely define the individually optimal arrival rate, which we shall denote by λ^e. (To avoid trivialities we shall assume that $\lim_{\lambda \to \infty} U'(\lambda) < \lim_{\lambda \to \infty} \pi(\lambda)$. The equilibrium conditions then have a unique solution since $\pi(\lambda)$ is strictly increasing and continuous.)

Equivalent equilibrium conditions are the following:

$$\pi(\lambda) - U'(\lambda) \geq 0$$
$$\lambda \geq 0$$

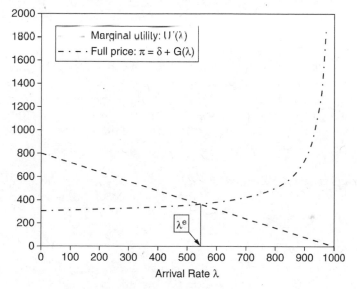

Figure 2.1 *Characterization of Equilibrium Arrival Rate*

$$\lambda(\pi(\lambda) - U'(\lambda)) = 0$$

Note that the equality constraint takes the form of a complementary-slackness condition. This form of the equilibrium conditions will facilitate comparison of individual optimization with social optimization.

For the present case, in which $\pi(\lambda) = \delta + G(\lambda)$, the equilibrium conditions (2.5) and (2.6) take the form

$$U'(\lambda) \leq \delta + G(\lambda) \text{ , and } \lambda \geq 0 \tag{2.7}$$
$$U'(\lambda) = \delta + G(\lambda) \text{ , if } \lambda > 0 \tag{2.8}$$

Figure 2.1 illustrates the case of a solution to the equilibrium condition (2.8).
  Equivalent equilibrium conditions are the following:

$$\delta + G(\lambda) - U'(\lambda) \geq 0$$
$$\lambda \geq 0$$
$$\lambda(\delta + G(\lambda) - U'(\lambda)) = 0$$

Note that if $U'(0) = \infty$, then the equilibrium conditions reduce to the single equation,

$$U'(\lambda) = \delta + G(\lambda) .$$

In this case the individually optimal arrival rate, λ^e, is the unique solution to this equation and $\lambda^e > 0$.

We have characterized a positive individually optimal arrival rate by equating the marginal utility to the full price of entering the system. Here is an informal motivation for this definition. (A formal justication in terms of the Nash equilibrium is contained in Section 2.3 on probabilistic joining rules.)

Suppose the current arrival rate is λ and one must decide whether to increase it to $\lambda + \Delta$. Think of this decision from the perspective of the increment in flow, Δ. An additional value per unit time, $U(\lambda + \Delta) - U(\lambda)$, will result from this increment in flow. On the other hand, the increment in flow will pay a price per unit time approximately equal to $\Delta \cdot \pi$. It will therefore be profitable for this increment in flow to add itself to the total flow if and only if

$$U(\lambda + \Delta) - U(\lambda) \geq \Delta \cdot \pi \,.$$

From this perspective, increments of flow will continue to add themselves to the total flow until an equilibrium, that is, a point of indifference, is reached, at which

$$U(\lambda + \Delta) - U(\lambda) = \Delta \cdot \pi \,.$$

Dividing both sides of this equality by Δ and letting $\Delta \to 0$ leads to (2.2).

Of course, this characterization depends on the assumption that the decision about increasing the flow is based on the benefits and costs to the increment in flow, without taking account of the effect of the increment on the costs incurred by the existing flow, λ. It is this effect – the *external effect* – that must be considered in order to find a value of the arrival rate that is optimal from the perspective of the total flow, that is, from the perspective of the collective of all customers. We turn our attention to this *socially optimal* flow in the next subsection.

2.1.2 Socially Optimal Arrival Rate

Now consider the decision problem of the system operator, who wishes to select an arrival rate that maximizes the average net benefit earned per unit time by the collective of all customers. We call such an arrival rate *socially optimal*. The problem is formulated as follows:

$$\max_{\{\lambda \geq 0\}} \ \mathcal{U}(\lambda) := U(\lambda) - \lambda G(\lambda) \tag{2.9}$$

(Since the toll is simply a transfer fee, it does not appear in the objective function for social optimality.)

The first-order necessary condition for an interior maximum is:

$$U'(\lambda) = G(\lambda) + \lambda G'(\lambda) \tag{2.10}$$

Let λ^s denote the optimal arrival rate for this problem. The assumed concavity of $U(\lambda)$ and convexity of $H(\lambda) = \lambda G(\lambda)$ imply that the maximum net benefit occurs at $\lambda^s = 0$, if $U'(0) < G(0)$. Otherwise, λ^s is the solution to the first-order necessary condition (2.10) and $\lambda^s \geq 0$.

Now suppose that the facility operator wishes to implement the socially optimal arrival rate by charging a toll and allowing the arriving customers, who are individual optimizers, to decide whether or not to enter the system.

In this case it follows from (2.7) and (2.10) that

$$\delta^s = \lambda^o G''(\lambda^s) \, . \tag{2.11}$$

That is, the optimal toll equals the external effect. If the system operator charges entering customers the toll δ^s, then $\lambda^e = \lambda^s$: the individually optimal arrival rate is also socially optimal.

Note that, by requiring that $G(0) < U'(0)$, one can guarantee the existence of a (unique) solution to (2.10) in $A = [0, \infty)$, so that the constraint $\lambda \geq 0$ can effectively be ignored. In particular, this is the case if $U'(0) = \infty$.

2.1.3 Facility Optimal Arrival Rate

Now consider the system from the point of view of a facility operator whose goal is to set a toll δ that will maximize its revenue. We call such a toll (and the associated arrival rate) *facility optimal*. Assuming that the arriving customers are individual optimizers, choosing a value δ for the toll will result in an arrival rate λ that (uniquely) satisfies the equilibrium conditions,

$$\begin{aligned} U'(\lambda) &\leq \delta + G(\lambda) \, , \text{ and } \lambda \geq 0 \\ U'(\lambda) &= \delta + G(\lambda) \, , \text{ if } \lambda > 0 \, , \end{aligned}$$

or equivalently,

$$\begin{aligned} \lambda(\delta + G(\lambda) - U'(\lambda)) &= 0 \\ \delta + G(\lambda) - U'(\lambda) &\geq 0 \\ \lambda &\geq 0 \end{aligned}$$

Thus, the facility optimization problem may be written in the following form:

$$\begin{aligned} \max_{\{\delta, \lambda\}} \quad & \lambda\delta \\ \text{s.t.} \quad & \lambda(\delta + G(\lambda) - U'(\lambda)) = 0 \\ & \delta + G(\lambda) - U'(\lambda) \geq 0 \\ & \lambda \geq 0 \end{aligned}$$

Subtracting the term $\lambda(\delta + G(\lambda) - U'(\lambda))$ (which equals zero by the first constraint) from the objective function and simplifying leads to the following equivalent formulation:

$$\begin{aligned} \max_{\{\delta, \lambda\}} \quad & \lambda U'(\lambda) - \lambda G(\lambda) \\ \text{s.t.} \quad & \delta \geq U'(\lambda) - G(\lambda) \\ & \delta = U'(\lambda) - G(\lambda) \, , \text{ if } \lambda > 0 \\ & \lambda \geq 0 \end{aligned}$$

To avoid technical difficulties we shall assume for now that $\lim_{\lambda \to 0} \lambda U'(\lambda) = 0$. (See below for further discussion of this point.)

The first two constraints now serve simply to define δ (nonuniquely), given

λ. Therefore, it suffices to solve the following problem with λ as the only decision variable:

$$\max_{\{\lambda \geq 0\}} \tilde{\mathcal{U}}(\lambda) := \lambda U'(\lambda) - \lambda G(\lambda) ,$$

and then choose δ to satisfy

$$\delta \geq U'(\lambda) - G(\lambda)$$
$$\delta = U'(\lambda) - G(\lambda) , \text{ if } \lambda > 0 .$$

Now if $G(0) \geq U'(0)$, then $G(\lambda) > U'(\lambda)$, for all $\lambda > 0$. In this case the facility-optimal arrival rate, λ^f, equals zero (along with the individually optimal arrival rate, λ^e, and the socially optimal arrival rate, λ^s). On the other hand, if $G(0) < U'(0)$, then the first-order necessary condition for a positive value of λ to be optimal for this problem is

$$U'(\lambda) + \lambda U''(\lambda) - G(\lambda) - \lambda G'(\lambda)) = 0 . \tag{2.12}$$

Note that (2.12) may not be sufficient for $\lambda > 0$ to be optimal, since the objective function may not be concave (or even unimodal), because $\lambda U'(\lambda)$ may not be concave.

Remark 3 Note that the objective function for facility optimization, $\tilde{\mathcal{U}}(\lambda) = \lambda U'(\lambda) - \lambda G(\lambda)$, takes the same form as the objective function for social optimization, but with a modified utility function, $\tilde{U}(\lambda) := \lambda U'(\lambda)$. This observation suggests that we can directly apply the results from our analysis of social optimization to the facility optimization problem (at least under the technical assumption that $\lim_{\lambda \to 0} \lambda U'(\lambda) = 0$). A crucial difference, however, is that the modified utility function, $\tilde{U}(\lambda)$, need not be concave (as we just observed). Indeed, it need not even be nondecreasing (see Section 2.1.3.5). So, to treat the facility optimal problem as a special case of the socially optimal problem, we would first have to extend the formulation and analysis of the latter to allow for utility functions that fail to be concave and nondecreasing. Instead of doing this, however, we prefer to deal directly with the facility optimization problem.

Recall that we allow $U'(0) = \infty$. In this case, $\lambda U'(\lambda)$ is undefined at $\lambda = 0$. However, the only reasonable value for the objective function, $\tilde{\mathcal{U}}(\lambda) = \lambda U'(\lambda) - \lambda G(\lambda)$, to assume at $\lambda = 0$ is zero, since $\lambda = 0$ corresponds to a decision on the part of the facility operator not to operate the facility at all. Indeed, it was to ensure the continuity of the objective function at $\lambda = 0$ that we made the technical assumption above that $\lim_{\lambda \to 0} \lambda U'(\lambda) = 0$. Other limits are possible, however, including $\lim_{\lambda \to 0} \lambda U'(\lambda) = \kappa$, $0 < \kappa \leq \infty$. (We shall consider this situation in detail when we return to the topic of facility optimality in Section 2.3.3.)

Let us briefly consider each of these possibilities in turn.

Case 1. Suppose $\lambda U'(\lambda) \to 0$ as $\lambda \to 0$. Then the value of λ that maximizes $\tilde{\mathcal{U}}(\lambda)$ may be $\lambda = 0$. This will be the case if and only if $\tilde{\mathcal{U}}(\lambda) \leq 0 = \tilde{\mathcal{U}}(0)$ for all $\lambda > 0$ (which is true if and only if $U'(0) \leq G(0)$). If this is not the case, then the maximum will occur at a positive value of λ which must satisfy the first-order necessary condition, (2.12). As noted above, if $\lambda U'(\lambda)$ is not concave, then this equation may have multiple solutions, some of which are local maxima or minima, and only one of which can be the global maximum.

Case 2. Suppose $\lambda U'(\lambda) \to \kappa$ as $\lambda \to 0$, where $0 < \kappa < \infty$. In this case, $\tilde{\mathcal{U}}(\lambda)$ has a discontinuity at $\lambda = 0$, since (by convention) $\tilde{\mathcal{U}}(0) = 0$, whereas $\tilde{\mathcal{U}}(0+) = \kappa > 0$. Now the maximum can no longer occur at $\lambda = 0$, but it may occur at $\lambda = 0+$. More precisely, if $\tilde{\mathcal{U}}(\lambda) < \kappa$ for all $\lambda > 0$, then $\sup_{\lambda>0} \tilde{\mathcal{U}}(\lambda) = \kappa$, but this supremum is not attained. Instead, the facility operator can attain a profit arbitrarily close to κ by choosing an arbitrarily small positive arrival rate λ. We shall use the notation, $\lambda^f = 0+$, as a shorthand for this property. On the other hand, if there exists a positive value of λ such that $\tilde{\mathcal{U}}(\lambda) \geq \kappa$, then (as in Case 1) the (positive) maximizing value of λ is a solution to the first-order necessary condition, (2.12).

Case 3. Suppose $\lim_{\lambda \to 0} \lambda U'(\lambda) = \infty$. In this case the facility optimization problem has an unbounded objective function, $\tilde{\mathcal{U}}(\lambda)$, which approaches ∞ as λ approaches zero. In other words a profit-maximizing facility operator can earn an arbitrarily large profit by charging an arbitrarily large toll, resulting in an arbitrarily small arrival rate. So again we have $\lambda^f = 0+$. Note that, by contrast, the individually and socially optimal arrival rates, λ^e and λ^s, still exist and are positive and the associated values of the objective function are still finite. In Section 2.3.3 we analyze facility optimization in the context of probabilistic joining rules, and there we are able to give a behavioral interpretation of the property, $\lim_{\lambda \to 0} \lambda U'(\lambda) = \infty$.

2.1.3.1 Comparison of Facility Optimal and Socially Optimal Arrival Rates

What is the relationship between λ^f and the socially optimal arrival rate, λ^s? The following theorem shows that $\lambda^f \leq \lambda^s$. In other words, a facility operator concerned only with maximizing the revenue received from tolls will choose a toll that results in fewer customers joining the system than is optimal from the point of view of the total welfare of all customers. This result is a direct consequence of the concavity of $U(\lambda)$ and is consistent with classical results from welfare economics.

The proof of this result depends on the following lemma, which is of independent interest. (In this section we shall again assume that $\lim_{\lambda \to \infty} \lambda U'(\lambda) = 0$, unless otherwise noted.)

Lemma 2.1 *Let $d(\lambda) := U(\lambda) - \lambda U'(\lambda)$, $\lambda \geq 0$. The function $d(\lambda)$ is nondecreasing in $\lambda \geq 0$ if (and only if) $U(\lambda)$ is concave in $\lambda \geq 0$.*

Proof Differentiating $d(\lambda)$ yields

$$
\begin{aligned}
d'(\lambda) &= U'(\lambda) - U'(\lambda) - \lambda U''(\lambda) \\
&= -\lambda U''(\lambda) .
\end{aligned}
$$

Since $\lambda \geq 0$, we conclude that $d'(\lambda) \geq 0$ if and only if $U''(\lambda) \leq 0$. ∎

Theorem 2.2 *The facility optimal arrival rate is bounded above by the socially optimal arrival rate: $\lambda^f \leq \lambda^s$. If $U'(0) - G(0) \leq 0$, then $\lambda^f = \lambda^s = 0$. If $U'(0) - G(0) > 0$, then $0 < \lambda^f \leq \lambda^s$. The inequality is strict if $U(\lambda)$ is strictly concave.*

Proof   First consider the case $U'(0) - G(0) \leq 0$. In this case $\lambda^s = 0$ (see Section 2.1.2). Let $\psi(\lambda) := U'(\lambda) - G(\lambda) - \lambda G'(\lambda)$ and note that $\psi(0) = U'(0) - G(0) \leq 0$. But $\psi(\lambda)$ is nonincreasing in λ and $\lambda U''(\lambda) \leq 0$ for all $\lambda \geq 0$. Hence

$$\tilde{\mathcal{U}}'(\lambda) = U'(\lambda) - G(\lambda) - \lambda G'(\lambda) + \lambda U''(\lambda) = \psi(\lambda) + \lambda U''(\lambda) \leq 0 \,,$$

for all $\lambda \geq 0$. It follows that $\tilde{\mathcal{U}}(\lambda)$ is nonincreasing in $\lambda \geq 0$ and hence is maximized by $\lambda = 0$. In other words, $\lambda^f = 0$.

Now consider the case $U'(0) - G(0) > 0$. In this case $\lambda^s > 0$ and is the unique solution to $\psi(\lambda) = 0$. But λ^f is also positive (since $\lambda^f = 0$ only if $U'(0) - G(0) \leq 0$) and therefore is a solution to

$$\psi(\lambda) = -\lambda U''(\lambda) \geq 0 \,.$$

(There may, of course, be multiple such solutions.) Since $\psi(\lambda)$ is nonincreasing and $\psi(\lambda^s) = 0$, we conclude that $0 < \lambda^f \leq \lambda^s$. If $U(\lambda)$ is strictly concave, then $-\lambda U''(\lambda) > 0$ for all $\lambda > 0$, which implies that $\lambda^f < \lambda^s$. ∎

Remark 4   Let $f(1, \lambda) := U(\lambda)$, $f(2, \lambda) := \lambda U'(\lambda)$. That is, $f(1, \lambda)$ and $f(2, \lambda)$ are the (gross) utility functions for social optimization and facility optimization, respectively. Then $d(\lambda) = U(\lambda) - \lambda U'(\lambda) = f(1, \lambda) - f(2, \lambda)$ and the condition (in Lemma 2.1) that $d(\lambda)$ is nondecreasing is equivalent to *submodularity* of the function $f(i, \lambda)$ in (i, λ). It then follows from basic properties of submodularity (see, e.g., Topkis [193], Heyman and Sobel [92]) that the maximizing value of λ is a nonincreasing function of i; in other words, the socially optimal arrival rate is larger than the facility optimal arrival rate. This property, which (as we have already indicated) is a familiar result in welfare economics, is thus seen (via Lemma 2.1) to be a basic and immediate consequence of the concavity of $U(\cdot)$.

We now give conditions and examples of utility functions $U(\lambda)$ for which the function $\lambda U'(\lambda)$ is concave. The condition for concavity is that the second derivative be nonpositive, that is,

$$\frac{d^2}{d\lambda^2} \lambda U'(\lambda) = 2U''(\lambda) + \lambda U'''(\lambda) \leq 0 \,, \text{ for all } \lambda \in [0, \Lambda] \,. \tag{2.13}$$

Let us now consider some examples.

2.1.3.2 Example 1: Logarithmic Utility Function

Suppose $U(\lambda) = a \ln(1 + b\lambda)$, where $a > 0$, $b > 0$. Then $U(0) = 0$ and for $\lambda > 0$ we have

$$U'(\lambda) \;\; = \;\; ab(1 + b\lambda)^{-1} > 0 \,,$$

$$U''(\lambda) = -ab^2(1+b\lambda)^{-2} < 0,$$
$$U'''(\lambda) - 2ab^3(1+b\lambda)^{-3} > 0.$$

Thus $U(\lambda)$ is nonnegative, strictly increasing, and strictly concave, and hence satisfies our general requirements for a utility function. Moreover,

$$\frac{d^2}{d\lambda^2}\lambda U'(\lambda) = -2ab^2(1+b\lambda)^{-2} + 2ab^3\lambda(1+b\lambda)^{-3}$$
$$= -2ab^2(1+b\lambda)^{-3} < 0,$$

so that condition (2.13) is satisfied and hence $\lambda U'(\lambda)$ is always (strictly) concave for this example.

2.1.3.3 Example 2: Power Utility Function

Suppose $U(\lambda) = a\lambda^b$, where $a > 0$, $0 < b < 1$. Then $U(0) = 0$ and for $\lambda > 0$ we have

$$U'(\lambda) = ab\lambda^{b-1} > 0,$$
$$U''(\lambda) = ab(b-1)\lambda^{b-2} < 0,$$
$$U'''(\lambda) = ab(b-1)(b-2)\lambda^{b-3} > 0.$$

Thus $U(\lambda)$ is nonnegative, strictly increasing, and strictly concave, and hence satisfies our general requirements for a utility function. Moreover,

$$\frac{d^2}{d\lambda^2}\lambda U'(\lambda) = 2ab(b-1)\lambda^{b-2} + \lambda ab(b-1)(b-2)\lambda^{b-3}$$
$$= ab(b-1)\lambda^{b-2}(2+b-2)$$
$$= ab^2(b-1)\lambda^{b-2} < 0,$$

so that condition (2.13) is satisfied and hence $\lambda U'(\lambda)$ is always (strictly) concave for this example. Note that for this example we have $U'(0) = \infty$, but $\lim_{\lambda\to0}\lambda U'(\lambda) = 0$.

2.1.3.4 Example 3: Exponential Utility Function

Suppose $U(\lambda) = a(1 - e^{-b\lambda})$, where $a > 0$, $b > 0$. Then $U(0) = 0$ and for $\lambda > 0$ we have

$$U'(\lambda) = abe^{-b\lambda} > 0,$$
$$U''(\lambda) = -ab^2e^{-b\lambda} < 0,$$
$$U'''(\lambda) = ab^3e^{-b\lambda} > 0.$$

Thus $U(\lambda)$ is nonnegative, strictly increasing, and strictly concave, and hence satisfies our general requirements for a utility function. In this case,

$$\frac{d^2}{d\lambda^2}\lambda U'(\lambda) = -2ab^2e^{-b\lambda} + ab^3\lambda e^{-b\lambda}$$
$$= ab^2e^{-b\lambda}(-2 + b\lambda),$$

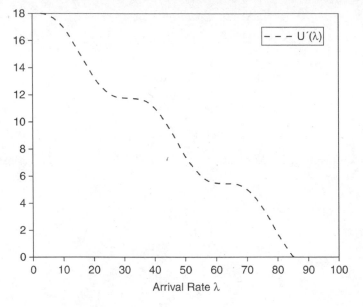

Figure 2.2 *Graph of the Function $U'(\lambda)$*

which is negative for $\lambda < 2b^{-1}$ and positive for $\lambda > 2b^{-1}$. Thus $\lambda U'(\lambda)$ is (strictly) concave over the interval $[0, 2b^{-1}]$ and strictly convex over the interval $[2b^{-1}, \infty)$. If the feasible set of values for λ is $[0, \Lambda]$ with $\Lambda \le 2b^{-1}$, then condition (2.13) is satisfied; otherwise, not. In particular, if $\Lambda = \infty$, that is, if λ has no finite upper bound, then condition (2.13) is not satisfied and $\lambda U'(\lambda)$ is not concave in $\lambda \in [0, \Lambda]$.

2.1.3.5 Example with Multiple Local Maxima

When the utility function $U(\lambda)$ fails to satisfy the condition that $\lambda U'(\lambda)$ be concave, the objective function for facility optimization may also fail to be concave. As a result, the first-order optimality conditions may not have a unique solution and there may be several local maxima.

To indicate what can happen qualitatively, suppose $U'(\lambda)$ takes the form of a "smoothed step function," as illustrated in Figure 2.2. In this case, Figure 2.3 illustrates the form taken by the function $\lambda U'(\lambda)$. As a result, the objective function, $\lambda U'(\lambda) - \lambda G(\lambda)$, for the profit-maximizing facility operator has a graph that looks like Figure 2.4. In this example, the objective function has two local maxima and one local minima, each of which is a solution to the necessary condition for optimality,

$$U'(\lambda) + \lambda U''(\lambda) - G(\lambda) - \lambda G'(\lambda)) = 0 \ .$$

Note that in this case $\lambda U'(\lambda)$ not only is not concave, but fails to be nondecreasing.

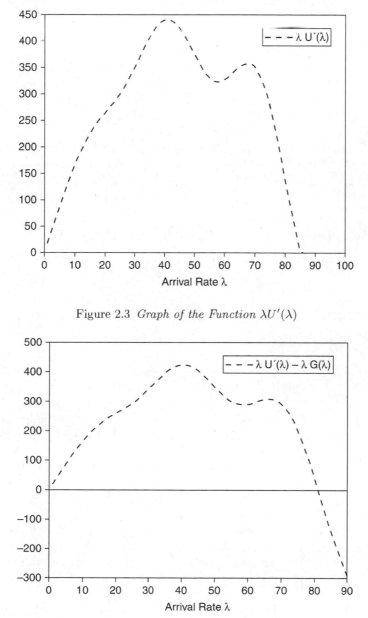

Figure 2.3 *Graph of the Function* $\lambda U'(\lambda)$

Figure 2.4 *Graph of the Objective Function:* $\lambda U'(\lambda) - \lambda G(\lambda)$

As we shall see, such a situation is particularly likely to occur when the arriving customers tend to be heterogeneous with respect to the value they place on being served. The extreme case is when they belong to a finite number of classes, with customers in the same class all sharing the same value and the values of different classes being distinct. In this case the smoothed step function, $U'(\lambda)$, pictured in Figure 2.2 is replaced by an actual step function. Since this form for $U'(\cdot)$ violates our assumption that $U(\cdot)$ is continuously differentiable, we do not yet have the machinery to deal with this model. We shall therefore pick up this thread again after we have generalized our model to allow for a discontinuous $U'(\cdot)$ (in the next section) and introduced a behavioral model based on probabilistic joining rules (in Section 2.3), into which context the multiclass model fits more naturally.

2.2 Generalizations of Basic Model

In this section we examine the effects of relaxing some of the assumptions of the basic model considered in the last section.

2.2.1 Nondifferentiable Utility Function

Suppose the utility function $U(\lambda)$ is concave but not differentiable. From the concavity of $U(\lambda)$ it follows that the derivative $U'(\lambda)$ exists a.e. on $[0, \infty)$. Moreover, $U(\lambda)$ has a right-hand and a left-hand derivative at all points $\lambda \in [0, \infty)$. We can extend the definition of $U'(\lambda)$ to all $\lambda \in [0, \infty)$ by letting $U'(\lambda)$ denote the right-hand derivative at λ. Since $U(\lambda)$ is concave, $U'(\lambda)$ is a nonincreasing function of λ which is *RCLL* (*Right-Continuous with Left Limits*). The left-hand derivative of $U(\lambda)$ is the left-limit of $U'(\lambda)$, denoted $U'(\lambda-)$ and we have

$$U'(\lambda-) \geq U'(\lambda) \,, \ \lambda \in [0, \infty) \,,$$

with equality at all points where $U(\lambda)$ is differentiable.

Figure 2.5 illustrates the behavior of the function $U'(\lambda)$ for the extreme case in which $U'(\lambda)$ is piecewise constant (i.e., a step function). As we shall see (cf. Section 2.3.4 below), this case corresponds to the situation when the arriving customers belong to a finite number of classes, with customers in the same class all sharing the same value and the values of different classes being distinct.

Now consider how the definition of the individually optimal arrival rate changes in this generalized setting. Again, our intuition is based upon interpreting $U'(\lambda)$ as the utility received by the marginal user when the arrival rate is λ. Suppose the full price equals π. As long as this marginal utility is greater than π, there is an incentive to increase λ. Thus, in the generalized setting, the individually optimal arrival rate associated with this particular full price π is characterized by the following condition:

$$U'(\lambda-) \geq \pi \geq U'(\lambda) \,. \tag{2.14}$$

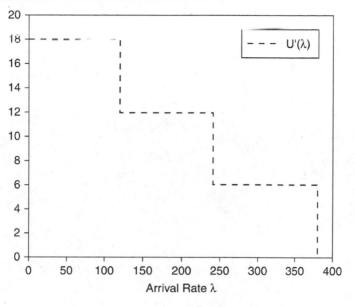

Figure 2.5 *Graph of the Function $U'(\lambda)$*

(Note that the case where the individually optimal rate equals zero is accommodated by defining $U'(0-) := \infty$.) If $\pi = \pi(\lambda)$ for a value of λ satisfying these conditions, then the system is in equilibrium: no individually optimizing customer acting unilaterally will have any incentive to deviate from its current action. In this case we have

$$U'(\lambda-) \geq \pi(\lambda) \geq U'(\lambda) . \tag{2.15}$$

This is the equilibrium condition which uniquely defines the individually optimal arrival rate, λ^e. The equilibrium conditions have a unique solution since $\pi(\lambda)$ is strictly increasing and continuous. (To avoid trivialities we again assume that $\lim_{\lambda\to\infty} U'(\lambda) < \lim_{\lambda\to\infty} \pi(\lambda)$.) Figures 2.6 and 2.7 illustrate the possibilities.

2.2.2 Upper Bound on Arrival Rate

Now suppose the feasible set for λ takes the form $A = [0, \Lambda]$, where $\Lambda < \infty$. As we suggested in a comment following the definition of the utility function $U(\cdot)$, we can accommodate a finite upper bound on λ by defining $U(\lambda) = U(\Lambda)$ (equivalently, $U'(\lambda) = 0$) for $\lambda \geq \Lambda$, at the possible cost of violating the assumption that $U(\cdot)$ is differentiable. Now that we have a mechanism for dealing with a nondifferentiable $U(\cdot)$, we can use this approach without additional conditions. In particular, we see that the system is in equilibrium

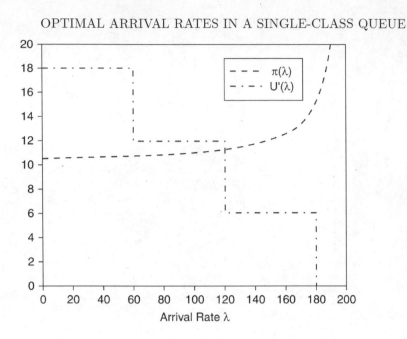

Figure 2.6 *Equilibrium Arrival Rate. Case 1:* $U'(\lambda-) > \pi(\lambda) > U'(\lambda)$

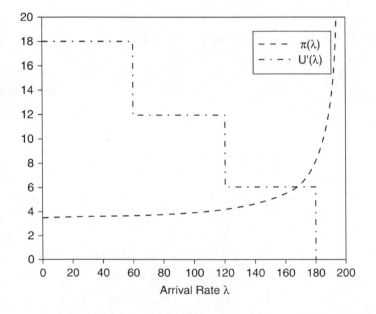

Figure 2.7 *Equilibrium Arrival Rate. Case 2:* $U'(\lambda-) = \pi(\lambda) = U'(\lambda)$

with $\lambda^e = \Lambda$ if and only if

$$\bar{U}'(\Lambda-) \geq \pi(\Lambda) \geq U'(\Lambda) = 0 . \qquad (2.16)$$

2.3 *GI/GI/1* Queue with Probabilistic Joining Rule

In this section we consider a special case of our general model, in which we are able to give a precise behavioral interpretation of the individually optimal allocation of flow. This interpretation is based on the assumption that we have individually optimizing customers with heterogeneous utilities (random rewards). The customers behave like players in a noncooperative game, and we show that the individually optimal arrival rate (as defined in Section 2.1.1) is in fact the arrival rate corresponding to the (unique) Nash equilibrium of this game.

We first provide an informal description of the behavior of the customers. Later, we shall give a more formal development.

We assume that there is a stream of *potential* arriving customers with a fixed rate Λ. Each potential customer may either join the system or balk. The customers who join the system constitute the *actual* arriving customers (or more simply the *arrivals*). Each potential arriving customer has an associated reward which it will receive if it joins the system and receives service.

Customers are self-optimizing. They cannot observe the congestion in the system, however, before deciding whether or not to join. Hence, they do not know the actual full price of admission before entering. We assume that customers form an estimate or prediction, $\hat{\pi}$, of the admission price, π, and base their join/balk decision on this predicted price. (We shall discuss possible mechanisms for forming this prediction presently.) A potential arriving customer, seeking to maximize its net benefit, has an incentive to join the system if its reward exceeds the predicted price, $\hat{\pi}$. If its reward is less than the predicted price, then the customer has an incentive to balk. If the reward equals the predicted price, then the customer will be indifferent between joining and balking. The resulting actual arrival (or entrance) rate λ will then be the product of Λ and the proportion of potential arrivals that enter the system.

A predicted price $\hat{\pi}$ is *individually optimal* (a Nash equilibrium) if no customer has an incentive to deviate unilaterally from the join/balk decision based on $\hat{\pi}$. This will be the case if and only if $\hat{\pi} = \pi = \pi(\lambda)$, that is, the predicted price equals the actual price.

Here is a formal description of the system. Consider a single-server queueing facility. Customers are served according to a *FIFO* queue discipline, with i.i.d. service times distributed as a generic random variable $\boldsymbol{S}$. The service rate is $\mu := (\mathrm{E}[\boldsymbol{S}])^{-1}$. Potential arrivals occur according to a renewal process with rate Λ. Suppose that each potential arriving customer places a certain value on entering the system and that the values of successive customers are independent and identically distributed random variables with distribution function, $F(r) = P\{\boldsymbol{R} \leq r\}$, $r \geq 0$. Thus the generic random variable $\boldsymbol{R}$ represents the value (or reward) associated with an arbitrary potential arriving customer. This reward becomes known (realized) at the instant of the customer's potential arrival, but not before.

A customer with realized reward r may either join the system (and receive r) or balk (and receive nothing). We assume that a customer with reward r (called a *type-r customer*) joins the system with probability $p(r)$, where $p(r), r \geq 0$, is a measurable function. These joining probabilities are decision variables. The decisions may be made by the customers themselves (individual optimality: see Section 2.3.1), by an agent operating on behalf of the collective of all customers (social optimality: see Section 2.3.2), or by the operator of the system (facility optimality: see Section 2.3.3). In each case the objective function is the appropriate one for the decision maker in question, as will become clear below.

Given a joining rule, $\{p(r), r \geq 0\}$, the unconditional probability that a potential arriving customer enters is $\int_0^\infty p(r)dF(r)$. The result is that customers enter the system according to a renewal process with mean rate

$$\lambda = \Lambda \int_0^\infty p(r)dF(r) . \tag{2.17}$$

We call λ the actual arrival rate or simply the arrival rate. The gross value or utility earned per unit time is given by

$$U = \Lambda \int_0^\infty rp(r)dF(r) .$$

Balanced against this value received is the (full) price of admission, π, which customers must pay for use of the system. In the models considered in this chapter, the full price of admission is the sum of the admission fee and the waiting cost, which is a function of λ. For the time being, however, we shall develop our results in the context of a general formulation, in which π is an arbitrary function of λ:

$$\pi = \pi(\lambda) .$$

(Unless otherwise noted, we assume that $\pi(\lambda)$ is continuous and strictly increasing in $\lambda \geq 0$.) This will facilitate our use of the results of this section in the analysis of more complicated models which we shall consider in later chapters. Later we shall return to the special case, $\pi = \delta + G(\lambda)$ and develop this case in more detail.

Among probabilistic joining rules, $p(r), r \geq 0$, we shall be particularly interested in *probabilistic reward-threshold* rules. A rule of this type is characterized by a threshold value x and a probability p such that a type-r customer will enter:

$$
\begin{aligned}
\text{with probability } p(r) &= 0 \text{ if } r < x ; \\
\text{with probability } p(r) &= p \text{ if } r = x ; \\
\text{with probability } p(r) &= 1 \text{ if } r > x .
\end{aligned}
$$

The joining rate associated with the rule (x, p) is given by

$$\lambda = \Lambda[\bar{F}(x) + p(\bar{F}(x-) - \bar{F}(x))] , \tag{2.18}$$

where $\bar{F}(x) := 1 - F(x)$. Note that distinct joining rules can have the same associated joining rate. For example, if $\bar{F}(x) = \bar{F}(x + h)$, the joining rules $(x, 0)$ and $(x + h, 0)$ both have associated joining rate λ, where

$$\lambda = \Lambda[\bar{F}(x)] = \Lambda[\bar{F}(x + h)] \;.$$

Similarly, if $\bar{F}(x-) = \bar{F}(x)$ (that is, if $\bar{F}$ is continuous at x), then all joining rules of the form (x, p) have the same joining rate, $\lambda = \Lambda \bar{F}(x)$. In both examples, the point is that changing the joining probability $p(r)$ for any or all $r \in B$, where B is a set of F-measure zero, has no effect on the arrival rate.

2.3.1 Individual Optimality

We first consider the system from the point of view of individually optimizing customers. That is, we assume that each arriving customer makes a decision whether or not to enter and that the decision is based only on the rewards and costs that accrue to that customer if it joins. In the language of game theory, the customers are *players* and the decision of each player takes the form of a *randomized strategy*, that is, a probability that the customer will join the system. We assume that all customers have the same information about the parameters of the system. It follows that all type-r customers will follow the same strategy; that is, each will choose the same probability, $p(r)$, of entering the system. Since there may be an uncountably infinite number of types of customers (one for each $r \in [0, \infty)$), we have an example of what is sometimes called a *nonatomic* game.

The resulting set of strategies, $\{p(r), r \geq 0\}$, therefore constitutes a probabilistic joining rule as defined above, with associated arrival rate λ given by (2.17). The full price paid for entrance, $\pi(\lambda)$, is a given continuous and strictly increasing function of λ. The *payoff* (expected net benefit) to a type-r customer who uses the randomized strategy $p(r)$ is therefore

$$p(r)(r - \pi(\lambda)) \;.$$

Borrowing again from the language of game theory, a joining rule (set of strategies) is said to be a *Nash equilibrium* if no player has any incentive to deviate from its strategy, assuming that all other players continue to follow their strategies. Thus, a joining rule, $\{p(r), r \geq 0\}$, will be a Nash equilibrium if and only if no customer of any type r can increase its payoff by choosing an entrance probability different from $p(r)$, assuming that the arrival rate remains equal to λ.

Now, since all customers have the same information about the system, they will all form the same estimate or prediction, denoted $\hat{\pi}$, of the price they will pay if they enter the system. Since the goal of a type-r customer is to maximize its expected net benefit, the joining rule based on $\hat{\pi}$ will be a probabilistic reward-threshold rule, (x, p), with $x = \hat{\pi}$ and $0 \leq p \leq 1$. That is, a type-r customer will join

i. with probability $p(r) = 0$, if $r < \hat{\pi}$;

ii. with probability $p(r) = p$, if $r = \hat{\pi}$;

iii. with probability $p(r) = 1$, if $r > \hat{\pi}$.

The arrival rate associated with this joining rule is

$$\lambda = \Lambda[\bar{F}(\hat{\pi}) + p(\bar{F}(\hat{\pi}-) - \bar{F}(\hat{\pi}))] \ . \qquad (2.19)$$

It follows that a joining rule will be a Nash equilibrium if and only if it is a probabilistic reward-threshold rule, $(\hat{\pi}, p)$, and the prediction of the entrance price coincides with the actual price, that is, $\hat{\pi} = \pi(\lambda)$, where λ is given by (2.19). To see this, consider first the case of a type-r customer with $r > \hat{\pi} = \pi(\lambda)$. The proposed strategy for such a customer is $p(r) = 1$, with expected net benefit equal to $r - \pi(\lambda) > 0$. If this customer were to deviate from this strategy by choosing a probability $0 \leq p < 1$ of joining, it would receive an expected net benefit equal to

$$p(r - \pi(\lambda)) < r - \pi(\lambda) \ ,$$

that is, a strictly smaller expected net benefit. Hence, such a customer has no incentive to deviate from its current strategy. Conversely, if such a customer were following a strategy $p(r)$ with $p(r) < 1$, it would find that it could increase its expected net benefit by joining with probability one. A similar argument shows that a type-r customer with $r < \hat{\pi} = \pi(\lambda)$ would also receive a strictly smaller expected net benefit by deviating from the proposed strategy, namely, $p(r) = 0$. Now consider a type-r customer with expected net benefit equal to $r - \pi(\lambda) = 0$. Such a customer currently joins with probability p, $0 \leq p \leq 1$. Deviating from this strategy would not change the expected net benefit: it would still be zero. Therefore, such a customer also has no incentive to deviate from its current strategy.[†]

These results are summarized in the following theorem.

Theorem 2.3 *A joining rule, $\{p(r), r \geq 0\}$, is a Nash equilibrium if and only if it is a probabilistic reward-threshold rule, $(\hat{\pi}, p)$, with*

$$\begin{aligned} \lambda &= \Lambda[\bar{F}(\hat{\pi}) + p(\bar{F}(\hat{\pi}-) - \bar{F}(\hat{\pi}))] \\ \hat{\pi} &= \pi(\lambda) \ . \end{aligned}$$

An arrival rate λ satisfying these equations is called an *individually optimal* (or *equilibrium*) arrival rate.

To relate these results to those derived in Section 2.1.1, we first define the inverse of the complementary cumulative distribution function $\bar{F}$ as follows:

$$\bar{F}^{-1}(y) := \sup\{x : \bar{F}(x) > y\} \ , \ 0 \leq y \leq 1 \ .$$

Note that $\bar{F}^{-1}(y)$ is a nonincreasing, right-continuous function of $y \in [0, 1]$, which takes values in $[0, \infty]$. The following lemma is immediate.

[†] It also has no incentive *not* to deviate. For this reason, one might call such a joining rule a *weak* Nash equilibrium.

Lemma 2.4 *The following are equivalent:*

$$y = \bar{F}(x) + p(\bar{F}(x-) - \bar{F}(x)) , \ 0 \leq p \leq 1$$
$$\bar{F}^{-1}(y-) \geq x \geq \bar{F}^{-1}(y) .$$

From this lemma and Theorem 2.3 we obtain the following equivalent characterization of the individually optimal arrival rate.

Theorem 2.5 *An arrival rate λ is an individually optimal arrival rate if and only if*

$$\bar{F}^{-1}\left(\frac{\lambda}{\Lambda}-\right) \geq \pi(\lambda) \geq \bar{F}^{-1}\left(\frac{\lambda}{\Lambda}\right) . \tag{2.20}$$

Recall that in the previous section we provided a characterization of the individually optimal arrival rate based on the derivative of the utility function. The next lemma provides the connection between the two characterizations and shows that in fact they are equivalent.

Lemma 2.6 *For all probabilistic reward-threshold rules, (x,p), the utility U is a function only of the associated arrival rate,*

$$\lambda = \Lambda[\bar{F}(x) + p(\bar{F}(x-) - \bar{F}(x))] , \tag{2.21}$$

and is given by

$$U = U(\lambda) := \int_0^\lambda \bar{F}^{-1}\left(\frac{\nu}{\Lambda}\right) d\nu . \tag{2.22}$$

Proof (Note that (2.21) is just (2.18), which we have rewritten here for convenience.) First observe that, for any probabilistic reward-threshold rule (x,p), U is independent of p and is given by

$$U = \lambda x + \Lambda \int_x^\infty \bar{F}(r)dr . \tag{2.23}$$

To see this, first note that

$$
\begin{aligned}
U &= \Lambda \int_0^\infty rp(r)dF(r) \\
&= \Lambda \left[p \cdot x(\bar{F}(x-) - \bar{F}(x)) + \int_{x+}^\infty rdF(r) \right] .
\end{aligned}
$$

Suppose $\bar{F}(x-) \neq \bar{F}(x)$. Solving (2.21) for p in terms of λ and x yields

$$p = \frac{\lambda - \Lambda\bar{F}(x)}{\Lambda(\bar{F}(x-) - \bar{F}(x))} .$$

Substituting this expression for p in the above equation for U, we obtain

$$
\begin{aligned}
U &= (\lambda - \Lambda\bar{F}(x))x + \Lambda \int_{x+}^\infty rdF(r) \\
&= \lambda x + \Lambda \int_x^\infty (r - x)dF(r) \\
&= \lambda x + \Lambda \int_x^\infty \bar{F}(r)dr .
\end{aligned}
$$

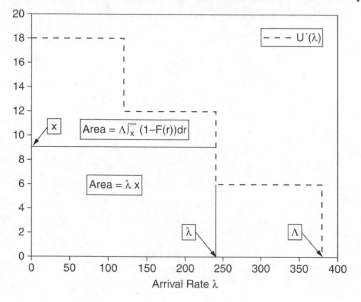

Figure 2.8 *Graphical Interpretation of $U(\lambda)$ as an Integral: Case 1*

On the other hand, if $\bar{F}(x-) = \bar{F}(x)$, then $\lambda = \Lambda\bar{F}(x)$ and

$$
\begin{aligned}
U &= \Lambda \int_x^\infty r dF(r) \\
&= \Lambda \left(x\bar{F}(x) + \int_x^\infty (r-x)dF(r) \right) \\
&= \lambda x + \Lambda \int_x^\infty \bar{F}(r)dr \ .
\end{aligned}
$$

But, for any $\lambda \in [0, \Lambda]$, it follows by a change of variable that

$$
\lambda x + \Lambda \int_x^\infty \bar{F}(r)dr = \int_0^\lambda \bar{F}^{-1}(\nu/\Lambda)d\nu = U(\lambda) \ , \tag{2.24}
$$

where x is any threshold value such that (2.21) holds for some p. Thus, U depends only on λ and is given by (2.22). This completes the proof of the lemma. ∎

Figures 2.8 and 2.9 illustrate (2.24) for the extreme case in which $U'(\lambda)$ is piecewise constant, which corresponds to a discrete reward distribution (see Section 2.3.4).

Theorem 2.7 *A probabilistic reward-threshold rule, (x,p), is an individually optimal joining rule if and only if its associated arrival rate, given by (2.21), satisfies the condition*

$$
U'(\lambda-) \geq \pi(\lambda) \geq U'(\lambda) \ . \tag{2.25}
$$

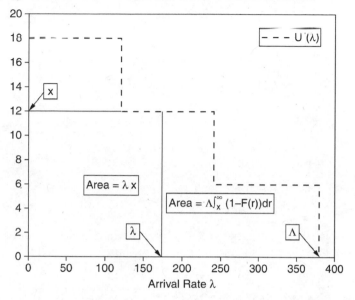

Figure 2.9 *Graphical Interpretation of $U(\lambda)$ as an Integral: Case 2*

Proof   Recall from the previous section that $U'(\lambda)$ is defined as the right-hand derivative of $U(\lambda)$, which exists for all $\lambda \in [0, \Lambda]$, provided that $U(\lambda)$ is concave. In the present context, in which $U(\lambda)$ is given by (2.22), concavity follows from the fact that $\bar{F}^{-1}(\nu/\Lambda)$ is nonincreasing in ν. Moreover,

$$U'(\lambda) = \bar{F}^{-1}(\frac{\lambda}{\Lambda})$$

$$U'(\lambda-) = \bar{F}^{-1}(\frac{\lambda}{\Lambda}-) .$$

But we showed in Theorem 2.5 that λ is an individually optimal arrival rate if and only if

$$\bar{F}^{-1}\left(\frac{\lambda}{\Lambda}-\right) \geq \pi(\lambda) \geq \bar{F}^{-1}\left(\frac{\lambda}{\Lambda}\right) .$$

It follows that λ is an individually optimal arrival rate if and only if (2.25) holds.   ∎

2.3.2 Social Optimality

Recall that the social optimization problem takes the form,

$$\max_{\{\lambda \geq 0\}} \mathcal{U}(\lambda) = U(\lambda) - \lambda G(\lambda) .$$

The socially optimal arrival rate, λ^s, is the unique solution to the *KKT* conditions,

$$
\begin{aligned}
G(\lambda) + \lambda G'(\lambda) - U'(\lambda) &\geq 0 \\
\lambda &\geq 0 \\
\lambda(G(\lambda) + \lambda G'(\lambda) - U'(\lambda)) &= 0 .
\end{aligned}
$$

When we relax the assumption that $U(\lambda)$ is differentiable, we can rewrite these conditions in equivalent form as

$$
U'(\lambda-) \geq \pi(\lambda) \geq U'(\lambda) ,
$$

where, in the case of social optimality,

$$
\pi(\lambda) = G(\lambda) + \lambda G'(\lambda) .
$$

Thus, as we have observed, the socially optimal arrival rate λ^s can be induced by charging customers (who are individual optimizers) the toll,

$$
\delta^s = \lambda^s G'(\lambda^s) .
$$

We have shown (cf. Theorems 2.3 and 2.7) that an alternate characterization of the individually optimal arrival rate corresponding to a given toll, δ, is

$$
\bar{F}^{-1}\left(\frac{\lambda}{\Lambda}-\right) \geq \pi(\lambda) \geq \bar{F}^{-1}\left(\frac{\lambda}{\Lambda}\right) ,
$$

or equivalently,

$$
\begin{aligned}
\lambda &= \Lambda[\bar{F}(\hat{\pi}) + p(\bar{F}(\hat{\pi}-) - \bar{F}(\hat{\pi}))] \\
\hat{\pi} &= \pi(\lambda)
\end{aligned}
$$

for some $0 \leq p \leq 1$, where $\pi(\lambda) = \delta + G(\lambda)$. It follows that the socially optimal arrival rate, λ^s, satisfies the following conditions:

$$
\begin{aligned}
\lambda &= \Lambda[\bar{F}(\hat{\pi}) + p(\bar{F}(\hat{\pi}-) - \bar{F}(\hat{\pi}))] \\
\hat{\pi} &= G(\lambda) + \lambda G'(\lambda) .
\end{aligned}
$$

In the case of a continuous reward distribution, F, these conditions simplify to

$$
\lambda = \Lambda \bar{F}(G(\lambda) + \lambda G'(\lambda)) .
$$

2.3.3 Facility Optimality

Recall that one possible formulation of the facility optimization problem is the following:

$$
\max_{\{\lambda \geq 0\}} \tilde{U}(\lambda) = \lambda(U'(\lambda) - G(\lambda)) .
$$

The facility optimal arrival rate, λ^f, is a solution (not necessarily unique) to the *KKT* conditions,

$$
G(\lambda) + \lambda G'(\lambda) - U'(\lambda) - \lambda U''(\lambda) \geq 0
$$

$$\lambda \geq 0$$
$$\lambda(G(\lambda) + \lambda G'(\lambda) - U'(\lambda) \quad \lambda U''(\lambda)) \;=\; 0 .$$

(assuming the first and second derivatives of $U(\lambda)$ exist). The optimality conditions can be written equivalently as

$$U'(\lambda-) \geq \pi(\lambda) \geq U(\lambda) ,$$

where, in the case of facility optimality,

$$\pi(\lambda) = -\lambda U''(\lambda) + G(\lambda) + \lambda G'(\lambda) .$$

Thus, as we have observed, the facility optimal arrival rate λ^f can be induced by charging customers (who are individual optimizers) the toll,

$$\delta^f = -\lambda^f U''(\lambda^f) + \lambda^f G'(\lambda^f) .$$

We have shown (cf. Theorems 2.3 and 2.7) that an alternate characterization of the individually optimal arrival rate corresponding to a given toll, δ, is

$$\bar{F}^{-1}\left(\frac{\lambda}{\Lambda}-\right) \geq \pi(\lambda) \geq \bar{F}^{-1}\left(\frac{\lambda}{\Lambda}\right) ,$$

or equivalently,

$$\lambda \;=\; \Lambda[\bar{F}(\hat{\pi}) + p(\bar{F}(\hat{\pi}-) - \bar{F}(\hat{\pi}))]$$
$$\hat{\pi} \;=\; \pi(\lambda)$$

for some $0 \leq p \leq 1$, where $\pi(\lambda) = \delta + G(\lambda)$. It follows that the facility optimal arrival rate, λ^f, satisfies the following conditions:

$$\lambda \;=\; \Lambda[\bar{F}(\hat{\pi}) + p(\bar{F}(\hat{\pi}-) - \bar{F}(\hat{\pi}))]$$
$$\hat{\pi} \;=\; -\lambda U''(\lambda) + G(\lambda) + \lambda G'(\lambda) .$$

In the case of a continuous reward distribution, F, these conditions simplify to

$$\lambda = \Lambda \bar{F}(-\lambda U''(\lambda) + G(\lambda) + \lambda G'(\lambda)) .$$

2.3.3.1 Heavy-Tailed Rewards

In Section 2.1.3 we noted that, when $\lambda U'(\lambda) \to \kappa$ as $\lambda \to 0$, where $0 < \kappa \leq \infty$, the facility optimal arrival rate may be $\lambda^f = 0+$. That is, the maximum of $\tilde{\mathcal{U}}(\lambda)$ may not be attained, but instead we have

$$\lim_{\lambda \to 0+} \tilde{\mathcal{U}}(\lambda) = \kappa .$$

In particular, when $\lambda U'(\lambda) \to \kappa = \infty$, the problem has an unbounded objective function and an arbitrarily large profit can be earned by charging an arbitrarily large toll, resulting in an arbitrarily small arrival rate.

How might such a situation arise in the case of a *GI/GI/1* queue with a probabilistic joining rule? First observe that we can alternatively formulate the profit-maximizing facility operator's problem as one of choosing a full price, π, to maximize the profit, $\lambda(\pi - G(\lambda))$, where λ satisfies the equilibrium equation, $U'(\lambda) = \pi$. (The corresponding toll is $\delta = \pi - G(\lambda)$.) In the case of a $GI/GI/1$ queue with a probabilistic joining rule, we have $U'(\lambda) = \bar{F}^{-1}(\lambda/\Lambda)$. Under the assumption that the reward distribution function, $F(x)$, is continuous and strictly increasing, the equilibrium equation, $U'(\lambda) = \pi$, is equivalent to

$$\lambda = \Lambda \bar{F}(\pi) \ .$$

Therefore, the objective function for facility optimization may be rewritten as

$$\Lambda \bar{F}(\pi)(\pi - G(\Lambda \bar{F}(\pi))) \ .$$

Thus we see that the condition that $\lim_{\lambda \to 0} \lambda U'(\lambda) = \kappa$ is equivalent to

$$\lim_{x \to \infty} x \bar{F}(x) = \kappa \ . \tag{2.26}$$

Note that, when $\kappa > 0$, (2.26) implies that $\bar{F}(x) \sim \kappa x^{-1}$ as $x \to \infty$, which in turn implies that the reward distribution function, F, has infinite mean.

A necessary (but not sufficient) condition for a distribution function F to satisfy (2.26) is that it be *heavy tailed*.

Definition Let X be a nonnegative random variable with distribution function $F(x) = P\{X \le x\}$, $x \ge 0$. The distribution F (or the random variable X) is said to have a *heavy tail* if

$$E\left[e^{\nu X}\right] = \int_0^\infty e^{\nu x} dF(x) = \infty \ ,$$

for all $\nu > 0$.

Since $\int_x^\infty e^{\nu y} dF(y) \ge e^{\nu x} \bar{F}(x)$ for all $x \ge 0$, we see that (2.26) implies that F has a heavy tail.

Example 1 Consider a Pareto reward distribution, in which

$$\bar{F}(x) = \left(\frac{\kappa}{\kappa + x}\right)^\alpha \ , \quad x \ge 0 \ ,$$

where $\alpha > 0$, $\kappa > 0$. It is easy to verify that

$$\lim_{x \to \infty} x \bar{F}(x) = \begin{cases} \infty \ , & \text{if } \alpha < 1 \\ \kappa \ , & \text{if } \alpha = 1 \\ 0 \ , & \text{if } \alpha > 1 \end{cases} \ .$$

Thus in particular we can conclude that $\lambda U'(\lambda) \to \infty$ as $\lambda \to 0$ when $\alpha < 1$. Therefore, in this case the facility optimal value of λ is $\lambda^f = 0+$. That is, the facility operator can obtain an arbitrarily large profit by selecting an arbitrarily small (but positive) arrival rate.

Example 2 Consider a reward distribution F with a tail that is *regularly varying* at ∞ of index $-\alpha$ (denoted $\bar{F} \in \mathcal{R}_{-\alpha}$); that is,

$$\lim_{x \to \infty} \frac{\bar{F}(tx)}{\bar{F}(x)} = t^{-\alpha} \ , \quad t > 0 \ ,$$

where $0 < \alpha < 1$. It can be shown that

$$\lim_{x \to \infty} x\bar{F}(x) = \infty \ .$$

It follows that $\lambda U'(\lambda) \to \infty$ as $\lambda \to 0$, so that again the facility optimal value of λ is $\lambda^f = 0+$.

It is interesting to examine the behavior of the function, $\lambda U'(\lambda)$, in more detail in the case of a Pareto reward distribution. Since $\bar{F}(x)$ is continuous and strictly decreasing in $x \geq 0$, we have $U'(\lambda) = \bar{F}^{-1}(\lambda/\Lambda)$ for all $\lambda \in [0, \Lambda]$, where Λ is the arrival rate of potential customers. We can easily evaluate $\bar{F}^{-1}(y)$ explicitly by solving the equation

$$y = \bar{F}(x) = \left(\frac{\kappa}{\kappa + x} \right)^\alpha \ ,$$

for x in terms of y, obtaining

$$x = \bar{F}^{-1}(y) = \kappa(y^{-\beta} - 1) \ ,$$

where $\beta := \alpha^{-1}$. Thus,

$$\lambda U'(\lambda) = \kappa\lambda \left[\left(\frac{\Lambda}{\lambda} \right)^\beta - 1 \right] \ .$$

To check whether $\lambda U'(\lambda)$ is nondecreasing and/or concave, we evaluate its first and second derivatives:

$$\frac{d}{d\lambda} (\lambda U'(\lambda)) = U'(\lambda) + \lambda U''(\lambda)$$

$$= -\kappa \left[(\beta - 1) \left(\frac{\Lambda}{\lambda} \right)^\beta + 1 \right]$$

$$\frac{d^2}{d\lambda^2} (\lambda U'(\lambda)) = \left(\frac{\kappa(\beta - 1)\beta}{\Lambda} \right) \left(\frac{\Lambda}{\lambda} \right)^{\beta+1} \ .$$

Thus we see that when $\beta > 1$ (that is, when $\alpha < 1$), the function $\lambda U'(\lambda)$ is neither nondecreasing nor concave. In fact, it is *strictly decreasing* and *strictly convex* over its entire domain, $\lambda \in [0, \Lambda]$. Figure 2.10 illustrates the behavior of $\lambda U'(\lambda)$ for the case of a Pareto reward distribution with $\kappa = 1$ and $\alpha = 1/2$, with $\Lambda = 400$.

Since $\lambda G(\lambda)$ is nondecreasing in λ, it follows that the objective function,

$$\tilde{\mathcal{U}}(\lambda) = \lambda U'(\lambda) - \lambda G(\lambda) \ ,$$

for facility optimization is strictly decreasing in $\lambda \in [0, \Lambda]$ when $\alpha < 1$. Thus, in this case, not only is $\lambda^f = 0+$, but there are no local maxima of $\tilde{\mathcal{U}}(\lambda)$ in $(0, \Lambda]$. Figure 2.11 displays the graph of $\tilde{\mathcal{U}}(\lambda)$ in the case of a steady-state *M/M/1* queue with $\Lambda = 400$, $\mu = 450$, a linear waiting cost with $h = 500$, and a Pareto reward distribution with $\kappa = 1$ and $\alpha = 1/2$.

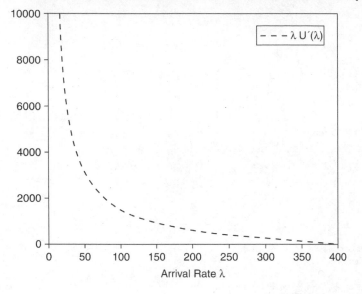

Figure 2.10 *Graph of $\lambda U'(\lambda)$: Pareto Reward Distribution ($\alpha < 1$)*

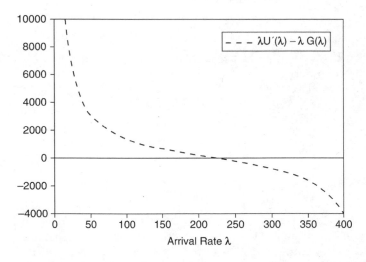

Figure 2.11 *Graph of $\tilde{\mathcal{U}}(\lambda)$:* M/M/1 *Queue with Pareto Reward Distribution ($\alpha < 1$)*

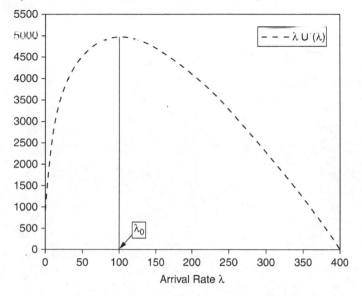

Figure 2.12 *Graph of* $\lambda U'(\lambda)$*: Pareto Reward Distribution* $(\alpha > 1)$

On the other hand, when $\beta < 1$ $(\alpha > 1)$:

$\lambda U'(\lambda)$ is concave and nondecreasing for $0 \le \lambda \le \lambda_0$;

$\lambda U'(\lambda)$ is concave and nonincreasing for $\lambda_0 \le \lambda \le \Lambda$;

where

$$\lambda_0 := \Lambda(1 - \beta)^{1/\beta} = \Lambda \left(\frac{\alpha - 1}{\alpha} \right)^{\alpha} .$$

Figure 2.12 illustrates the behavior of $\lambda U'(\lambda)$ in the case of a Pareto reward distribution with $\kappa = 50$ and $\alpha = 2$, with $\Lambda = 400$. In this case we have $\lambda_0 = 100$.

Since $H(\lambda) = \lambda G(\lambda)$ is convex (by assumption), the objective function, $\tilde{\mathcal{U}}(\lambda) = \lambda U'(\lambda) - \lambda G(\lambda)$, is concave when $\beta < 1$ $(\alpha > 1)$. Hence the facility optimal arrival rate, λ^f, is the unique solution to the first-order optimality condition,

$$U'(\lambda) + \lambda U''(\lambda) = G(\lambda) + \lambda G'(\lambda) ,$$

which in this case reduces to

$$\kappa \left[(1 - \beta) \left(\frac{\Lambda}{\lambda} \right)^{\beta} - 1 \right] = G(\lambda) + \lambda G'(\lambda) .$$

Note that $\tilde{\mathcal{U}}(\lambda)$ is decreasing for $\lambda \ge \lambda_0$ and therefore $\lambda^f \le \lambda_0$. An interesting implication of this property is that we have an upper bound on the facility optimal arrival rate which is completely independent of the waiting-cost function, $G(\lambda)$. In particular, even in a system with no congestion effects or

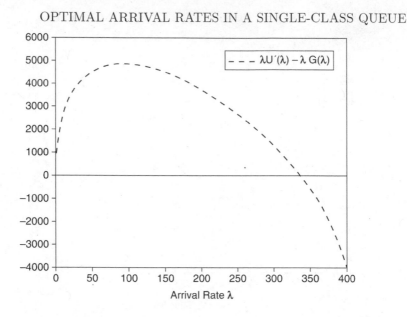

Figure 2.13 *Graph of* $\tilde{\mathcal{U}}(\lambda)$: M/M/1 *Queue with Pareto Reward Distribution* ($\alpha > 1$)

a system in which the customers are completely insensitive to congestion, a profit-maximizing facility operator will always set the arrival rate at or below λ_0. In other words, a customer whose reward from being served is below a certain threshold will always be denied service, regardless of the waiting-cost function.

Figure 2.13 displays the graph of $\tilde{\mathcal{U}}(\lambda)$ in the case of a steady-state M/M/1 queue with $\Lambda = 400$, $\mu = 450$, a linear waiting cost with $h = 500$, and a Pareto reward distribution with $\kappa = 50$ and $\alpha = 2$.

What is the significance of these results? Of course, in the real world, a reward distribution with infinite support, let alone a heavy tail, does not exist. But heavy-tailed distributions can be a useful way of capturing extreme situations, at least as a first approximation. Let us consider a facility that serves a population of customers whose rewards from being served follow a heavy-tailed distribution F satisfying

$$\lim_{x \to \infty} x\bar{F}(x) = \infty \ .$$

Such a reward distribution has the property that, for a small fraction of the population (the *elite* customers, if you will), the value of the service provided is much larger than for the rest of the population. Confronted by such a situation, a profit-maximizing facility operator will find it possible to earn an extremely large profit by charging an extremely high admission fee, thereby restricting service to the elite customers and denying service to everyone else. As a result, the time interval between successive admissions of customers to the system will be extremely large, but the reward received when an elite customer arrives and is admitted will more than compensate for this long interval. To put it more graphically, the facility operator will find that profit is maximized by reserving service for the likes of Warren Duffett and Bill Gates and ignoring the rest of us.

We leave it to the reader to decide if this model is a realistic approximation of the behavior of service facilities in the real world. It is interesting to note that the Pareto distribution was introduced in the early twentieth century by Wilfred Pareto (Pareto [153]) as a model for the observed distribution of incomes in the population of an industrialized society. Modern economists have confirmed that income distributions continue to tend to be heavy tailed. If one makes a further (innocuous?) assumption – that the utility gained from a service is proportional to one's income – then we have an empirical confirmation that the phenomenon,

$$\lim_{x \to \infty} x \bar{F}(x) \to \kappa , \ 0 < \kappa \leq \infty ,$$

is at least plausible in the real world.

2.3.4 Discrete Reward Distributions: Multiple Reward Classes

A discrete reward distribution reflects the situation when the arriving customers are distinctly heterogeneous with respect to the value they place on being served. The extreme case is when they belong to a finite number of classes, with customers in the same class all sharing the same value and the values of different classes being distinct.

Multi-class queues are an important subject of research and worthy of a chapter to themselves (see Chapter 4). When the classes are heterogeneous with respect to their delay costs as well as their service values, then analysis is considerably more complicated and a separate treatment is called for. In particular, it is appropriate to consider priority disciplines, in which classes of customers that are more sensitive to delay receive preferential service. We shall deal with these issues at length in Chapter 4.

But when the customers of different classes are homogeneous with respect to delay costs and the queue discipline is *FIFO* (or, more generally, class independent), we can use the techniques which we have developed in this chapter, treating the system as a special case of our general model for a single-class queue, with an appropriately chosen value function, $U(\lambda)$.

We begin by reconsidering the model with a piecewise linear concave, non-decreasing utility function, $U(\lambda)$ (introduced above in Section 2.2.1), showing how such a utility function can arise from a multiclass *GI/GI/1* queueing system with a probabilistic joining rule.

Let $r_1 < r_2 < \ldots < r_m$, $0 = \nu_{m+1} < \nu_m < \nu_{m-1} < \ldots < \nu_1$, and suppose $U(\lambda)$ takes the following (piecewise-linear) form:

$$U(\lambda) = r_m \lambda , \ 0 \leq \lambda \leq \nu_m$$
$$U(\lambda) = U(\nu_m) + r_{m-1}(\lambda - \nu_m) , \ \nu_m \leq \lambda \leq \nu_{m-1}$$

$$\vdots$$

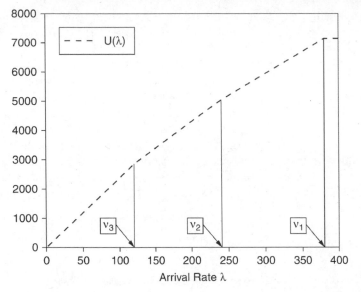

Figure 2.14 $U(\lambda)$ *for Three-Class Example*

$$U(\lambda) \;=\; U(\nu_2) + r_1(\lambda - \nu_2) \,,\; \nu_2 \le \lambda \le \nu_1$$
$$U(\lambda) \;=\; U(\nu_1) = \sum_{i=1}^{m} r_i(\nu_i - \nu_{i+1}) \,,\; \nu_1 \le \lambda \,.$$

Figure 2.14 illustrates the graph of U. Between breakpoints, ν_{i+1} and ν_i, one can increase λ by admitting customers with reward r_i (class-i customers). Customers of class i are arriving at rate $\Lambda_i = \nu_i - \nu_{i+1}$ (with $\nu_{m+1} = 0$). When the supply of such customers is exhausted (that is, when λ reaches the breakpoint $\nu_i = \Lambda_m + \ldots + \Lambda_i$), in order to increase λ further one must admit customers with the next lower reward, r_{i-1} (class-$(i-1)$ customers).

As the graph in Figure 2.14 clearly shows, the utility function U corresponds to a situation in which the marginal utility of the arrival rate λ is piecewise constant. The corresponding marginal utility function, $U'(\lambda)$, is given by

$$U'(\lambda) \;=\; r_m \,,\; 0 \le \lambda < \nu_m$$
$$U'(\lambda) \;=\; r_{m-1} \,,\; \nu_m \le \lambda < \nu_{m-1}$$
$$\vdots$$
$$U'(\lambda) \;=\; r_1 \,,\; \nu_2 \le \lambda < \nu_1$$
$$U'(\lambda) \;=\; 0 \,,\; \nu_1 \le \lambda \,.$$

The graph of $U'(\lambda)$ is illustrated in Figure 2.15.

It follows from the analysis in Section 2.3 that $U'(\lambda) = \bar{F}^{-1}(\lambda/\Lambda)$, where the system under study is a *GI/GI/1* queue with a discrete reward distribution,

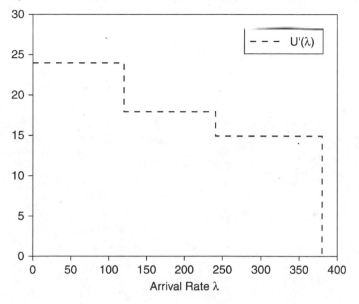

Figure 2.15 *U'(λ) for Three-Class Example*

$F(x) = \mathrm{P}\{\boldsymbol{R} \le x\}$ given by

$$
\begin{aligned}
F(x) &= 0 \,,\, 0 \le x < r_1 \\
F(x) &= p_1 \,,\, r_1 \le x < r_2 \\
F(x) &= p_1 + p_2 \,,\, r_2 \le x < r_3 \\
&\vdots \\
F(x) &= p_1 + p_2 + \ldots + p_{m-1} \,,\, r_{m-1} \le x < r_m \\
F(x) &= p_1 + p_2 + \ldots + p_{m-1} + p_m = 1 \,,\, r_m \le x
\end{aligned}
$$

Here $\Lambda = \nu_1$ and

$$
p_i = \mathrm{P}\{\boldsymbol{R} = r_i\} = \frac{\Lambda_i}{\Lambda} = \frac{\nu_i - \nu_{i+1}}{\nu_1} \,,\, i = 1, \ldots, m \,.
$$

This reward distribution corresponds to a situation in which potential customers arrive at rate Λ, the rewards of successive customers are i.i.d., and the probability that a potential customer has reward equal to r_i (that is, belongs to class i) is p_i. When the potential arrivals are from a Poisson process with parameter Λ (an *M/GI/1* queue), another interpretation is that potential customers of class i arrive from a Poisson process with parameter Λ_i, $i = 1, \ldots, m$, and that these Poisson processes are mutually independent.

What do optimal arrival rates look like in the case of a discrete reward distribution? In the case of individual optimality, we already observed that the equilibrium condition can be written in either of the following equivalent forms:

$$U'(\lambda-) \geq \pi(\lambda) \geq U'(\lambda) ,$$

or equivalently,

$$\lambda = \Lambda[\bar{F}(\hat{\pi}) + p(\bar{F}(\hat{\pi}-) - \bar{F}(\hat{\pi}))]$$
$$\hat{\pi} = \pi(\lambda)$$

for some $0 \leq p \leq 1$, where $\pi(\lambda) = \delta + G(\lambda)$. In the case of a discrete reward distribution, these equivalent conditions hold if and only if

$$r_{i+1} > \pi(\lambda) \geq r_i , \tag{2.27}$$

$$\lambda = \sum_{k=i+1}^{m} \Lambda_k + p \cdot \Lambda_i , \tag{2.28}$$

for some $i \in \{1, 2, \ldots, m\}$ and $0 \leq p \leq 1$. In other words, in equilibrium customers in classes k with $r_k > \pi(\lambda)$ (in this case classes $m, m-1, \ldots, i+1$) join with probability one, customers in a class k with $r_k = \pi(\lambda)$ (in this case possibly class i) join with probability p, $0 \leq p \leq 1$, and customers in classes k with $r_k < \pi(\lambda)$ (in this cases classes $i-1, \ldots, 1$) join with probability zero.

By appropriate choice of $\pi(\lambda)$, we obtain individually optimal or socially optimal solutions. If

$$\pi(\lambda) = G(\lambda) + \delta ,$$

then the solution to the equilibrium conditions, (2.27), (2.28), is individually optimal for the given toll, δ. If

$$\pi(\lambda) = G(\lambda) + \lambda G'(\lambda) ,$$

then the solution to the equilibrium conditions, (2.27), (2.28), is socially optimal.

Now consider facility optimization. Recall that, for a differentiable utility function, $U(\lambda)$, a positive facility optimal arrival rate is a solution to the equation,

$$U'(\lambda) + \lambda U''(\lambda) = G(\lambda) + \lambda G'(\lambda) .$$

Hence, by setting the full price, $\pi(\lambda)$, equal to $G(\lambda) + \lambda G'(\lambda) - \lambda U''(\lambda)$ and charging the toll, $\delta = \pi(\lambda) - G(\lambda) = \lambda G'(\lambda) - \lambda U''(\lambda)$, one can induce facility-optimal behavior on the part of individually optimizing customers. In the case of a discrete reward distribution, however, $U''(\lambda)$ does not exist so this approach does not work. Hence, a different approach is called for.

First note that the facility optimal arrival rate is a solution to the following optimization problem:

$$\max_{\{\lambda \geq 0\}} \tilde{\mathcal{U}}(\lambda) = \lambda(U'(\lambda) - G(\lambda)) .$$

In the case of a discrete reward distribution, the function $\lambda U'(\lambda)$ takes the form illustrated in Figure 2.16.

In this example (Case 1), the objective function, $\tilde{\mathcal{U}}(\lambda)$, takes the form illustrated in Figure 2.17. (For this example, we have assumed that the service

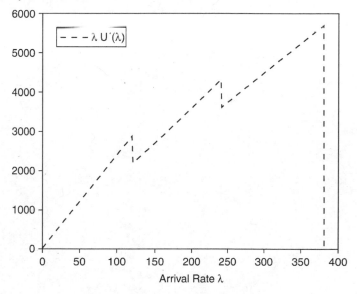

Figure 2.16 $\lambda U'(\lambda)$ *for Three-Class Example*

facility is a steady-state *M/M/1* queue with $\mu = 450$ and a linear waiting cost function with $h = 1000$.) The facility optimal arrival rate is $\lambda^f = \nu_2 = \Lambda_3 + \Lambda_2$, which corresponds to admitting all customers of class 3 and class 2, while rejecting all customers of class 1. This is done by charging the facility optimal toll,

$$\delta^f = r_2 - G(\lambda) \,,$$

which leads to the full price,

$$\pi(\lambda) = G(\lambda) + \delta^f = r_2 \,.$$

Thus, the facility optimal arrival rate λ^f satisfies the equilibrium equations, (2.27), (2.28), corresponding to this choice of toll and the full price, $\pi(\lambda)$, with $p = 1$.

The other possibility (Case 2) for a facility optimal arrival rate is illustrated in Figure 2.18, in which the parameters, ν_i, $i = 1, 2, 3$, have been perturbed slightly. (The other parameters – r_i, $i = 1, 2, 3$, μ, and h – are unchanged.) In this case, the facility optimal arrival rate is $\lambda^f = \nu_2 + p \cdot \Lambda_1 = \Lambda_3 + \Lambda_2 + p \cdot \Lambda_1$, which corresponds to admitting all customers of class 3 and class 2, while admitting each customer of class 1 with probability p, where $0 < p < 1$. This is done by charging the facility optimal toll,

$$\delta^f = r_1 - G(\lambda) \,,$$

which leads to the full price,

$$\pi(\lambda) = G(\lambda) + \delta^f = r_1 \,.$$

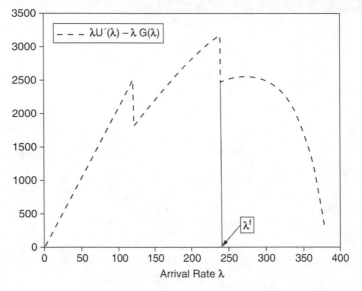

Figure 2.17 $\tilde{\mathcal{U}}(\lambda)$ *for Three-Class Example (Case 1)*

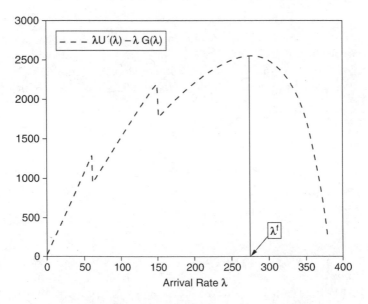

Figure 2.18 $\tilde{\mathcal{U}}(\lambda)$ *for Three-Class Example (Case 2)*

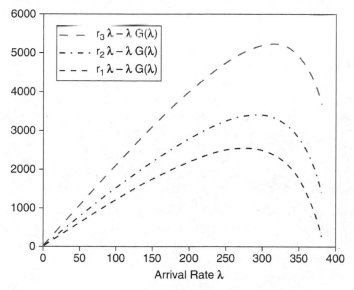

Figure 2.19 $\mathcal{U}_i(\lambda)$, $i = 1, 2, 3$, for Three-Class Example

Thus, the facility optimal arrival rate λ^f satisfies the equilibrium equations, (2.27), (2.28), corresponding to this choice of toll and the full price, $\pi(\lambda)$, with $0 < p < 1$.

Note that in both Cases 1 and 2, there are three local maxima of the objective function, $\tilde{\mathcal{U}}(\lambda)$: two at the break-points, ν_3 and ν_2, between classes, and one strictly between the break-points, ν_2 and ν_1. In general, the facility optimization problem with a discrete reward distribution will have local maxima at $\nu_m, \nu_{m-1}, \ldots, \nu_{i+1}$, and at some $\lambda \in (\nu_{i+1}, \nu_i]$, for some $i \in M$. There is no way of identifying which of these local maxima is the global maximum without evaluating the objective function at each.

Remark 5 Another way of looking at the case of a discrete reward distribution is to look first at the net utility function, $\mathcal{U}_i(\lambda)$, for a problem in which all customers belong to the single class i and there is no upper bound on λ. That is,

$$\mathcal{U}_i(\lambda) := r_i \cdot \lambda - \lambda G(\lambda) \ .$$

(This is the objective function for both social and facility optimization, since the reward is deterministic.) For the three-class example we have been considering, the functions, $\mathcal{U}_i(\lambda)$, $i = 1, 2, 3$, are displayed in Figure 2.19. Obviously, with no upper bound on the arrival rate in class m, an optimal solution will admit only customers of class m (since $r_m > r_i$, $i \neq m$). The optimal value of λ in this case will be the optimal arrival rate for the problem with deterministic reward (linear utility function) (first considered in Section 1.2 of Chapter 1) with $r = r_m$, that is, the unique solution to

$$r_m = G(\lambda) + \lambda G'(\lambda)$$

(assuming $r_m \geq G(0)$). The introduction of an upper bound, Λ_i, on the arrival rate of each class i creates the break-points, $\nu_i = \sum_{k=i}^{m} \Lambda_k$, at which the net utility function, $\tilde{\mathcal{U}}(\lambda)$, for facility optimization "drops down" from the graph of $\mathcal{U}_i(\lambda)$ to $\mathcal{U}_{i-1}(\lambda)$. As we saw in the three-class example, the placement of the break-points determines which of the local maxima will be the global maximum.

2.3.4.1 Infinite Number of Reward Classes

We can extend this model to a discrete-reward distribution with an infinite number of classes. That is, the reward $\boldsymbol{R}$ has a probability mass function, $p_i := \mathrm{P}\{\boldsymbol{R} = r_i\}$, where $0 < r_1 < r_2 < \ldots < r_m < \ldots$, and $\sum_{i=1}^{\infty} p_i = 1$.

Of particular interest in this case are discrete reward distributions with a heavy tail. We shall start with the Peter and Paul distribution, which is based on the St. Petersburg paradox (cf. Embrechts et al. [61]). The model is as follows. Peter tosses a fair coin. If the first head appears on the k-th toss, then Peter pays Paul 2^k dollars. Since the probability that the first head appears on the k-th toss is 2^{-k}, $k = 1, 2, \ldots$, the c.d.f. for amount of money paid to Paul is $F(x) = \sum_{k:2^k \leq x} 2^{-k}$. Suppose we use this as the c.d.f. for the reward received by an entering customer in our queueing model. Then for $x = 2^n$ we have

$$
\begin{aligned}
2^n \bar{F}(2^n) &= 2^n \left(1 - \sum_{k=1}^{n} 2^{-k} \right) \\
&= 2^n 2^{-n} = 1 \, ,
\end{aligned}
$$

for all $n \geq 1$ and thus, a fortiori, $\lim_{n \to \infty} 2^n \bar{F}(2^n) = 1$. Note also that for all $2^n \leq x < 2^{n+1}$ we have $x \bar{F}(x) = x 2^{-n}$, so that $\lim_{x \uparrow 2^{n+1}} x \bar{F}(x) = 2$. It follows that

$$\liminf_{x \to \infty} x \bar{F}(x) = 1 \, , \qquad \limsup_{x \to \infty} x \bar{F}(x) = 2 \, .$$

By appropriate modifications of the reward that Paul receives if the first head occurs on the n-th trial, we can construct examples with a range of finite nonzero limits for $x \bar{F}(x)$, as well as examples in which the limit is 0 or ∞.

Example 1 Suppose $r_n = 2^{n+k}$ and (as before) $p_n = \mathrm{P}\{\boldsymbol{R} = r_n\} = 2^{-n}$, $n \geq 1$, where k is a positive integer. Then $2^{n+k} \bar{F}(2^{n+k}) = 2^{n+k} 2^{-n} = 2^k$ for all $n \geq 1$. In this case, for all $2^{n+k} \leq x < 2^{n+1+k}$ we have $x \bar{F}(x) = x 2^{-n}$, so that $\lim_{x \uparrow 2^{n+1+k}} x \bar{F}(x) = 2^{k+1}$. It follows that

$$\liminf_{x \to \infty} x \bar{F}(x) = 2^k \, , \qquad \limsup_{x \to \infty} x \bar{F}(x) = 2^{k+1} \, .$$

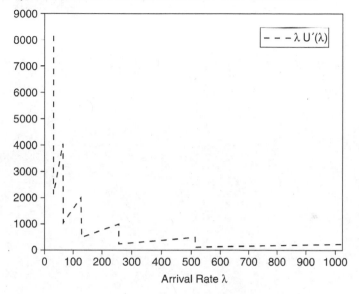

Figure 2.20 $\lambda U'(\lambda)$ *for Example 3*

Example 2 Suppose $r_n = 2^{n/2}$ and (as before) $p_n = \mathrm{P}\{\boldsymbol{R} = r_n\} = 2^{-n}$, $n \geq 1$. Then $2^{n/2}\bar{F}(2^{n/2}) = 2^{n/2}2^{-n} = 2^{-n/2}$ for all $n \geq 1$. In this case, for all $2^{n/2} \leq x < 2^{(n+1)/2}$ we have $x\bar{F}(x) = x2^{-n}$, so that $\lim_{x\uparrow 2^{(n+1)/2}} x\bar{F}(x) = 2^{-(n-1)/2}$. It follows that

$$\limsup_{x\to\infty} x\bar{F}(x) = \lim_{n\to\infty} 2^{-(n-1)/2} = 0 \ .$$

Example 3 Suppose $r_n = 2^{2n}$ and (as before) $p_n = \mathrm{P}\{\boldsymbol{R} = r_n\} = 2^{-n}$, $n \geq 1$. Then $2^{2n}\bar{F}(2^{2n}) = 2^{2n}2^{-n} = 2^n$ for all $n \geq 1$. In this case, for all $2^{2n} \leq x < 2^{2(n+1)}$ we have $x\bar{F}(x) = x2^{-n}$, so that $\lim_{x\uparrow 2^{2(n+1)}} x\bar{F}(x) = 2^{n+2}$. It follows that

$$\liminf_{x\to\infty} x\bar{F}(x) = \lim_{n\to\infty} 2^n = \infty \ .$$

Figure 2.20 illustrates the behavior of the function $\lambda U'(\lambda)$ in the case of Example 3 with $\Lambda = 1024$. In this example, the objective function, $\tilde{\mathcal{U}}(\lambda)$, takes the form illustrated in Figure 2.21. (For this example, we have assumed that the service facility is a steady-state *M/M/1* queue with $\mu = 1200$ and a linear waiting cost function with $h = 1200$.)

In particular, Example 3 provides another instance in which F has a heavy tail and $\lim_{x\to\infty} x\bar{F}(x) = \infty$.

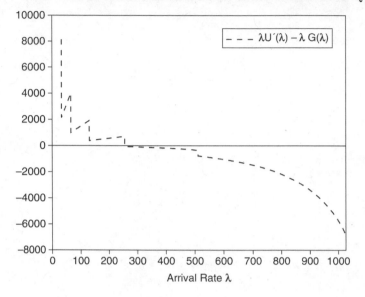

Figure 2.21 $\tilde{\mathcal{U}}(\lambda)$ *for Example 3*

2.4 Uniform Value Distribution: Stability

In this section we consider a system in which the value of service has a uniform distribution, $F(r) = r/a$, $0 \le r \le a$, $F(r) = 1$, $r > a$. In this case $U'(\lambda) = a(1 - \lambda/\Lambda)$, $0 \le \lambda \le \Lambda$. The service facility is an *M/M/1* queue with linear delay cost.

The individually optimal arrival rate λ^e satisfies

$$U'(\lambda) = G(\lambda) + \delta ,$$

which in this case reduces to

$$a(1 - \lambda/\Lambda) = h/(\mu - \lambda) + \delta , \tag{2.29}$$

where δ is the toll charged for admission. (We assume that $\delta + h/\mu < a$. Otherwise no customer has an incentive to join the system.) Since $a(1 - \lambda/\Lambda)$ is strictly decreasing and $h/(\mu - \lambda)$ is strictly increasing in λ, λ^e is the unique solution in $[0, \mu)$ to (2.29). We can interpret $S(\lambda) := \delta + h/(\mu - \lambda)$ and $D(\lambda) := a(1 - \lambda/\Lambda)$ as supply and demand curves, respectively, so that λ^e is the point at which the supply and demand curves intersect, as illustrated in Figure 2.22. We can use (2.29) to derive an explicit expression for λ^e in terms of the problem parameters. First we note that (2.29) is equivalent to

$$a'(1 - \lambda/\Lambda') = h/(\mu - \lambda) , \tag{2.30}$$

where $a' := a - \delta$, and $\Lambda' := [(a - \delta)/a]\Lambda$. Thus λ^e is a root of the quadratic equation

$$a'(\Lambda' - \lambda)(\mu - \lambda) = h\Lambda' .$$

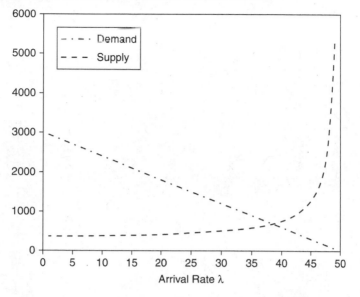

Figure 2.22 *Supply and Demand Curves: Uniform Value Distribution*

The two roots of this equation are

$$\lambda = (\Lambda' + \mu \pm [(\Lambda' + \mu)^2 - 4\Lambda'(\mu h/a')]^{1/2})/2 \ .$$

Since $\lambda^e < \mu$, the desired root is the one with the minus sign. That is,

$$\lambda^e = (\Lambda' + \mu - [(\Lambda' + \mu)^2 - 4\Lambda'(\mu h/a')]^{1/2})/2 \ . \tag{2.31}$$

2.4.1 Stability of the Equilibrium Arrival Rate

In economic theory a supply-demand equilibrium may be stable or unstable. Roughly speaking, an equilibrium is stable if the system tends to return to equilibrium after a perturbation. To give a precise interpretation of stability and instability in the present context, let us consider a dynamic version of our problem. The service rate μ is fixed, as before, but now the system operates over a succession of time periods, labeled $n = 0, 1, \ldots$. During time period n the arrival rate is fixed at a particular value λ_n and each time period is assumed to last long enough for the system to attain steady state (approximately). The arrival rate λ_n induces a cost per customer equal to $h/(\mu - \lambda_n) + \delta$ during period n. The customers react collectively to this perceived cost by adjusting the arrival rate during the next period so as to equate this cost with the marginal value received per unit time. That is, λ_{n+1} is the unique solution in $[0, \mu)$ of the equation $a(1 - \lambda_{n+1}/\Lambda) = h/(\mu - \lambda_n) + \delta$, or, equivalently,

$$a'(1 - \lambda_{n+1}/\Lambda') = h/(\mu - \lambda_n) \ . \tag{2.32}$$

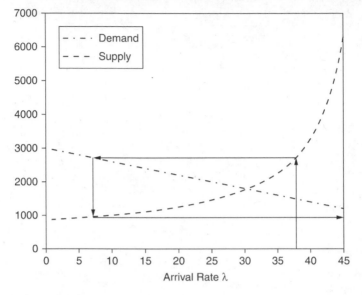

Figure 2.23 *An Unstable Equilibrium*

If $\lambda_{n+1} = \lambda_n$, then $\lambda_n = \lambda^e$ and the individually optimal arrival rate has been attained. Otherwise, the process repeats itself, with n replaced by $n + 1$. If $\lambda_n \to \lambda^e$ as $n \to \infty$, for every starting value λ_0, $0 \leq \lambda_0 \leq \lambda^e$, then we call the individually optimal arrival rate λ^e *stable*. Otherwise, it is called *unstable*. (In economic theory and the theory of dynamical systems, our definition of stability corresponds to *global* stability. By contrast, *local* stability would refer to the situation where convergence occurs only for λ_0 in some neighborhood of λ^e. Our next result, however, shows that λ_n converges to λ^e for some λ_0 if and only if it converges to λ^e for every λ_0, $0 \leq \lambda_0 \leq \Lambda$. That is, local stability and global stability coincide for our problem.)

The following theorem is proved in Stidham [187].

Theorem 2.8 *The individually optimal arrival rate λ^e is stable if and only if $\mu > \Lambda' = [(a - \delta)/a]\Lambda$.*

Hence, for any starting value λ_0, when $\mu < \Lambda'$ the sequence $\{\lambda_n, n \geq 0\}$ diverges (Figure 2.23), whereas when $\mu > \Lambda'$, the sequence $\{\lambda_n, n \geq 0\}$ converges to λ^e, as illustrated in Figure 2.24. The "cobweb" diagrams of Figures 2.23 and 2.24 are familiar from microeconomic theory, but this sort of analysis has not often been applied to queueing models.

An implication of Theorem 2.8 is that a service facility may not have a stable individually optimal arrival rate if the service capacity μ is too low relative to the "adjusted" maximal arrival rate, $\Lambda' = [(a - \delta)/a]\Lambda$. Put another way, a particular service capacity μ may give rise to a stable individually optimal arrival rate when the toll δ is high, but not when it is low.

In our discussion so far, we have allowed the toll δ to be arbitrary (subject

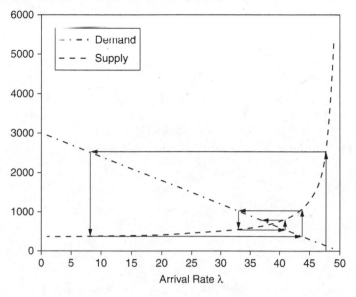

Figure 2.24 *Convergence to a Stable Equilibrium*

to $\delta + h/\mu < a$). In the social optimization problem, however, the system operator's objective is to maximize the net benefit,

$$U(\lambda) - h\lambda/(\mu - \lambda) ,$$

over $0 \leq \lambda < \mu, \lambda \leq \Lambda$. In the case of an interior maximum, this objective is achieved by setting the toll equal to the external effect (cf. equation (2.11)),

$$\delta^s = h\lambda^s/(\mu - \lambda^s)^2 , \qquad (2.33)$$

where λ^s, the socially optimal arrival rate, is the unique solution to the optimality condition (2.10), which in this case reduces to

$$a(1 - \lambda/\Lambda) = h\mu/(\mu - \lambda)^2 . \qquad (2.34)$$

We would like to see if the necessary and sufficient condition for steady state, $\mu > \Lambda'$, is satisfied when $\delta = \delta^s$. The following theorem is also proved in Stidham [187].

Theorem 2.9 *The optimal arrival rate λ^s is a stable equilibrium with $\delta = \delta^s = h\lambda^s/(\mu - \lambda^s)^2$ if and only if $\mu > [\Lambda + (h\Lambda/a)^{1/2}]/2$.*

Recall that we have assumed that $h/\mu < a$, in order to ensure that $\lambda^s > 0$. Suppose $\Lambda \leq h/a$. Then since $\mu > h/a$, we have $\mu > [\Lambda + (h\Lambda/a)^{1/2}]/2$ ($> \Lambda$), so that λ^s is a stable equilibrium. Now suppose $\Lambda > h/a$. Then $\mu > [\Lambda + (h\Lambda/a)^{1/2}]/2$ implies that $\mu > h/a$; that is, the steady-state condition is stronger than the condition for $\lambda^s > 0$. In this case, if

$$h/a < \mu \leq [\Lambda + (h\Lambda/a)^{1/2}]/2 ,$$

then the optimal arrival rate λ^s is strictly positive, but an unstable equilibrium with $\delta = \delta^s$.

2.5 Power Criterion

We have presented a model with net benefit as the objective to be maximized, arguing that this objective is a reasonable mechanism for capturing the trade-off between throughput and delay (waiting time). As an alternative, consider the following performance measure (to be maximized):

$$P(\lambda) := \frac{\lambda}{W(\lambda)} \ . \tag{2.35}$$

We call $P(\lambda)$ the *power* of the arrival rate, λ. Note that $P(\lambda)$ is the ratio of a benefit (the throughput, λ) to a cost (the waiting time, $W(\lambda)$). Benefit/cost ratios have frequently been proposed as a decision-making tool for reconciling conflicting objectives, but their use is not without controversy and pitfalls. (See Remark 7 below.)

Let the optimal arrival rate for the power criterion be denoted by λ^p. That is, λ^p solves the problem:

$$\max_{\{\lambda:0\leq\lambda<\mu\}} P(\lambda) \ . \tag{2.36}$$

Suppose $W(\lambda)$ is differentiable and satisfies conditions (1) – (3), which we repeat here for convenience:

1. $W(\lambda)$ is strictly increasing in $0 \leq \lambda < \mu$;
2. $W(\lambda) \uparrow \infty$ as $\lambda \uparrow \mu$;
3. $W(0) = 1/\mu$.

Since $P(\lambda) > 0$ for all $0 < \lambda < \mu$ and $P(\lambda) \to 0$ as $\lambda \to 0$ and as $\lambda \to \mu$, an optimal solution always lies in the interior of the feasible region: $0 < \lambda^p < \mu$. The necessary condition for an interior solution is, of course, that $P'(\lambda) = 0$ for $\lambda = \lambda^p$. Since

$$P'(\lambda) = \frac{W(\lambda) - \lambda W'(\lambda)}{(W(\lambda))^2} \ ,$$

$P'(\lambda) = 0$ if and only if

$$W(\lambda) - \lambda W'(\lambda) = 0 \ . \tag{2.37}$$

Remark 6 It follows from equation (2.37) that maximum power is achieved at a point where the *external effect* equals the *internal effect* for the case of a linear waiting-cost function, $G(\lambda) = h \cdot W(\lambda)$. Although this characteristic is interesting, its economic significance is not clear.

An equivalent form for the necessary condition for optimality is

$$\frac{W(\lambda)}{\lambda} = W'(\lambda) \ ;$$

that is, the average waiting time per unit of flow equals the marginal waiting time. That this should be the case is also clear from the fact that, since λ_p maximizes $P(\lambda) = \lambda/W(\lambda)$, it also minimizes $W(\lambda)/\lambda$, the average waiting cost per unit of arrival rate.

Is the necessary condition for maximum power also sufficient? If $P(\lambda)$ were concave, then the answer would be "yes." Without additional conditions on $W(\lambda)$, however, we cannot conclude that $P(\lambda)$ is concave. (It *is* concave in the special case of an $M/GI/1$ queue, in which

$$W(\lambda) = \frac{1}{\mu} + \frac{\lambda\beta}{\mu(\mu - \lambda)} \, , \text{ where } \beta := \frac{\mathrm{E}[\boldsymbol{S}^2]}{2(\mathrm{E}[\boldsymbol{S}])^2} \, .$$

See Exercise 1 at the end of this section.)

Although $P(\lambda)$ is not concave in general, if we make the additional assumption that $W(\lambda)$ is strictly convex, we can show that $P(\lambda)$ is unimodal, which implies that the necessary condition (2.37) has a unique solution and is in fact sufficient for maximum power.

Theorem 2.10 *Suppose that $W(\lambda)$ is differentiable and strictly convex and satisfies conditions (1) – (3). Then λ^p is the unique solution in $[0, \mu)$ to the optimality condition (2.37).*

Proof We have already observed that λ^p must satisfy (2.37). To verify unimodality of $P(\lambda)$, note that strict convexity of $W(\lambda)$ implies that the numerator of $P'(\lambda)$, namely, $W(\lambda) - \lambda W'(\lambda)$, is strictly decreasing in λ. Therefore, $P'(\lambda) > 0$ for $\lambda < \lambda^p$, $P'(\lambda) = 0$ for $\lambda = \lambda^p$, and $P'(\lambda) < 0$ for $\lambda > \lambda^p$. Thus, $P(\lambda)$ is strictly increasing for $\lambda < \lambda^p$ and strictly decreasing for $\lambda > \lambda^p$; that is, $P(\lambda)$ is unimodal and its maximizer, λ^p, is therefore the unique solution to (2.37). ∎

The assumption that $W(\lambda)$ is strictly convex is satisfied in many queueing models, including the $M/GI/1$ queue. (See Exercise 2 at the end of this section.)

Figure 2.25 graphically illustrates these facts. Since λ^p maximizes $\lambda/W(\lambda)$, it also minimizes $W(\lambda)/\lambda$, which is the slope of the chord connecting the origin to the point $(\lambda, W(\lambda))$ on the graph of $W(\lambda)$. But this slope is minimized over $\lambda \in [0, \mu)$ when the chord is tangent to the graph of $W(\lambda)$, that is when $W(\lambda)/\lambda = W'(\lambda)$. (In Figure 2.25, $\alpha := \min_{\lambda \in [0,\mu)} W(\lambda)/\lambda = 1/P(\lambda^p)$.)

It is interesting to compare the arrival rate, λ^p, dictated by the power criterion, to the system-optimal and individually optimal arrival rates, λ^s and λ^e, respectively, for the model based on deterministic reward and linear waiting cost. In particular, for an $M/M/1$ queue, note that $\lambda^p = \mu/2$, so that $\lambda^p = \lambda^s = \mu - \sqrt{\mu h/r}$ if and only if $h/r = \mu/4$, and $\lambda^p = \lambda^e = \mu - h/r$ if and only if $h/r = \mu/2$.

Remark 7 The power criterion takes the form of a *benefit-cost ratio*. It is well known in economic theory that maximizing the ratio of benefit to cost can lead to decisions that are inconsistent with intuition and common sense in certain problem contexts. This is particularly true when, as in the present

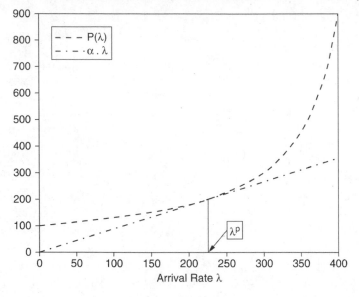

Figure 2.25 *Graphical Illustration of Power Maximization*

context, the benefits and costs do not accrue to the same entity. For note that λ is a measure of benefit (per unit time) to the system operator or (in the case of social optimality) to the collective of all customers, whereas $W(\lambda)$ is a cost incurred by *each* customer. By contrast, a benefit-cost ratio in which both benefit and cost are measured per unit time and accrue to the same entity would be

$$\frac{\text{throughput per unit time}}{\text{waiting cost per unit time}} = \frac{\lambda}{L(\lambda)} = \frac{\lambda}{\lambda W(\lambda)} = \frac{1}{W(\lambda)},$$

which is maximized by setting $\lambda = 0$: a not-very-interesting result.

2.5.0.1 Exercises

1. Show that $P(\lambda)$ is strictly concave in the case of a steady-state $M/GI/1$ queue.

2. Show that $W(\lambda)$ is strictly convex in the case of a steady-state $M/GI/1$ queue.

2.5.1 Generalized Power Criterion

The definition of power (2.35) implies a specific trade-off between throughput and delay. Different system operators, however, may have different relative sensitivities to throughput and delay. In order to accommodate such differences, several authors have proposed a generalized power criterion, in which the original definition of power is replaced by

$$P(\lambda) := \frac{\lambda^{\alpha}}{W(\lambda)} . \qquad (2.38)$$

Here the parameter $\alpha > 0$ is a weighting factor that may be used by the system manager to achieve a desired trade-off between throughput and average waiting time.

As in the special case, $\alpha = 1$, already considered above, we have $P(\lambda) > 0$ for all $0 < \lambda < \mu$ and $P(\lambda) \to 0$ as $\lambda \to 0$ and as $\lambda \to \mu$. Hence, an optimal solution always lies in the interior of the feasible region: $0 < \lambda^p < \mu$. The necessary condition for an interior solution is $P'(\lambda) = 0$ for $\lambda = \lambda^p$. Since

$$P'(\lambda) = \frac{\lambda^{\alpha-1}}{(W(\lambda))^2} \left(\alpha W(\lambda) - \lambda W'(\lambda) \right) ,$$

$P'(\lambda) = 0$ if and only if

$$\alpha W(\lambda) - \lambda W'(\lambda) = 0 . \qquad (2.39)$$

Remark 8   The maximum generalized power is achieved at a point where the *external effect* equals α times the *internal effect* for the case of a linear waiting-cost function, $G(\lambda) = h \cdot W(\lambda)$. Thus, in choosing the parameter α in the generalized power criterion, we are in effect choosing the ratio of external effect to internal effect for the case of a linear waiting-cost function.

Again, we would like to know if and when the necessary condition (2.37) is also sufficient. Note that if $W(\lambda)$ is convex and $\alpha < 1$, then $\alpha W(\lambda) - \lambda W'(\lambda)$ is strictly decreasing in λ and hence (2.37) has a unique solution, $\lambda = \lambda^p$. When $\alpha > 1$, however, we cannot establish sufficiency without further making assumptions about $W(\lambda)$.

As an example, consider the special case of an *M/M/1* queue with service rate μ, for which

$$P(\lambda) = \lambda^{\alpha}(\mu - \lambda) ,$$

so that

$$P'(\lambda) = \lambda^{\alpha-1}(\alpha\mu - (\alpha + 1)\lambda) ,$$

and hence the necessary condition $P'(\lambda) = 0$ has a unique solution,

$$\lambda = \left(\frac{\alpha}{\alpha + 1} \right) \mu , \qquad (2.40)$$

which is *ipso facto* the optimal arrival rate, λ^p. Note that in this special case the necessary condition always has a unique solution, regardless of the value of α, even though $P(\lambda)$ is not concave unless $\alpha \leq 1$.

2.5.2 Generalized Power Criterion as Special Case of Net-Benefit Criterion

Note that maximizing $P(\lambda)$, as defined in (2.38), is equivalent to maximizing

$$\ln(P(\lambda)) = \alpha \ln(\lambda) - \ln(W(\lambda)) . \qquad (2.41)$$

This objective function is in the general form of our net-benefit model, with logarithmic utility function

$$U(\lambda) = \alpha \ln(\lambda) , \qquad (2.42)$$

and waiting cost per unit time

$$H(\lambda) = \lambda G(\lambda) = \ln(W(\lambda)) . \qquad (2.43)$$

Thus the waiting cost per customer takes the form

$$G(\lambda) = \frac{\ln(W(\lambda))}{\lambda} .$$

Since $\ln(\cdot)$ is concave, increasing, and differentiable, the utility function defined by (2.42) satisfies our default conditions. The default assumptions for the waiting-cost function, $G(\lambda) = H(\lambda)/\lambda$, are that it be differentiable and strictly increasing. This will be true if and only if

$$\ln(W(\lambda)) < \frac{\lambda W(\lambda)}{W'(\lambda)} , \ \lambda \geq 0 . \qquad (2.44)$$

Hence, under condition (2.44), the (generalized) power criterion may rightly be considered as a special case of our net-benefit criterion.

Note, however, that the assumption that $H(\lambda) = \lambda G(\lambda)$ is convex, which we needed for many of our results in the net-benefit model, will not necessarily hold. Convexity of $H(\lambda) = \ln(W(\lambda))$ is equivalent to log-convexity of $W(\lambda)$ – a stronger property than just convexity of $W(\lambda)$. As an example in which $W(\lambda)$ is log-convex, consider again the *M/M/1* queue with service rate μ, in which

$$\ln(W(\lambda)) = \ln((\mu - \lambda)^{-1}) = -\ln(\mu - \lambda) ,$$

which is convex in λ.

Note that, since $H(0) = 0$, convexity of $H(\lambda)$ (that is, strict log-convexity of $W(\lambda)$) implies that $G(\lambda) = H(\lambda)/\lambda$ is strictly increasing, that is, condition (2.44) holds.

Thus, if $W(\lambda)$ is log-convex, all of our default assumptions hold for the net-benefit maximization problem (2.41) equivalent to the power-maximization problem. The question remains:

> Viewed from the perspective of the net-benefit criterion, how realistic are the specific utility and waiting-cost functions implied by the power criterion (cf. (2.42) and (2.43))?

The requirement of a logarithmic utility function is, of course, very specific, but it does satisfy our default assumptions for utility functions. The assumption that $H(\lambda) = \ln(W(\lambda))$ is quite restrictive, however. Note that it implies that $G(\lambda) = f(W(\lambda), \lambda)$, where

$$f(w, \lambda) := \frac{\ln w}{\lambda} , \ w > 0 , \ 0 < \lambda < \mu .$$

Thus, for a fixed λ, a customer's sensitivity to delay is a function only of its *average* waiting time, $W(\lambda)$. Higher moments of the waiting time arc irrelevant. Moreover, the marginal sensitivity *decreases* as the average waiting time increases (since the logarithm is a concave function). Finally, for a fixed w, the sensitivity to delay *decreases* as λ increases. Again, these assumptions are quite restrictive and even counterintuitive.

Our conclusion is that, before adopting the power criterion, one should be aware of its unorthodox behavioral implications, when viewed through the lens of standard utility theory.

2.6 Bidding for Priorities

Consider the following variant of the individual optimization problem. Instead of being charged a fixed toll, δ, to enter the system, individually optimizing customers submit bids to the system controller and are served in decreasing order of the size of their bids. In this section we consider a special case of this model.

We consider a steady-state $M/M/1$ queue with service rate μ, linear utility function, $U(\lambda) = r \cdot \lambda$, and linear waiting cost, with $h = $ cost per unit time spent waiting in the system. That is, each customer who joins receives a fixed reward r and incurs a waiting cost at rate h per unit time while in the system. Now, however, those customers who decide to join may offer a payment to purchase priority. A joining customer is placed in the queue ahead of all those with smaller payments, including the customer in service. That is, the customer in service is preempted when a higher-paying customer joins. The preempted customer's service resumes where it left off at the next time instant when no customers who made larger payments are in the system. As usual we assume that the arriving customers are not able to observe the system state (that is, the number of customers present and the payments they made), but instead base their joining and payment decisions on knowledge of the system parameters, μ, r, and h.

We shall characterize the individually optimal joining probability and the associated distribution of payments and compare the resulting arrival rate to the socially optimal arrival rate, λ^s. As before, λ^s maximizes social welfare, that is, the overall expected net benefit per unit time, $B(\lambda) = \lambda(r - hW(\lambda))$. Like the congestion toll discussed previously, the payments of customers act simply as transfer payments; the facility operator collects the payments and then redistributes them in equal shares among all potential arrivals. Thus they have no effect on social welfare. Note also that, since the service-time distribution is exponential, the remaining service time of each customer in the system has the same (exponential) distribution. Hence, the order of service has no effect on $W(\lambda)$, the waiting time in the system averaged over all joining customers, which is therefore given by

$$W(\lambda) = \frac{1}{\mu - \lambda},$$

as in the case of a *FIFO* queue discipline. To simplify the analysis, let us assume that $r > h/\mu$, so that λ^s is the (unique) solution in $[0, \mu)$ to the first-order optimality condition $B'(\lambda) = 0$, or, equivalently,

$$r = h(W(\lambda) + \lambda W'(\lambda)) = \frac{h\mu}{(\mu - \lambda)^2} . \tag{2.45}$$

Now let a denote the joining probability and $\{F(x), x \geq 0\}$ the cumulative distribution function of the payment offered by a joining customer. Assume for now that $F(0) = 0$ and that $F(x)$ is continuous and strictly increasing on an interval $[0, x_{max}]$, where $x_{max} > 0$ and $F(x_{max}) = 1$. (It turns out that these assumptions are without loss of generality, as we shall see presently.)

We seek an individually optimal pair, $(a, \{F(x), x \geq 0\})$. First we shall consider the equilibrium behavior of those customers who join the system, for a given (arbitrary) value of the arrival rate, λ. The governing property of a Nash equilibrium – that no customer shall have an incentive to deviate unilaterally – requires that the expected net benefit should be the same for all such customers. Let $W_x(\lambda)$ denote the expected waiting time in the system of a customer who joins and pays x. The expected net benefit for such a customer is $r - (x + hW_x(\lambda))$. It follows that, for some constant K,

$$r - (x + hW_x(\lambda)) = K , \text{ for all } x \in [0, x_{max}] . \tag{2.46}$$

In particular,

$$r - hW_0(\lambda) = K . \tag{2.47}$$

Now let us evaluate $W_0(\lambda)$. Recall that λ is the rate at which customers are joining the system. Note first that a joining customer who pays zero goes to the end of the queue and stays there until its service is completed. (The assumption that $F(0) = 0$ ensures that the probability is zero that another customer offers a zero payment.) If there are n customers present when it joins, then its waiting time in the system is distributed as the sum of $n + 1$ busy periods, generated by the arrivals during the service of each of the n customers as well as its own service. Each of these busy periods has expected duration $1/(\mu - \lambda)$ and the expected number of customers present at the arrival (using *PASTA*) is $L(\lambda) = \lambda/(\mu - \lambda)$. Thus

$$
\begin{aligned}
W_0(\lambda) &= \left(\frac{\lambda}{\mu - \lambda} + 1 \right) \frac{1}{\mu - \lambda} \\
&= \frac{\mu}{(\mu - \lambda)^2} .
\end{aligned} \tag{2.48}
$$

To derive x_{max} and $\hat{F}(x)$, first note that (2.46), (2.47), and (2.48) imply that

$$W_x(\lambda) = W_0(\lambda) - \frac{x}{h} = \frac{\mu}{(\mu - \lambda)^2} - \frac{x}{h} . \tag{2.49}$$

In particular, a customer who pays $x = h[\mu/(\mu - \lambda)^2 - 1/\mu]$ waits for $W_x(\lambda) = 1/\mu$ time units. But no matter what its payment, a joining customer must wait for at least $1/\mu$ time units, since this is its expected time in service. Therefore, no customer will have an incentive to pay more than this amount and we conclude that

$$x_{max} = h\left(\frac{\mu}{(\mu - \lambda)^2} - \frac{1}{\mu}\right). \tag{2.50}$$

(Note that this is just the cost of the time spent in the queue by a customer who pays zero, which is the maximal waiting cost that can be saved by making a positive payment.)

Now the same argument that we used to derive an explicit expression for $W_0(\lambda)$ can be applied to yield

$$W_x(\lambda) = \frac{\mu}{(\mu - \lambda(1 - F(x)))^2}, \tag{2.51}$$

for an arbitrary arrival rate λ and payment distribution $F(x)$. This is because a customer who pays x must wait for all customers present who have paid more than x and for the busy periods generated by all arrivals during their service times who pay more than x. The arrival rate of such customers is $\lambda(1 - F(x))$. It follows from (2.51) and (2.49) that

$$\begin{aligned}
\lambda(1 - F(x)) &= \mu - \sqrt{\frac{\mu}{W_x(\lambda)}} \\
&= \mu - \sqrt{\frac{h\mu}{\frac{h\mu}{(\mu - \lambda)^2} - x}}. \tag{2.52}
\end{aligned}$$

Now let us turn our attention to λ^e, the equilibrium value of λ. In order to have a Nash equilibrium for all customers – those who join and those who do not – we must have $K = 0$, since the net benefit of a customer who does not join is zero. Therefore, from (2.47) it follows that λ^e must satisfy

$$r = hW_0(\lambda).$$

It follows from this result and equations (2.45) and (2.48) that $\lambda^s = \lambda^e$.

We conclude that the individual customers, given the option of paying for preemptive priority and then left to their own devices, attain the socially optimal arrival rate. This result is in sharp contrast to what happens in the system without priority payments, in which the individually optimal arrival rate is strictly greater than the socially optimal arrival rate and it is necessary to impose a congestion toll in order to induce the customers to behave optimally.

The following is an intuitive explanation for this result. It follows from (2.46) that λ^e satisfies

$$x + hW_x(\lambda^e) = hW_0(\lambda^e), \text{ for all } x \in [0, x_{max}]. \tag{2.53}$$

Now since the customer who pays zero goes to the end of the queue and stays there until its service is completed, this customer imposes no external effect (in the form of additional waiting costs) on other customers. That is, it bears the entire social cost (internal plus external effect) of its decision to join. On the other hand, since the service times are exponentially distributed and the queue discipline is preemptive resume, the order of service has no effect on the social welfare. Thus, the social cost caused by each customer who joins is independent of its payment and therefore equals $hW_0(\lambda)$. Now a customer who pays x waits $W_x(\lambda)$ time units and the costs it incurs contribute $hW_x(\lambda)$ to the social cost. The difference, $h[W_0(\lambda) - W_x(\lambda)]$ represents the external effect imposed by this customer on other customers. But (2.53) implies that this expression equals x, so that in equilibrium the amount paid by each joining customer is exactly equal to its external effect, which explains why the customers' individually optimizing behavior is socially optimal.

2.6.1 Explicit Solution for Distribution of Priority Payments

Equation (2.52) gives an explicit expression for the equilibrium distribution of priority payments when the arrival rate equals λ. If the arrival rate is also in equilbrium – that is, if $\lambda = \lambda^e = \lambda^s$ – then it follows from (2.45) and (2.52) that $F(x) \equiv \hat{F}(x)$, and

$$\hat{F}(x) = \frac{\sqrt{\frac{h\mu}{r-x}} - \sqrt{\frac{h\mu}{r}}}{\mu - \sqrt{\frac{h\mu}{r}}} \ , \ x \in [0, x_{max}] \ . \tag{2.54}$$

Note that $\hat{F}(0) = 0$ and $\hat{F}(x)$ is continuous and strictly increasing on $[0, x_{max}]$ (as we assumed). Moreover, it follows from equations (2.50) and (2.45) that

$$x_{max} = r - h/\mu \ .$$

The equilibrium bid distribution, $\hat{F}(x)$, is graphed in Figure 2.26, for the case $r = 50$, $h = 1$, $\mu = 1$. It is interesting to note that approximately 80% of the bids are within 20% of the maximum bid, x_{max}.

2.7 Endnotes

Section 2.1

The basic model for arrival-rate selection introduced in this section is a generalization of the simple *M/M/1* model with deterministic reward and linear waiting costs discussed in Section 1.2 of Chapter 1.

The assumption that utility (or value) is a concave function of the arrival rate pervades the literature on optimization of queues. Mendelson [143] developed a queueing model for a computer facility in a firm, in which the aggregate value of the facility to the firm is assumed to be a concave function of the number of jobs arriving (and processed) per unit time. (See also Dewan and Mendelson [54] and Stidham [187].) Mendelson cites Sharpe [176] as a source for this assumption in the context of models for computer facilities. Decreasing marginal utility (i.e., a concave utility function) is also a standard assumption in classical decision theory and microeconomics. In models for communication networks, it is often assumed that each user or application

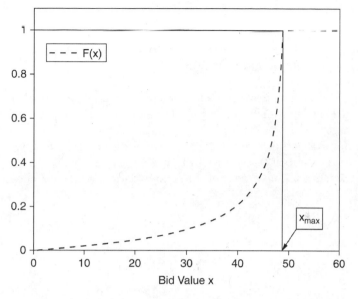

Figure 2.26 *Graph of Equilibrium Bid Distribution*

has a utility function which is concave in its allocated bandwidth (i.e., the number of messages/packets/bytes transmitted per unit time). Shenker [179] uses the term *elastic* demand to refer to such a user or application. See also Kelly [105]. (But note that "elastic" in this context does not mean the same thing as in classical microeconomics.)

The distinction between individual and social optimization was introduced in the context of control of an *M/M/1* queue with deterministic reward and linear waiting cost in a seminal paper by Naor [151]. As mentioned in the endnotes to Chapter 1, Edelson and Hildebrand [59] studied the same model in a design setting and showed that the socially optimal and facility optimal arrival rates coincide. They did not explicitly compare individual and social optimization, however. Stidham [185] made this comparison and gave the explicit expression for the external effect. To the best of my knowledge, the material on facility optimal arrival rates and the comparison with socially optimal rates for a general concave utility function is new.

Section 2.3

In models for control of queues, the assumption that potential arriving customers have i.i.d. random rewards was introduced by Miller [147] and Cramer [47] and extended by Lippman [132], Lippman and Stidham [131], Stidham [183], and Johansen and Stidham [100]. Random rewards in the context of design models have a shorter history. The connection between a concave utility (value) function and random rewards was pointed out by Dewan and

Mendelson [144], under restrictive assumptions on the reward distribution. The extension in this section to a general reward distribution is new.

The analysis of reward distributions with heavy tails and the implications for facility optimization have not previously appeared in the literature, as far as I know.

A good reference on heavy tails is Embrechts et al. [61]. Our definition of a heavy-tailed distribution coincides with theirs; there are other definitions in the literature.

For a proof that $\lim_{x \to \infty} x\bar{F}(x) = \infty$ for a distribution F with regularly varying tail (Example 2), see Proposition A3.8 on p. 568 in [61].

For a reference on the Peter and Paul distribution, see Embrechts et al [61].

Section 2.4

The material in this section is based on Stidham [187].

Section 2.5

Kleinrock [114] introduced the concept of *power* as a measure of the trade-off between throughput and delay in communication systems. Power and generalized power have continued to be popular in the communications literature.

Section 2.6

The model and results in this section are from Hassin [85], with additional material from Stidham [188].

CHAPTER 3

Dynamic Adaptive Algorithms: Stability and Chaos

This chapter focuses on dynamic algorithms for finding the equilibrium or optimal arrival rates and prices in a single-facility, single-class queueing system. We introduced this topic in Chapter 2, proposing there a simple dynamic algorithm for adjusting the arrival rate at discrete time points and showing that it either converges to the equilibrium arrival rate (*stability*) or diverges (*instability*), depending on the values of the economic and structural parameters in the model. In this chapter we expand on this subject and consider a variety of dynamic algorithms in discrete and continuous time.

After introducing the basic model (in Section 3.1), we extend (in Section 3.2) the discrete-time, dynamic-system algorithm of Chapter 2 to allow *adaptive expectations*, in which customers predict the future price based on a convex combination of the current price and the previous prediction. We give conditions for local and global stability of the equilibrium flow allocation. We show that the algorithm can lead to chaotic behavior when the equilibrium is unstable. That is, the price and arrival rate can follow aperiodic orbits, which appear to be completely random. Section 3.2.1 presents the basic results concerning stability, instability, and chaotic behavior of the sequence of prices and arrival rates in the multi-period model. Most of these results consist of applications of theorems from dynamic-systems theory to our model. These results are then illustrated in the context of a steady-state $M/GI/1$ queue with uniform value distribution in Section 3.2.2. We exhibit explicit ranges of parameter values in which the system is stable, unstable, or exhibits chaotic behavior, and present a numerical example (an $M/M/1$ queue). We summarize the results for the basic discrete-time dynamic model in Section 3.2.3.

In Section 3.3 we briefly discuss variants of the discrete-time algorithm in which customers predict future arrival rates rather than prices and/or the predictions are based on current arrival rates instead of prices.

Sections 3.4 and 3.5 introduce and analyze continuous-time dynamic adaptive algorithms, in which the prices or arrival rates evolve continuously according to a differential rather than a difference equation. Here we use the concept of Lyapunov functions from the theory of dynamical systems to show that the algorithms converge to the equilibrium price or arrival rate, regardless of the values of the system parameters. In particular, chaotic behavior is not possible.

3.1 Basic Model

Our basic model is the single-facility model we introduced in Chapter 2. For convenience we summarize the model assumptions here. We consider a service facility defined by the following ingredients:

- the arrival rate λ – the average number of customers actually entering the system per unit time (a decision variable);

- the (gross) utility $U(\lambda)$ received per unit time when the arrival rate is λ;

- the average waiting cost per customer, $G(\lambda)$;

- the admission fee or toll, δ, paid by each entering customer.

The set of feasible values for λ is denoted A. Our default assumption will be that $A = [0, \infty)$. The utility function $U(\lambda)$ is strictly increasing, differentiable, and strictly concave in $\lambda \in A$. The waiting-cost function $G(\lambda)$ takes values in $[0, \infty]$ and is nondecreasing and differentiable in $\lambda \in A$. The sum of the toll and the waiting cost constitutes the *full price* of admission: $\pi = \delta + G(\lambda)$.

Our primary interest is in the equilibrium arrival rate and the equilibrium full price for individually optimizing customers, assuming a given, fixed toll (or in some cases a time-varying toll). As usual we will be able to extend our results to social and facility optimality by assuming that the facility operator can set the toll to achieve these objectives.

Customers are self-optimizing. A potential arriving customer, seeking to maximize its net benefit, has an incentive to join the system if its service value exceeds the admission price. Customers cannot observe the congestion in the system, however, before deciding whether or not to join. Hence, they do not know the waiting cost, $G(\lambda)$, nor the admission price, $\pi = \delta + G(\lambda)$, before entering. We assume that customers form an estimate or prediction, $\hat{\pi}$, of the admission price, π, and base their join/balk decision on this predicted price. (We shall discuss possible mechanisms for forming this prediction presently.) Thus, a potential arriving customer joins the system if and only if its service value exceeds the predicted price, $\hat{\pi}$.

The resulting arrival rate, denoted by $\lambda(\hat{\pi})$, is therefore the unique solution in $[0, \infty)$ to the conditions,

$$U'(\lambda) \leq \hat{\pi} ,$$
$$U'(\lambda) = \hat{\pi} , \text{ if } \lambda > 0 .$$

As a result, the actual full price experienced by a joining customer is $\pi = \delta + G(\lambda(\hat{\pi}))$. If $\hat{\pi} = \pi$ – that is, if the predicted price equals the actual price – then π is a Nash equilibrium (for individually optimizing customers) because no customer has an incentive to deviate unilaterally from the prediction $\hat{\pi} = \pi$ (and the associated join/balk decision based on it).

The equilibrium price (denoted π^e) thus satisfies the equation

$$\pi^e = \delta + G\left(\lambda(\pi^e)\right) . \tag{3.1}$$

For $\lambda = \lambda(\pi^e)$, we therefore have (by the definition of $\lambda(\hat{\pi})$)

$$U'(\lambda) \ \leq \ \delta + G(\lambda) \,, \tag{3.2}$$
$$U'(\lambda) \ = \ \delta + G(\lambda) \,, \text{ if } \lambda > 0 \,. \tag{3.3}$$

It follows that $\lambda(\pi^e) = \lambda^e$, the unique solution to the equilibrium equations for an individually optimal arrival rate (cf. Section 2.1.1 of Chapter 2).

3.2 Discrete-Time Dynamic Adaptive Model

Now, following the approach introduced in Chapter 2, we consider a dynamic model in which the service facility operates over a succession of discrete time periods labeled $n = 0, 1, \ldots$.* Let $\hat{\pi}^{(n)}$ denote the predicted price for period n, $n = 0, 1, \ldots$. The resulting arrival rate in period n is given by $\lambda^{(n)} = \lambda(\hat{\pi}^{(n)})$. The arrival rate $\lambda^{(n)}$ in turn induces the actual price $\pi^{(n)} = \delta + G(\lambda^{(n)})$. Observing this price, the customers then collectively form a prediction $\hat{\pi}^{(n+1)}$ of the price in the next period, $\pi^{(n+1)}$, using a convex combination of the current price $\pi^{(n)}$ and the previous estimate $\hat{\pi}^{(n)}$, viz.

$$\hat{\pi}^{(n+1)} = (1 - \omega)\hat{\pi}^{(n)} + \omega\pi^{(n)}, \tag{3.4}$$

where $\omega \in (0, 1]$.

The prediction (3.4) is an exponential-smoothing forecast of $\pi^{(n+1)}$, based on past prices $\pi_0, \ldots, \pi^{(n)}$, under what is known as *adaptive expectations* in the economics literature. In the boundary case, $\omega = 1$, we have $\hat{\pi}^{(n+1)} = \pi^{(n)}$ and hence

$$\lambda^{(n+1)} = \lambda(\pi^{(n)}) = \lambda(\delta + G(\lambda^{(n)})) \,,$$

which is the general form of the difference equation governing the dynamic model that we studied in Chapter 2. This is generally referred to as *static expectations* in the economics literature. In the general case of adaptive expectations ($\omega \in (0, 1]$) we shall find it more convenient to work with a recursion in the predicted prices such as (3.4)) than with a recursion in the arrival rates.

Remark 1   The static-expectations model is naive in several regards. For one it assumes that each customer ignores the presence of others acting in the same manner. During periods when arrivals are frequent, customers predict that the high waiting cost will persist and thus they tend to elect not to enter during the next period, producing a period of low utilization with a corresponding low waiting cost. This then triggers a surge of arrivals in the subsequent period as customers overreact to the low price. Since there is no learning process involved, this pattern tends to repeat itself in periodic swings about the equilibrium.

In an attempt to rectify these shortcomings, the adaptive expectations model allows customers to learn from and correct for past errors. By rewriting

---

* The material in this section is taken from Rump and Stidham [170], in which one may find proofs of the results, which are quite technical. The reader who is interested in a brief verbal summary of the results is referred to Section 3.2.3.

(3.4) as

$$\hat{\pi}^{(n+1)} - \hat{\pi}^{(n)} = \omega(\pi^{(n)} - \hat{\pi}^{(n)}) , \tag{3.5}$$

one readily sees the dampening effect that adjustments have on the observed prediction error $\pi^{(n)} - \hat{\pi}^{(n)}$ in period n. As we shall see later, however, this quantitative dampening may be accompanied by a qualitative *destabilization*, in which the swings about the equilibrium are wildly erratic and apparently random (*chaotic*).

Since each successive forecast, $\hat{\pi}^{(n)}$, seeks to predict the price, $\pi^{(n)} = \delta + G(\lambda(\hat{\pi}^{(n)}))$, we can view this dynamic pricing process as an equilibrium-seeking pricing algorithm governed by the first-order nonlinear difference equation,

$$\hat{\pi}^{(n+1)} = \hat{\Pi}(\hat{\pi}^{(n)}) , \tag{3.6}$$

where

$$\hat{\Pi}(\hat{\pi}) \ := \ (1 - \omega)\hat{\pi} + \omega\Pi(\hat{\pi}) , \tag{3.7}$$

$$\Pi(\hat{\pi}) \ := \ \delta + G(\lambda(\hat{\pi})) . \tag{3.8}$$

Consider a fixed point solution of the map $\hat{\Pi}(\cdot)$, that is, a value of $\hat{\pi}$ such that

$$\hat{\pi} = \hat{\Pi}(\hat{\pi}) .$$

It follows from (3.7) and (3.8) that

$$\hat{\pi} = \Pi(\hat{\pi}) = \delta + G(\lambda(\hat{\pi})) ,$$

that is, $\hat{\pi}$ is also a fixed point of the map $\Pi(\cdot)$ and, by (3.1), an equilibrium price, that is, $\hat{\pi} = \pi^e$. Note that $\Pi(\hat{\pi})$ is nonincreasing, since we have assumed that $U'(\lambda)$ is nondecreasing and $G(\cdot)$ is nondecreasing. Thus, the equilibrium price, π^e, is the unique fixed point of both Π and $\hat{\Pi}$.

Returning to the dynamic algorithm, observe that, based on the comparison between the value of service and the prediction $\hat{\pi}^{(n)}$ of the price in period n, the customers adjust their service demand (arrival rate) for that period. The quantity of service demanded, that is, the resulting arrival rate $\lambda^{(n)}$ is given by

$$\lambda^{(n)} = \lambda(\hat{\pi}^{(n)}) = D^{-1}(\hat{\pi}^{(n)}) .$$

Here $D^{-1}(\cdot) := \lambda(\cdot)$ may be interpreted as the inverse demand curve from standard economic theory. Using this relation we can rewrite the price during period n as a function of the predicted price for period n, as follows:

$$\pi^{(n)} = \delta + G(\lambda(\hat{\pi}^{(n)})) = S(D^{-1}(\hat{\pi}^{(n)})) , \tag{3.9}$$

where $S(\cdot) := \delta + G(\cdot)$ may be interpreted as the supply curve. When the system is in economic equilibrium, the predicted and actual prices coincide, so that

$$\pi^{(n)} = S(D^{-1}(\pi^{(n)})) ,$$

or, equivalently,

$$S^{-1}(\pi^{(n)}) = D^{-1}(\pi^{(n)}) .$$

That is, supply equals demand.

Remark 2 Note that the behavior of the service facility is *demand driven* by the customers' decisions, as opposed to the production markets frequently analyzed in the economic literature, which are *supply driven*. In a supply-driven market the quantity *supplied* to the market $S^{-1}(\hat{\pi}^{(n)})$ is based on the *supplier's* price predictions $\hat{\pi}^{(n)}$. This quantity then determines demand, and the price set by the market, $\pi^{(n)} = D(S^{-1}(\hat{\pi}^{(n)}))$, is the marginal value at the demand level.

3.2.1 Stability, Instability, and Chaos

We now study the asymptotic behavior of the dynamical system (3.6) about the unique fixed-point equilibrium. In particular, we seek to develop conditions for equilibrium stability, i.e., asymptotic convergence of the algorithm to the equilibrium. More precisely, the equilibrium price, π^e, is called *locally stable* (or simply *stable*) if $\hat{\pi}^{(n)} \to \pi^e$ for all $\hat{\pi}^{(0)}$ in a neighborhood of π^e. It is called *globally stable* if the algorithm converges for all starting prices, $\hat{\pi}^{(0)}$. We shall also study the dynamics about an unstable equilibrium, i.e., the behavior of the algorithm when it diverges.

In this section and the immediately following section, we shall focus attention on single-server queues in which the waiting cost (and other induced measures) depend on an additional service-capacity parameter, μ. In applications to classical queueing models, such as the *M/GI/1* queue operating in steady state (cf. the model in Section 3.2.2), μ is the service rate of the single server (the reciprocal of the average service time). To denote explicitly the dependence on μ, we shall often append μ as a subscript to the economic and structural measures and write, for example, $G_\mu(\cdot)$, $\Pi_\mu(\cdot)$, $\hat{\Pi}_\mu(\cdot)$, π_μ, and $\hat{\pi}_\mu$.

We shall also restrict our attention to the special case where the utility function, $U(\lambda)$, is induced by a probabilistic reward-threshold joining rule on the part of customers with i.i.d. service values with distribution function, $F(r) = P\{R \le r\}$, $r \ge 0$. (See Section 2.3 of Chapter 2.)

We make the following assumptions:

- for each fixed $\mu > 0$, $G_\mu(\lambda)$ is strictly increasing in $\lambda \in [0, \mu)$, $G_\mu(0) > 0$, $\lim_{\lambda \uparrow \mu} G_\mu(\lambda) = \infty$, and $G_\mu(\lambda) = \infty$ for $\lambda \ge \mu$;

- for each fixed $\lambda \ge 0$, $G_\mu(\lambda)$ is strictly decreasing in $\mu > \lambda$;

- $G_\mu(\cdot, \cdot) \in \mathcal{C}^3$ as a function of (λ, μ) for $0 \le \lambda < \mu$;

- F is strictly increasing and $F \in \mathcal{C}^3$ in its support, $\mathcal{R} = [d, a]$.

These assumptions are satisfied, for example, in the steady-state *M/GI/1* model with linear waiting cost which we introduce in the following section. (The last two assumptions are not required for all our results, but are imposed here to avoid technicalities and simplify the exposition.)

Theorem 3.1 *A noninterior price equilibrium, $\pi_\mu^e \notin (d, a)$, is globally stable.*

Moreover, global convergence to the equilibrium is monotonic after the first step.

In light of this result, we shall henceforth focus attention on the case of an interior equilibrium, $\pi_\mu^e \in (d, a)$. We make two assumptions to ensure that this is the case. In order to have $\pi_\mu^e < a$, we require $\delta + G_\mu(0) < a$. (That is, there is an incentive for at least some customers to join the facility.) This implies $\delta < a$ and imposes a lower bound, $\underline{\mu}$, on the allowable service rate, since $G_\mu(\lambda)$ is decreasing in μ. (If the maximal service value a is infinite, then $\underline{\mu} := 0$.) Likewise, in order to have $\pi_\mu^e > d$, we require that $\delta + G_\mu(\Lambda) > d$, so that the incentive is not so great that all potential arrivals decide to join. For $\delta < d$ and $\mu > \Lambda$, this requirement imposes an upper bound, $\bar{\mu}$, on the allowable service capacity, since $G_\mu(\Lambda)$ is decreasing in μ. Otherwise, this upper bound is not required, and we let $\bar{\mu} = \infty$.

Theorem 3.2 *There exists a $\mu_g \geq 0$ such that, if $\mu > \mu_g$, then π_μ^e is a globally stable equilibrium under both static and adaptive expectations ($\omega \in (0, 1]$).*

Remark 3   This is a *sufficiency* condition. It does not imply that the equilibrium is unstable when $\mu < \mu_g$. It may still be stable but not *globally* so.

From now on we shall concern ourselves with local behavior of the algorithm (3.6) in a neighborhood of an interior equilibrium. It is well known that the equilibrium is locally stable (unstable) if $|\frac{\partial}{\partial \hat{\pi}} \hat{\Pi}_\mu(\pi_\mu^e)| < 1$ ($|\frac{\partial}{\partial \hat{\pi}} \hat{\Pi}_\mu(\pi_\mu^e)| > 1$). The nonincreasing nature of $\Pi_\mu(\hat{\pi})$ guarantees $\frac{\partial}{\partial \hat{\pi}} \hat{\Pi}_\mu(\hat{\pi}) < 1$ for all $\hat{\pi}$, yielding the following result.

Theorem 3.3 *Let $v := \frac{\partial}{\partial \pi} \Pi_\mu(\pi_\mu^e)$.*

- *If $v > (\omega - 2)/\omega$, then the equilibrium π_μ^e is stable.*
- *If $v < (\omega - 2)/\omega$, then π_μ^e is unstable.*
- *If $v = (\omega - 2)/\omega$, then the test is inconclusive.*

It is interesting to observe that the size of the stability region grows as the smoothing coefficient ω decreases. For $\omega = 1$ the stability region is at its smallest. In other words, stability under static expectations implies stability under adaptive expectations, but not vice versa. Therefore, an unstable equilibrium under static expectations may be stable under adaptive expectations.

We now turn our attention to the region of *instability*. The change from stability to instability occurs at parameter values $\hat{\mu}$ – called *bifurcation values* – for which the stability test in Theorem 3.3 is inconclusive.

Remark 4   As indicated in Remark 3, the μ_g found in Theorem 3.2 need not be a bifurcation value. This is not the case in the *M/M/1* example studied in Chapter 2 for $\omega = 1$. In that model there is a single bifurcation value $\hat{\mu}$, and the equilibrium is globally stable if and only if $\mu > \hat{\mu}$.

The uncertainty about the equilibrium stability at these bifurcation values is partially removed by the following theorem which makes use of the so called Schwarzian derivative:

$$\sigma := -2\frac{\partial^3}{\partial\hat{\pi}^3}\hat{\Pi}_\mu(\pi_\mu^e) - 3\left(\frac{\partial^2}{\partial\hat{\pi}^2}\hat{\Pi}_\mu(\pi_\mu^e)\right)^2.$$

Theorem 3.4 *Suppose* $\frac{\partial}{\partial\hat{\pi}}\hat{\Pi}_\mu(\pi_\mu^e) = -1$, *i.e.*, $\frac{\partial}{\partial\pi}\Pi_\mu(\pi_\mu^e) = (\omega - 2)/\omega$.
- *If* $\sigma < 0$, *then the equilibrium* π_μ^e *is stable.*
- *If* $\sigma > 0$, *then* π_μ^e *is unstable.*

Let $\pi_{\hat{\mu}}^e$ denote the fixed-point equilibrium of the map $\hat{\Pi}_{\hat{\mu}}$ corresponding to a bifurcation value $\hat{\mu}$. The point $(\pi_{\hat{\mu}}^e, \hat{\mu})$ is called a *bifurcation point* since it is common that the stable equilibrium will "split apart" there. More formally, the map often undergoes a *period doubling*, in which stability passes from the equilibrium fixed point π_μ^e (period one) to a pair of period-two points $\pi_\mu^{(1)} < \pi_\mu^e$ and $\pi_\mu^{(2)} > \pi_\mu^e$ for which $\hat{\Pi}_\mu(\pi_\mu^{(1)}) = \pi_\mu^{(2)}$ and $\hat{\Pi}_\mu(\pi_\mu^{(2)}) = \pi_\mu^{(1)}$. In other words, the equilibrium π_μ^e becomes unstable, and the process cycles between two stable fixed points of the two-fold composition $\hat{\Pi}_\mu^2 = \hat{\Pi}_\mu \circ \hat{\Pi}_\mu$. The precise nature of a period-doubling bifurcation depends on the Schwarzian derivative as well as $\eta := \frac{\partial\hat{\Pi}_\mu}{\partial\mu}\frac{\partial^2\hat{\Pi}_\mu}{\partial\hat{\pi}^2} + 2\frac{\partial^2\hat{\Pi}_\mu}{\partial\hat{\pi}\partial\mu}$, as specified in the following theorem:

Theorem 3.5 *If a bifurcation point* $(\pi_{\hat{\mu}}^e, \hat{\mu})$ *satisfies* $\eta \neq 0$ *and* $\sigma \neq 0$, *then it is a period-doubling bifurcation. For* $\eta > 0$ *and* $\sigma < 0$ *we have the following behavior when* μ *lies in a neighborhood of* $\hat{\mu}$: *the equilibrium* π_μ^e *is stable for* $\mu \geq \hat{\mu}$ *and becomes unstable for* $\mu < \hat{\mu}$ *where a stable cycle of period-two points emerges. Reversing the sign of* η *reverses the stability of the equilibrium, while reversing the sign of* σ *reverses the stability of the period-two cycle. Also, reversing the sign of only one of either* η *or* σ *reverses the side of* $\hat{\mu}$ *on which the cycle lies.*

As stated in Theorem 3.5, the emerging period-two fixed points $\hat{\pi}_\mu^{(1)}$ and $\hat{\pi}_\mu^{(2)}$ of $\Pi_\mu^2(\hat{\pi})$ are stable only for parameters μ in a neighborhood of $\hat{\mu}$. Outside this neighborhood it is possible that the period-two points, themselves being fixed points of the map $\hat{\Pi}_\mu^2(\hat{\pi})$, undergo a period-doubling bifurcation as well when $|(\partial/\partial\hat{\pi})\hat{\Pi}_\mu^2(\hat{\pi})| = 1$. The chain rule implies that both points must bifurcate at the same time since the slope of $\hat{\Pi}_\mu^2(\hat{\pi})$ at each point is $(\partial/\partial\hat{\pi})\hat{\Pi}_\mu^2(\hat{\pi}_\mu^{(1)}) = (\partial/\partial\hat{\pi})\hat{\Pi}_\mu^2(\hat{\pi}_\mu^{(2)}) = (\partial/\partial\hat{\pi})\hat{\Pi}_\mu(\hat{\pi}_\mu^{(1)}) \cdot (\partial/\partial\hat{\pi})\hat{\Pi}_\mu(\hat{\pi}_\mu^{(2)})$.

Since the map $\hat{\Pi}_\mu(\cdot) = \Pi_\mu(\cdot)$ is nonincreasing under static expectations ($\omega = 1$), this additional bifurcation cannot occur in this setting. In fact, the only possible unstable behavior in this case is divergence or a period-2 cycle. We can readily see this by choosing without loss of generality an initial predicted price $\hat{\pi}^{(0)} > \pi_\mu^e$. Monotonicity implies $\hat{\pi}^{(1)} = P(\hat{\pi}^{(0)}) < P(\pi_\mu^e) = \pi_\mu^e$ and $\hat{\pi}^{(2)} = P(\hat{\pi}^{(1)}) > P(\pi_\mu^e) = hpi_\mu^e$. Induction then implies that convergence, a 2-cycle or divergence depends solely on whether $\hat{\pi}^{(2)} < \hat{\pi}^{(0)}$, $\hat{\pi}^{(2)} = \hat{\pi}^{(0)}$, or $\hat{\pi}^{(2)} > \hat{\pi}^{(0)}$, respectively. Due to this limited behavior under static expectations we focus in this chapter on the adaptive-expectations model where $\omega \in (0, 1)$.

The nature of this new bifurcation will again be determined by Theorem 3.5, with $\hat{\Pi}_\mu$ replaced by $\hat{\Pi}_\mu^2$ If the conditions in Theorem 3.5 are met, each branch of the formerly stable two-cycle will become unstable and split into two new stable branches, creating a stable four-cycle. In general, this period-doubling scenario can continue ad infinitum over a range of μ, creating regions where price predictions follow a stable cycle of period $2^n, n = 0, 1, \ldots$.

Typically, at a limit point of the sequence of period-doubling bifurcation points of the parameter μ, the price predictions will begin to oscillate in an aperiodic fashion, a telling sign of what has been coined *deterministic chaos*. In fact, periodic cycles of arbitrary order k (including aperiodic cycles) will become visible. In a somewhat weaker sense, the existence of such cycles of arbitrary order k (including aperiodic cycles), though possibly not attracting, is called *chaos in the Li/Yorke sense*. Such chaotic maps display a *sensitive dependence on initial conditions* for which (i) every aperiodic trajectory comes arbitrarily close to every other trajectory yet must eventually move away, and (ii) every aperiodic cycle must eventually move away from any periodic cycle. The condition for Li/Yorke chaos simply relies on the existence of a sequence of iterates $\hat{\pi}_0$, $\hat{\pi}_1 = \hat{\Pi}_\mu(\hat{\pi}_0)$, $\hat{\pi}_2 = \hat{\Pi}_\mu(\hat{\pi}_1)$ and $\hat{\pi}_3 = \hat{\Pi}_\mu(\hat{\pi}_2)$ satisfying $\hat{\pi}_2 < \hat{\pi}_1 < \hat{\pi}_0 \leq \hat{\pi}_3$ (or $\hat{\pi}_2 > \hat{\pi}_1 > \hat{\pi}_0 \geq \hat{\pi}_3$). For the case of linear waiting costs and bounded service values, we can form the following Li/Yorke theorem for our model.

Theorem 3.6 *Suppose that service values are bounded above by $a < \infty$ and that the waiting costs are linear, i.e., $G_\mu(\lambda) = hW_\mu(\lambda)$. Li/Yorke chaos exists under adaptive expectations provided the average delay (waiting time in the queue), $D_\mu(\lambda) = W_\mu(\lambda) - 1/\mu$, satisfies*

$$D_\mu(\lambda_2) \geq (\varpi + 1)(\underline{\mu}^{-1} - \mu^{-1}), \tag{3.10}$$

where $\varpi := (1 + (1 - \omega)^2)/(1 - \omega) > 2$, $\lambda_2 := \Lambda \bar{F}(a - \omega h(\underline{\mu}^{-1} - \mu^{-1}))$ and $\underline{\mu} := h/(a - \delta)$.

The form of the Li/Yorke condition in Theorem 3.6 motivates the following observations.

Remark 5   For a particular service capacity μ, Li/Yorke chaos exists for sufficient variability in the customers' job sizes. This arises from the fact that the mean queueing time in a system increases in the variability of the service times.

Remark 6   The existence of Li/Yorke chaos depends on the shape of the customer-value distribution, $F(\cdot)$. A distribution which is skewed to the left, where more customers receive high rewards close to a than low rewards near d, lends itself to chaos more than a uniform distribution. This occurs because, for a particular price prediction $\hat{\pi}_2$, a greater proportion of customers will have a value that exceeds $\hat{\pi}_2$ than in the uniform case. Thus, the resulting arrival rate $\lambda_2 = \Lambda \bar{F}(\hat{\pi}_2)$ and the associated expected queueing delay will both be greater. For the same reason a uniform distribution realizes chaos more often than a right-skewed distribution.

The period-doubling route to chaos has been shown to occur universally in a variety of one-dimensional, noninvertible maps often called "generally quadratic" in the sense that they have one critical point and a Schwarzian derivative that is negative. In our adaptive-expectations model, the map $\hat{\Pi}_\mu(\cdot)$ often contains a minimum. This is a consequence of the fact that $\Pi_\mu(\hat{\pi})$ is decreasing in $\hat{\pi}$, but asymptotic to $\pi + hW_\mu(0)$. Thus as $\hat{\pi}$ increases, the increasing linear term $(1 - \omega)\hat{\pi}$ comes to dominate in the map $\hat{\Pi}(\cdot)$. Thus, we expect that in general our model will exhibit period-doubling behavior towards a chaotic regime.

In the next section we shall examine the $M/GI/1$ queueing model with uniform value distribution. In this setting we shall explore the issues covered in this section, namely, equilibrium stability, bifurcation of the equilibrium, and the rise of Li/Yorke chaos.

3.2.2 Example: M/GI/1 Queue

In this section we consider a population of customers with linear waiting-cost functions and service values that are distributed uniformly on the interval $[d, a]$. Potential arrivals come from a Poisson process with mean rate Λ. For a price prediction $\hat{\pi}$, therefore, customers enter the system according to a Poisson process with mean rate λ given by

$$\lambda = \Lambda \bar{F}(\hat{\pi}) = \begin{cases} \Lambda & , \ 0 \leq \hat{\pi} \leq d \\ \Lambda(a - \hat{\pi})/(a - d) & , \ d \leq \hat{\pi} \leq a \\ 0 & , \ \hat{\pi} \geq a \end{cases} . \tag{3.11}$$

For the special case $\hat{\pi} = \delta$ we denote the resulting arrival rate by $\Lambda' := \Lambda(a - \delta)/(a - d)$. The sizes of customer service requirements are i.i.d. random variables, distributed as S with first and second moments $E[S] = 1$ and $E[S^2] = 2\beta$, respectively, where $\beta \geq 1/2$ is a given parameter. In other words, without loss of generality, we measure work requirements in units of the mean requirement. For notational compactness we often use an alternative parameter, $\beta' := 1 - \beta \leq 1/2$, to capture the variation in these requests.

If $\lambda < \mu$, then the facility behaves as an $M/GI/1$ queue in steady state. In other words, in our dynamic-system model, we assume that the arrival rate remains fixed throughout each time period and a period lasts long enough for the system to attain steady state (approximately).

By the Pollaczek-Khintchine formula the average waiting times in the system and queue are then, respectively,

$$W_\mu(\lambda) = (\mu - \beta'\lambda)[\mu(\mu - \lambda)]^{-1}, \qquad \text{and} \tag{3.12}$$

$$D_\mu(\lambda) = (1 - \beta')\lambda[\mu(\mu - \lambda)]^{-1}. \tag{3.13}$$

The expected delay is infinite whenever $\mu \leq \lambda \leq \Lambda$, which corresponds to price predictions $\hat{\pi} \leq \hat{\underline{\pi}} := \bar{F}^{-1}(\mu/\Lambda) = a - (a - d)\mu/\Lambda$.

An interior equilibrium arrival rate, $\lambda = \lambda_\mu^e \in (0, \Lambda)$, satisfies the equilibrium equation (3.3), which now takes the form

$$a - (a - d)\lambda/\Lambda = \delta + h(\mu - \beta'\lambda)[\mu(\mu - \lambda)]^{-1} .$$

By subtracting δ from both sides and dividing by $a - \delta > 0$, this becomes

$$1 - \lambda/\Lambda' = \underline{\mu}(\mu - \beta'\lambda)[\mu(\mu - \lambda)]^{-1} .$$

Thus, λ_μ^e is a root of the quadratic equation $(\Lambda' - \lambda)(\mu - \lambda) = \Lambda'(\underline{\mu}\mu^{-1})(\mu - \beta'\lambda)$, which we rewrite as

$$\lambda^2 + (\kappa_\mu - 2\mu)\lambda + \Lambda'(\mu - \underline{\mu}) = 0, \tag{3.14}$$

with

$$\kappa_\mu := \mu - \Lambda' + \beta'\Lambda'(\underline{\mu}\mu^{-1}) .$$

The two roots of (3.14) are $\lambda = (2\mu - \kappa_\mu \pm [\kappa_\mu^2 + 4\tau]^{1/2})/2$, where $\tau := (1 - \beta')\Lambda'\underline{\mu} > 0$. Since $\lambda_\mu^e < \mu$, the desired root is the smaller one, namely,

$$\lambda_\mu^e = (2\mu - \kappa_\mu - [\kappa_\mu^2 + 4\tau]^{1/2})/2 . \tag{3.15}$$

An expression of the equilibrium price in terms of the equilibrium arrival rate (3.15) is then given by

$$\pi_\mu^e = \bar{F}^{-1}(\lambda_\mu^e/\Lambda) = a - (a - d)\lambda_\mu^e/\Lambda . \tag{3.16}$$

3.2.2.1 Stability

For a uniform reward distribution, $\bar{F}(\cdot)$, the (sufficient) stability condition in Theorem 3.3 takes the form

$$(\omega - 2)/\omega \quad < \quad \frac{\partial}{\partial \hat{\pi}}\Pi_\mu(\pi_\mu^e) = h\Lambda\bar{F}'(\pi_\mu^e)\frac{\partial}{\partial \lambda}W_\mu(\lambda_\mu^e)$$

$$= -\Lambda'\underline{\mu}\frac{\partial}{\partial \lambda}W_\mu(\lambda_\mu^e). \tag{3.17}$$

At a bifurcation point, the inequality in (3.17) is replaced with an equality, producing the equation

$$\omega\Lambda'\underline{\mu}\frac{\partial}{\partial \lambda}W_\mu(\lambda_\mu^e) = 2 - \omega. \tag{3.18}$$

We now investigate the conditions of Theorem 3.4 in order to ascertain the behavior at a bifurcation point, i.e., when (3.18) is satisfied. From the conditions of Theorem 3.4, the following result implies that the bifurcation point is a stable equilibrium.

Lemma 3.7 *The Schwarzian derivative is negative at the bifurcation points.*

By virtue of Theorems 3.3 and 3.4, we have the following necessary and sufficient stability condition for our $M/GI/1$ setting:

$$(\mu - \lambda_\mu^e)^2 \geq \tau\omega/(2 - \omega). \tag{3.19}$$

The condition (3.19) is satisfied for $\lambda_\mu^e < \mu$ if and only if

$$\lambda_\mu^e \leq \mu - [\tau\omega/(2 - \omega)]^{1/2} . \tag{3.20}$$

Using the closed form expression (3.15) for λ_μ^e in this stability condition (3.20), we obtain the following result.

Lemma 3.8 *The equilibrium* λ_μ^e *is stable if and only if* $\kappa_\mu \geq \kappa$, *where*

$$\kappa := -2(1 - \omega)[\tau/(\omega(2 - \omega))]^{1/2} < 0.$$

Lemma 3.8 is now used to establish a stability threshold for the admission fee δ.

Theorem 3.9 *For the M/GI/1 queue with uniform distribution of service value and linear waiting costs, the interior equilibrium (3.16) is stable if and only if* $\delta \geq \hat\delta$, *where*

$$\hat\delta := a - (\mu - \kappa)/\Lambda'' - \beta'h/\mu < a ,$$

with $\Lambda'' := \Lambda/(a - d)$.

Since $\mu > 0$, the stability condition $\kappa_\mu \geq \kappa$ of Lemma 3.8 can be written $\mu\kappa_\mu \geq \mu\kappa$, or simply as the quadratic condition $\hat\phi(\mu) \geq 0$, where $\hat\phi(\mu) := \mu^2 - (\Lambda' + \kappa)\mu + \beta'\Lambda'\underline\mu$. If real-valued roots for $\hat\phi(\mu)$ exist, then the bifurcation values are given by

$$\hat\mu_\pm := (\Lambda' + \kappa \pm [(\Lambda' + \kappa)^2 - 4\beta'\Lambda'\underline\mu]^{1/2})/2 . \tag{3.21}$$

The convexity of $\hat\phi(\mu)$ then implies that $\hat\phi(\mu) \geq 0$ if and only if $\mu \leq \hat\mu_-$ or $\mu \geq \hat\mu_+$, providing the following necessary and sufficient stability conditions on the service capacity μ.

Theorem 3.10 *For the M/GI/1 queue with uniform distribution of service value and linear waiting costs, the interior equilibrium (3.16) is stable if and only if* $\mu \notin (\hat\mu_-, \hat\mu_+)$.

3.2.2.2 Period Doubling

We would like to now characterize the behavior about these bifurcation points. Under adaptive expectations, we would like to know if a period doubling occurs at the bifurcation point from which a stable 2-cycle emerges.

Lemma 3.11 *The bifurcation values* $\hat\mu_- < \hat\mu_+$ *satisfy*

1. $\hat\mu_-^2 < \beta'\Lambda'\underline\mu < \hat\mu_+^2$.

2. $\frac{\partial}{\partial\mu}\lambda_\mu^e \big|_{\mu=\hat\mu_-} > 1$ *and* $\frac{\partial}{\partial\mu}\lambda_\mu^e \big|_{\mu=\hat\mu_+} < 1$.

Theorem 3.12 *For the* M/GI/1 *queue with uniform distribution of service value and linear waiting costs, the bifurcation values* $\hat{\mu}_- < \hat{\mu}_+$ *are supercritical period-doubling bifurcation points.*

3.2.2.3 Chaotic Dynamics

Supercritical period doubling under adaptive expectations eventually leads to a region of Li/Yorke chaos as specified in the following result.

Theorem 3.13 *Using* $\varpi = (1 + (1 - \omega)^2)/(1 - \omega)$, *define* $\tilde{\mu}_\pm := (\omega \Lambda'/2)(1 \pm \varphi^{1/2}(\omega))$ *where* $\varphi(\omega) := 1 - (s/\omega)(\varpi + \beta')(\varpi + 1)^{-1}$ *and* $s := 4\mu/\Lambda'$. *Then a service rate* μ *in the interval* $[\tilde{\mu}_-, \tilde{\mu}_+]$ *induces Li/Yorke chaos about an interior equilibrium.*

We now examine how likely such chaotic regions are to exist. The Li/Yorke capacities $\tilde{\mu}_\pm$ are real-valued if and only if $\varphi(\omega) \geq 0$. Substitution of $\varpi = (1 + (1 - \omega)^2)/(1 - \omega)$ yields

$$\varphi(\omega) = 1 - (s/\omega)(\omega^2 - (2 + \beta')\omega + (2 + \beta'))/(\omega^2 - 3\omega + 3).$$

Since $\omega^2 - 3\omega + 3 > 0$ for all ω, the condition $\varphi(\omega) \geq 0$ is equivalent to $\tilde{\varphi}(\omega) \geq 0$, where

$$
\begin{aligned}
\tilde{\varphi}(\omega) &:= \omega(\omega^2 - 3\omega + 3)\varphi(\omega) \\
&= \omega^3 - (3 + s)\omega^2 + (3 + (2 + \beta')s)\omega - (2 + \beta')s.
\end{aligned}
$$

The function $\tilde{\varphi}(\omega)$ is continuous on the interval $[0, 1]$ taking values $\tilde{\varphi}(0) = -(2 + \beta')s$, $\tilde{\varphi}(s) = (1 - \beta')s(1 - s)$ and $\tilde{\varphi}(1) = 1 - s$. Moreover, $\tilde{\varphi}(\omega)$ is concave on $[0, 1]$ since $\tilde{\varphi}''(\omega) = -2(s + 3(1 - \omega)) < 0$ due to the fact that $\omega < 1 < 1 + s/3$.

For $\beta' > -2$ (i.e., small variation in service requests), $\tilde{\mu}_\pm \in \Re$ for sufficiently large ω provided $s < 1$. To see why, notice that $\tilde{\varphi}(0) < 0$ whereas $\tilde{\varphi}(s), \tilde{\varphi}(1) > 0$. Thus, there exists a root of $\tilde{\varphi}(\omega)$, $\omega_s \in (0, s)$, whereby $\tilde{\mu}_\pm \notin \Re$ for $\omega \in (0, \omega_s)$, and $\tilde{\mu}_\pm \in \Re$ for $\omega \in [\omega_s, 1)$.

For $\beta' \leq -2$ (i.e., large variation in service requests), we have $\tilde{\varphi}(0) \geq 0$. In the case of equality, $\tilde{\varphi}(\omega)$ is increasing at 0 since $\tilde{\varphi}'(0) = 3 + (2 + \beta')s = 3 > 0$. Otherwise, $\beta' < -2$ gives $\tilde{\varphi}(0) > 0$. In either case, continuity ensures $\tilde{\varphi}(\omega) \geq 0$ for all sufficiently small $\omega > 0$. For $s \leq 1$ we have $\tilde{\varphi}(1) \geq 0$ so that concavity ensures $\tilde{\varphi}(\omega) \geq 0$, i.e., $\tilde{\mu}_\pm \in \Re$, on the entire interval $\omega \in (0, 1)$. Otherwise, $s > 1$ implies $\tilde{\varphi}(1) < 0$ so that $\exists\, 0 < \omega_s < 1$ such that $\tilde{\mu}_\pm \in \Re$ for $\omega \in (0, \omega_s]$, and $\tilde{\mu}_\pm \notin \Re$ for $\omega \in (\omega_s, 1)$.

Remark 7   From a system-management perspective, it is interesting to note that the chaotic behavior made possible when $s < 1$ corresponds, upon expansion of s, to low admission fees, $\delta < \tilde{\delta}$, where $\tilde{\delta} := a - (2/\Lambda)[h(a - d)]^{1/2}$. Thus, imposing a larger fee reduces the possibility of chaos. In fact, for $\delta \geq \tilde{\delta}$, i.e., $s \geq 1$, we can guarantee no possibility of chaos for all $\omega \in (0, 1)$ when the variability of jobs is small. To wit, suppose $\beta \leq 2$, i.e., $\beta' \geq 0$. Now if $s \geq 1$ then $\tilde{\varphi}(\omega)$ is increasing on $\omega \in (0, 1)$ by concavity and the fact that $\tilde{\varphi}'(1) = \beta's \geq 0$. Thus, $\tilde{\varphi}(\omega) < \tilde{\varphi}(1) = 1 - s \leq 0$, for all $\omega \in (0, 1)$.

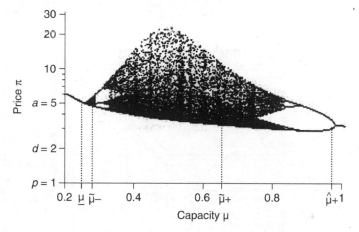

Figure 3.1 *Period-Doubling Bifurcations*

3.2.2.4 Stability, Bifurcation and Chaos: An Illustration

This section illustrates the asymptotic behavior of the price/arrival-rate dynamics under adaptive expectations. For simplicity, we consider the case of an $M/M/1$ queue ($\beta = 2$). Since $\beta' = 0$, the lower value $\hat{\mu}_- = 0 < \underline{\mu}$ in (3.21) can be ignored. The remaining bifurcation value $\hat{\mu}_+$ is given by

$$\hat{\mu}_+ = \Lambda' + \kappa = \Lambda' - 2(1-\omega)[\Lambda'\underline{\mu}/(\omega(2-\omega))]^{1/2} \leq \Lambda' < \Lambda. \qquad (3.22)$$

Clearly if $\Lambda' \leq \underline{\mu}$, then the equilibrium is stable for all service capacities $\mu > 0$. We therefore restrict attention to the more interesting case in which $\Lambda' > \underline{\mu}$.

To illustrate the asymptotic behavior of prices we rescale cost and time units by setting $h = 1$ and $\Lambda = 1$, respectively, and choose lower and upper bounds on customer values equal to $d = 2$ and $a = 5$, respectively. We also let $\omega = 0.7$.

The bifurcation diagram in Figure 3.1 displays attracting price values $\hat{\pi}$ over a range of service rates μ. The bifurcation diagram was constructed by iterating the mapping $\hat{\Pi}_\mu$ for 200 fixed service rates in the discretized interval from 0.2 to $\Lambda = 1$. The plotted points at each fixed service rate are the sequence of prices experienced after a transient period of 5000 iterations.

As expected, prices converge to a single point, namely, the stable equilibrium π^e_μ, when the capacity μ is at least $\hat{\mu}_+ \approx 0.97$. A period-doubling bifurcation of the equilibrium occurs as μ is decreased below $\hat{\mu}_+$. For capacities μ just below $\hat{\mu}_+$, prices alternate between a stable 2-cycle. Near $\mu = 0.86$ we see the 2-cycle split into a 4-cycle, which then in turn splits into an 8-cycle near $\mu = 0.82$. A period-doubling cascade then ensues as the service rate parameter is decreased further. The band of Li/Yorke chaos begins for $\mu \leq \tilde{\mu}_+ \approx 0.65$ and continues until μ drops below $\tilde{\mu}_- \approx 0.28$. Notice the large window of stable period-3 prices on the left side of this chaotic region for $\mu \geq \tilde{\mu}_-$. The presence of such a 3-cycle is a well-known sign of a chaotic mapping [130].

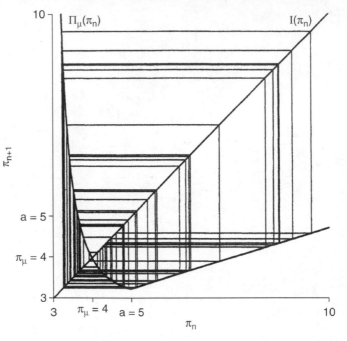

Figure 3.2 *Chaotic Cobweb*

In terms of the arrival rate, Figure 3.1 reveals the following dichotomy: points lying below $\hat{\pi} = a$ correspond to a positive arrival rate, whereas those above indicate no arrivals. Recall from Theorem 3.1 that violating the assumption $\mu > \underline{\mu}$ produces a stable noninterior arrival-rate equilibrium $\lambda_{\mu}^e = 0$, as indicated by the single curve of prices above a for $\mu \leq \underline{\mu} = 1/4$.

Now consider the particular service capacity $\mu = 2/3$ which lies just outside the interval of Li/Yorke chaos. This capacity yields $\kappa_{\mu} = \mu - \Lambda' = -2/3$ and $\tau = \Lambda'\mu = 1/3$, which produce via (3.15) and (3.16), respectively, an arrival rate equilibrium $\lambda_{\mu}^e = 1/3$ and corresponding price equilibrium $\pi_{\mu}^e = 4$.

Figure 3.2 depicts an economic cobweb diagram about this equilibrium $\pi_{\mu}^e = 4$ when $\mu = 2/3$. The cobweb connects each price iterate $\hat{\pi}^{(n)}$ with its mapping $\hat{\pi}^{(n+1)} = \hat{\Pi}_{\mu}(\hat{\pi}^{(n)})$ by reflecting $\hat{\pi}^{(n+1)}$ back to the $\hat{\pi}^{(n)}$-axis via the identity map $I(\cdot)$. Initial transient behavior has been removed and yet the evolution of price iterates has not stabilized into a small order cycle. The wild trajectory exhibits what can be described as an apparently "chaotic cobweb."

To illustrate how this behavior affects the arrival rate process, Figure 3.3 displays a histogram of arrival rates determined from a sequence of 100,000 price iterates at the capacity $\mu = 2/3$. The arrival-rate region $(0, \mu)$ is divided into 200 intervals to show the frequency at which the process visits the entire region. Low-degree periodic arrivals would tend to fill in just a few intervals unlike the observed dense distribution in Figure 3.3. A large spike, nearly 40%

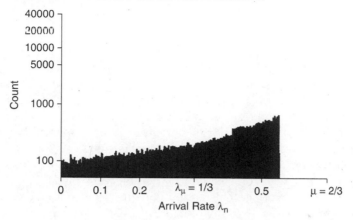

Figure 3.3 *Arrival Rate Distribution*

of all iterates, occurs at $\lambda = 0$. This corresponds to the frequent transitions to prices above the maximum reward a in the bifurcation diagram of Figure 3.1 when $\mu = 2/3$.

3.2.3 *Summary of Results*

We have examined the dynamic behavior of a simple input-pricing mechanism for self-optimizing customers using a service facility. At the equilibrium price and arrival rate, the marginal value of service equals the price: the sum of the admission fee and the customer waiting cost. We have used a variant of the discrete-time dynamical-systems model of Section 2.4.1 of Chapter 2, in which customers base their future join/balk decisions on their previous experience of congestion. Unlike that model, in which customers expect (predict) that the next period's price will be the same as in the current period, we have adopted a model with *adaptive expectations* from economic theory, in which the next period's price prediction is a convex combination of the predicted price and the actual observed price for the current period. This simple prediction constitutes an exponential smoothing forecast based on all previously observed prices.

The problem is formulated in terms of a dynamical system governed by the map, $\hat{\Pi}_\mu(\cdot)$, which maps the (predicted) price in the current period onto the (predicted) price in the next period. (Here the parameter μ is a measure of the service capacity.) The equilibrium price π^e_μ is the unique fixed point of this map. The equilibrium price is globally (locally) stable if the sequence of prices (iterates of the map $\hat{\Pi}_\mu(\cdot)$) converges to the equilibrium from every possible initial price (from every initial price in a neighborhood of π^e_μ).

In this context we have shown that there exists a region of service capacities which create an instability in the sense that the system does not return to the equilibrium after a perturbation. We have established that using static expectations creates a larger instability region than using adaptive expectations.

This implies that stability under static expectations ensures stability under adaptive expectations.

The inherent dampening of the amplitude of oscillations under adaptive expectations is accompanied by very different behavior inside the instability region. Under static expectations the loss of stability involves an immediate and complete destabilization, in which arrival rates diverge toward an extreme cycle involving an empty and/or full system. The initial loss of equilibrium stability under adaptive expectations, however, is accompanied by the emergence of a small, stable 2-cycle about the equilibrium.

In the literature on dynamical systems the emergence of a stable 2-cycle at the point of change from equilibrium stability to instability is called a *period-doubling* bifurcation. This period doubling may continue for a convergent sequence of values of the parameter μ, as stability passes to a 2-cycle, then a 4-cycle, then an 8-cycle, etc., eventually leading to a regime of *chaotic* behavior, in which the trajectory of prices is aperiodic and apparently completely random.

Focusing attention on a $M/GI/1$ model with uniform value distribution, we have shown that equilibrium does, in fact, undergo a period-doubling bifurcation and can become chaotic in the sense of Li and Yorke. That is, every aperiodic trajectory of prices comes arbitrarily close to every other trajectory yet must eventually move away. Thus the system displays *sensitive dependence on initial conditions*. Hence, the quantitative stabilization under adaptive expectations brings with it a qualitative destabilization, too, characterized by apparently random fluctuations about the equilibrium. From the perspective of a service-facility manager we show that raising the admission fee above a threshold reduces and possibly eliminates the presence of this chaotic instability. We also establish another threshold on the admission fee which, when exceeded, ensures the stability of the equilibrium.

Relaxing the assumption of uniformity in the customer value distribution, we determined from Theorem 3.6 that the possibility of this chaotic behavior increases with the variation in job size as well as the negative skewness of the service-value distribution. This latter result means that for two value distributions with the same upper bound on customer value, the one with a greater mass towards higher service values is more likely to create chaotic behavior.

3.3 Discrete-Time Dynamic Algorithms: Variants

In this section we describe some variants of the discrete-time algorithm introduced and studied in Section 3.2.

Recall that in the model of Section 3.2 the individually optimizing customers form an estimate or prediction, $\hat{\pi}$, of the admission price, $\pi = G(\lambda) + \delta$, and base their join/balk decision on this predicted price. A potential arriving customer joins the system if and only if its service value exceeds the predicted price, $\hat{\pi}$. The resulting arrival rate equals $\Lambda \bar{F}(\hat{\pi})$. If $\hat{\pi} = \pi$ (equivalently, if $\Lambda\bar{F}(\hat{\pi}) = \lambda$) then the system is in equilibrium and λ is the individually optimal arrival rate.

In the discrete-time algorithm introduced in Section 3.2, the predicted prices were updated sequentially according to the difference equation,

$$\hat{\pi}^{(n+1)} = (1 - \omega)\hat{\pi}^{(n)} + \omega\pi^{(n)} , \qquad (3.23)$$

where $\pi^{(n)} = G(\lambda^{(n)}) + \delta$ and $\lambda^{(n)} = \lambda(\hat{\pi}^{(n)}).$[†] The difference equation (3.23) embodies an exponential smoothing of the previous predicted prices.

As an alternative to (3.23), one could consider an exponential smoothing of predictions of the arrival rate, rather than the price. Let $\hat{\lambda}^{(n)}$ denote the prediction of the arrival rate at the nth step of the algorithm, $n = 0, 1, \ldots$ and suppose that the predicted arrival rates are updated sequentially according to the difference equation,

$$\hat{\lambda}^{(n+1)} = (1 - \omega)\hat{\lambda}^{(n)} + \omega\lambda^{(n)} , \qquad (3.24)$$

where again $\lambda^{(n)} = \lambda(\hat{\pi}^{(n)})$ but now $\hat{\pi}^{(n)} = G(\hat{\lambda}^{(n)}) + \delta$.

(In the extreme case where $\omega = 1$, we have

$$\hat{\lambda}^{(n+1)} = \lambda^{(n)} = \lambda(\hat{\pi}^{(n)}) = \lambda(G(\hat{\lambda}^{(n)}) + \delta) ,$$

and this variant is equivalent to the original algorithm introduced in Section 3.2.)

Note that in this version of the algorithm, stage n consists of the following steps:

1. given the predicted arrival rate at stage n, the predicted price is calculated: $\hat{\pi}^{(n)} = G(\hat{\lambda}^{(n)}) + \delta$;

2. given the predicted price, the actual arrival rate is calculated: $\lambda^{(n)} = \lambda(\hat{\pi}^{(n)})$;

3. the predicted arrival rate at stage $n+1$ is calculated by smoothing, based on the predicted and actual arrival rates at stage n.

By contrast, in the original version of the algorithm, stage n consists of the following steps:

1. given the predicted price at stage n, the actual arrival rate is calculated: $\lambda^{(n)} = \lambda(\hat{\pi}^{(n)})$;

2. given the actual arrival rate, the actual price is calculated: $\pi^{(n)} = G(\lambda^{(n)}) + \delta$;

3. the predicted price at stage $n + 1$ is calculated by smoothing, based on the predicted and actual prices at stage n.

In both versions, the predicted price determines the customers' joining behavior and therefore also the actual arrival rate in stage n. The difference lies in how the predicted price is calculated: directly by smoothing in the original

---

[†] Recall that $\lambda(\hat{\pi})$ is defined as the unique solution to the equilibrium condition, $U'(\lambda) = \hat{\pi}$. In the case where the utility function, $U(\lambda)$, is generated by potential arrivals at rate Λ who join with probability $\bar{F}(\hat{\pi})$, we have $\lambda(\hat{\pi}) = \Lambda\bar{F}(\hat{\pi})$.

version, versus indirectly in the second version by basing the predicted price on the predicted arrival rate, which is calculated by smoothing. Of course, in the extreme case when $\omega = 1$ (no smoothing), the two versions are equivalent.

In each of these variants, the prediction of a variable (price or arrival rate) is formed by smoothing the differences between the actual and predicted values of the variable in question. In the original version,

$$\hat{\pi}^{(n+1)} - \hat{\pi}^{(n)} = \omega(\pi^{(n)} - \hat{\pi}^{(n)}) . \tag{3.25}$$

In the second version,

$$\hat{\lambda}^{(n+1)} - \hat{\lambda}^{(n)} = \omega(\lambda^{(n)} - \hat{\lambda}^{(n)}) . \tag{3.26}$$

In the original version, for example, the effect of the smoothing is to increase (decrease) the predicted price in proportion to the amount by which the actual price is larger than (smaller than) the predicted price, and similarly for the second version (with price replaced by arrival rate).

An alternative is to let the change in the predicted price be proportional to the discrepancy between the predicted and actual values of the arrival rate:

$$\hat{\pi}^{(n+1)} - \hat{\pi}^{(n)} = \omega(\hat{\lambda}^{(n)} - \lambda^{(n)}) , \tag{3.27}$$

where in this context

$$\begin{aligned} \hat{\pi}^{(n)} &= G(\hat{\lambda}^{(n)}) + \delta , \\ \lambda^{(n)} &= \lambda(\hat{\pi}^{(n)}) . \end{aligned}$$

That is, $\lambda^{(n)}$, the actual arrival rate, is the arrival rate that results from the predicted price $\hat{\pi}^{(n)}$, while $\hat{\lambda}^{(n)}$, the predicted arrival rate, is the unique value of λ that would generate a price equal to the predicted price $\hat{\pi}^{(n)}$. In other words, it is the arrival rate that is consistent (from the point of view of individual optimization) with the price prediction $\hat{\pi}^{(n)}$.

Note that in this smoothing the predicted price *increases* (*decreases*) at stage $n + 1$ if the actual arrival rate is *less than* (*greater than*) the predicted arrival rate at stage n. This is consistent with the inverse relationship between the price and the arrival rate.

Similarly, one could let the change in the predicted arrival rate be proportional to the discrepancy between the predicted and actual values of the price:

$$\hat{\lambda}^{(n+1)} - \hat{\lambda}^{(n)} = \omega(\hat{\pi}^{(n)} - \pi^{(n)}) , \tag{3.28}$$

where in this context

$$\begin{aligned} \pi^{(n)} &= G(\hat{\lambda}^{(n)}) + \delta , \\ \hat{\lambda}^{(n)} &= \lambda(\hat{\pi}^{(n)}) . \end{aligned}$$

That is, $\pi^{(n)}$, the actual price, is the price that results from the predicted arrival rate, $\hat{\lambda}^{(n)}$ (which, in this setting, is the arrival chosen by the individually optimizing customers at stage n), while $\hat{\pi}^{(n)}$ is the unique value of π that would generate an arrival rate equal to the predicted arrival rate $\hat{\lambda}^{(n)}$.

In other words, it is the price that is consistent (from the point of view of individual optimization) with the customers' choice, $\lambda = \hat{\lambda}^{(n)}$.

Similarly to the previous variant, the predicted arrival rate *increases* (*decreases*) at stage $n+1$ if the actual price is *less than* (*greater than*) the predicted price at stage n.

Note that (3.28) is equivalent to

$$\hat{\lambda}^{(n+1)} - \hat{\lambda}^{(n)} = \omega \left(U'(\hat{\lambda}^{(n)}) - [G(\hat{\lambda}^{(n)}) + \delta] \right) , \tag{3.29}$$

since $U'(\hat{\lambda}^{(n)}) = \hat{\pi}^{(n)}$. Thus we see that in this variant the predicted arrival rate increases (decreases) at stage $n + 1$ if the marginal value is greater than (less than) the full price at stage n, in other words, if the demand curve, $D(\lambda) = U'(\lambda)$, lies above (below) the supply curve, $S(\lambda) = G(\lambda) + \delta$, at $\lambda = \hat{\lambda}^{(n)}$.

3.4 Continuous-Time Dynamic Adaptive Algorithms

In the dynamic algorithms discussed so far in this chapter, (predicted) arrival rates and prices change at discrete points in time and their evolution is governed by difference equations. As we have indicated, such algorithms may provide a model for systems in which users make decisions about system parameters such as arrival rates and tolls on a time scale which is significantly longer than the time scale on which the queue lengths and waiting times evolve. In particular, they may be an appropriate model for systems in which a steady state is reached (at least approximately) between successive adjustments to the system parameters.

By contrast, when parameter adjustments can take place on the same time scale as the evolution of the measures of congestion, continuous-time dynamic algorithms may be appropriate. They are the subject of this section.

We assume the same basic model as in our discussion of discrete-time algorithms and we review it here for the reader's convenience. The arrival rate λ is a decision variable and the feasible region for λ is a given set $A \subseteq [0, \infty)$. Unless otherwise noted, we assume that $A = [0, \infty)$. There is a utility function, $U(\lambda)$, which measures the average gross value received per unit time by entering customers as a function of the arrival rate λ.

For a given λ, $G(\lambda)$ denotes the average waiting cost of a job. In addition to incurring the waiting cost $G(\lambda)$, an entering customer may have to pay an admission fee (or toll) δ. The sum of the toll and the waiting cost $-\delta + G(\lambda)$ – constitutes the full price of admission.

As usual, our basic model will assume individually optimizing customers. Socially or facility optimal solutions may be implemented by charging entering customers an appropriate toll. Recall the Nash equilibrium conditions for an individually optimal solution:

$$U'(\lambda) \leq G(\lambda) + \delta ; \tag{3.30}$$

$$U'(\lambda) = G(\lambda) + \delta , \text{ if } \lambda > 0 . \tag{3.31}$$

Now suppose the arrival rate and toll evolve in continuous time. Let $\lambda(t)$ denote the arrival rate at time $t \geq 0$. Let $\delta(t)$ denote the toll charged per entering customer at time $t \geq 0$. Let ω be a positive constant. Consider the differential equation

$$\frac{d}{dt}\lambda(t) = \omega \left(U'(\lambda(t)) - [G(\lambda(t)) + \delta(t)]\right) . \tag{3.32}$$

We interpret (3.32) as follows. The facility operator charges a toll $\delta(t)$ per entering customer. Equation (3.32) corresponds to a rate-control algorithm in which the flow is adjusted in proportion to the current difference between the marginal utility and the full price (the waiting cost plus the toll) of entering the system. If this difference is positive (that is, if the marginal utility exceeds the full price), then the flow is increased; if negative then the flow is decreased. (If the current value of $\lambda(t)$ is zero, it is understood that the rate-control algorithm keeps this flow at zero in the latter case rather than reducing it further and making it negative.)

It is therefore reasonable to conjecture that this algorithm will converge to a value of λ satisfying the Nash equilibrium conditions, (3.30), (3.31), provided that $\delta(t) \to \delta$.

Note that this version of the continuous-time dynamic algorithm is the continuous-time analogue of the fourth variant of the discrete-time algorithm (cf. (3.29)) governed by the difference equations,

$$\hat{\lambda}^{(n+1)} - \hat{\lambda}^{(n)} = \omega \left(U'(\hat{\lambda}^{(n)}) - [G(\hat{\lambda}^{(n)}) + \delta]\right) ,$$

In the continuous-time setting, however, we are allowing the toll to be time varying and we are eliminating the "hat" on the arrival rate, even though it is appropriate to think of $\lambda(t)$ as a predicted arrival rate, in the sense that it is a prediction (or estimate) of an equilibrium arrival rate.

The stability analysis of the dynamic algorithm in the continuous-time case differs somewhat from the discrete-time case in both the analysis and the results. In particular, we shall see that, in contrast to the discrete-time case, chaotic behavior cannot occur in continuous time.

The specific behavior of the algorithm depends on how we choose the toll $\delta(t)$. If $\delta(t) = \delta$ (fixed toll), then we shall see that the algorithm converges to the individually optimal arrival rate associated with the fixed toll δ. If we choose $\delta(t)$ equal to the external effect at time t, the algorithm converges (not surprisingly) to the socially optimal arrival rate. Finally, we show that the algorithm converges to a solution to the optimality conditions for the facility optimization problem when $\delta(t)$ is equal to the dynamic counterpart of the facility optimal toll. In this case, however, since there may be more than one solution to the optimality conditions, the algorithm might not converge to the global optimum for the facility optimization problem.

We shall find it convenient to consider the case of social optimization first, since the cases of individual and facility optimization are both analyzed by exploiting the equivalence to a social optimization problem with a transformed utility or waiting-cost function.

3.4.1 Algorithm with Time-Varying Toll: Social Optimality

Consider the differential equation (3.32) with

$$\delta(t) = \lambda(t)G'(\lambda(t)) . \tag{3.33}$$

That is,

$$\frac{d}{dt}\lambda(t) = \omega\left(U'(\lambda(t)) - [G(\lambda(t)) + \lambda(t)G'(\lambda(t))]\right) . \tag{3.34}$$

In this case, the toll $\delta(t)$ equals the external effect of the flow through the system at time t, so that the full price at time t equals the marginal cost of waiting. As a result the r.h.s. of the differential equation is proportional to the difference between the marginal utility and the marginal cost of waiting. At a (positive) *stationary point* of the dynamical system corresponding to the differential equation, this difference equals zero, which is precisely the necessary and sufficient condition for a (positive) socially optimal arrival rate. Thus we may reasonably conjecture that the dynamical system will converge to the socially optimal arrival rate.

To prove this result formally, let $\mathcal{U}(\lambda) := U(\lambda) - \lambda G(\lambda)$, the net utility per unit time when the arrival rate equals λ. We have

$$\mathcal{U}'(\lambda) = U'(\lambda) - [G(\lambda) + \lambda G'(\lambda)] . \tag{3.35}$$

Consider $\mathcal{U}(\lambda(t))$, where $\lambda(t)$ satisfies the differential equation (3.32) with $\delta(t)$ given by (3.33). From (3.35), (3.32), and (3.33) it follows that

$$\frac{d}{dt}\mathcal{U}(\lambda(t)) = \mathcal{U}'(\lambda(t)) \cdot \frac{d}{dt}\lambda(t) = \omega\left(U'(\lambda(t)) - [G(\lambda(t)) + \lambda G'(\lambda)]\right)^2 ,$$

from which we see that $\mathcal{U}(\lambda(t))$ is strictly increasing over time unless the quantity in parentheses on the last line equals zero, that is, unless $\lambda(t)$ is a solution to the first-order conditions. It follows that the algorithm converges (with strictly increasing aggregate utility) to the unique solution to the first-order conditions, which, as we have noted, is the socially optimal arrival rate (and a Nash equilibrium for individually optimizing customers when charged the the socially optimal toll).

A practical implementation of this algorithm requires that the system operator be able to learn enough about the customer's waiting-cost functions to come up with a reasonable approximation of the external effect.

See Chapter 8 for extensions of this algorithm to multiclass networks of queues.

3.4.2 Algorithm with Fixed Toll: Individual Optimality

Consider the dynamical system governed by the differential equation (3.32), but now suppose that the toll charged each entering customer is a fixed quantity, δ. The differential equation is now

$$\frac{d}{dt}\lambda(t) = \omega\left(U'(\lambda(t)) - [G(\lambda(t)) + \delta]\right) . \tag{3.36}$$

As we already observed, a stationary point λ of this system, which satisfies (3.30), (3.31), is a Nash equilibrium for individually optimizing customers when entering customers are charged the toll δ. In fact, it is the unique Nash equilibrium corresponding to the toll δ and the dynamic algorithm (3.36) converges globally to this equilibrium.

To prove the latter two assertions, define a new net utility function, $\hat{\mathcal{U}}$, as follows:

$$\hat{\mathcal{U}}(\lambda) := U(\lambda) - \int_0^\lambda (G(\nu) + \delta)d\nu .$$

It is easily seen that $\hat{\mathcal{U}}(\lambda)$ is concave in λ under our assumptions. (In fact, the waiting-cost function, $G(\lambda)$, need not be convex, just nondecreasing.) Thus $\hat{\mathcal{U}}(\lambda)$ has a unique global maximum, which is the unique solution to the *KKT* conditions. But in this case the *KKT* conditions are equivalent to the above conditions for a Nash equilibrium.

The new net utility function $\hat{\mathcal{U}}(\cdot)$ is a Lyapunov function for the dynamical system governed by the differential equation (3.36), as can be seen by the following argument. Consider $\hat{\mathcal{U}}(\lambda(t))$, where $\lambda(t)$ satisfies the differential equation (3.36). We have

$$\begin{aligned}\frac{d}{dt}\hat{\mathcal{U}}(\lambda(t)) &= \mathcal{U}'(\lambda(t)) \cdot \frac{d}{dt}\lambda(t) \\ &= \omega\left(U'(\lambda(t)) - [G(\lambda(t)) + \delta]\right)^2 ,\end{aligned}$$

from which we see that $\hat{\mathcal{U}}(\lambda(t))$ is strictly increasing over time unless the quantity in parentheses on the last line equals zero, that is, unless $\lambda(t)$ is a solution to the first-order conditions. It follows that the algorithm converges (with strictly increasing aggregate utility) to the unique solution to the first-order conditions, which, as we have noted, is a Nash equilibrium for individually optimizing customers when charged the fixed toll δ.

We have also shown that the stationary point to which the algorithm converges maximizes the aggregate transformed utility, $\hat{\mathcal{U}}(\lambda)$. But this objective function does not have an obvious economic interpretation.

Of course, if one sets the fixed toll at each resource equal to the external effect at that resource *at the globally optimal flow allocation for Problem (P)*, then the Nash equilibrium to which the algorithm globally converges will be that globally optimal flow allocation. But implementing this version of the algorithm requires that the system manager know the globally optimal flow allocation and the associated external effect. In many, if not most applications, this may be an unrealistic assumption. By contrast, the algorithm with a time-varying toll equal to the external effect associated with the *current usage* of the system can be implemented without foreknowledge of the optimal solution.

3.4.3 Algorithm with Time-Varying Toll: Facility Optimality

Consider the differential equation (3.32) with

$$\delta(t) = \lambda(t)G'(\lambda(t)) - \lambda(t)U''(\lambda(t)) . \tag{3.37}$$

Substituting for $\delta(t)$ in (3.32) yields the differential equation

$$\frac{d}{dt}\lambda(t) = \omega\left(U'(\lambda(t)) - [G(\lambda(t)) + \lambda(t)G'(\lambda(t)) - \lambda(t)U''(\lambda(t))]\right) . \tag{3.38}$$

At a positive stationary point, $\lambda > 0$, of this system, we have

$$U'(\lambda) + \lambda U''(\lambda) - [G(\lambda) + \lambda G'(\lambda)] = 0 , \tag{3.39}$$

which is the necessary *KKT* condition for $\lambda > 0$ to be be a facility optimal arrival rate (cf. equation (2.12) in Chapter 2). To show that the dynamic algorithm defined by the differential equation (3.38) converges to such a stationary point, we shall once again make use of the equivalence, noted in Remark 3 in Chapter 2 between the facility optimization problem and a social optimization problem with modified utility function, $\tilde{U}(\lambda) := \lambda U'(\lambda)$. For this transformed social optimization problem the associated dynamic algorithm is governed by the differential equation

$$\frac{d}{dt}\lambda(t) = \omega\left(\tilde{U}'(\lambda(t)) - [G(\lambda(t)) + \lambda(t)G'(\lambda(t))]\right) , \tag{3.40}$$

which is equivalent to the differential equation (3.38). Define a new net utility function, $\hat{\mathcal{U}}$, as follows:

$$\tilde{\mathcal{U}}(\lambda) := \tilde{U}(\lambda) - \lambda G(\lambda) .$$

Note that

$$\begin{aligned}\tilde{\mathcal{U}}'(\lambda) &= \tilde{U}'(\lambda) - G(\lambda) + \lambda G'(\lambda) \\ &= U'(\lambda) + \lambda U''(\lambda) - G(\lambda) + \lambda G'(\lambda) .\end{aligned} \tag{3.41}$$

The function $\hat{\mathcal{U}}(\cdot)$ is a Lyapunov function for the dynamical system governed by the differential equation (3.40), as can be seen by the following argument. Consider $\tilde{\mathcal{U}}(\lambda(t))$, where $\lambda(t)$ satisfies the differential equation (3.38). We have

$$\begin{aligned}\frac{d}{dt}\tilde{\mathcal{U}}(\lambda(t)) &= \tilde{\mathcal{U}}'(\lambda(t)) \cdot \frac{d}{dt}\lambda(t) \\ &= \omega\left(U'(\lambda(t)) + \lambda(t)U''(\lambda(t)) - [G(\lambda(t)) + \lambda(t)G'(\lambda(t))]\right)^2 ,\end{aligned}$$

from which we see that $\tilde{\mathcal{U}}(\lambda(t))$ is strictly increasing over time unless the quantity in parentheses on the last line equals zero, that is, unless $\lambda(t)$ is a solution to the necessary condition (3.39) for a facility optimal arrival rate. It follows that the algorithm converges (with strictly increasing aggregate utility) to a solution to (3.39) (which, as we have noted, is a Nash equilibrium for individually optimizing customers when charged the toll $\delta = \lambda G'(\lambda) - \lambda U''(\lambda)$). However, since the modified utility function, $\tilde{U}(\lambda)$, need not be concave (as we observed) the solution to (3.39) to which the algorithm converges need not be *globally* optimal for the facility optimization problem.

3.5 Continuous-Time Dynamic Algorithm: Variants

Just as in the discrete-time case, we can construct variants of the continuous-time dynamic algorithm by choosing (i) whether the arrival rate or the price evolves according to a differential equation, and (ii) whether the right-hand-side of the differential equation involves differences in arrival rates or differences in prices. To illustrate, we shall consider one such variant and leave the analysis of other variants to the reader.

Consider the continuous-time analogue of the second variant of the discrete-time model of Section 3.3, in which the predicted arrival rate $\hat{\lambda}^{(n)}$ is updated at discrete stages $n = 0, 1, \ldots$, by a smoothing of the previous predicted and actual arrival rates. The governing difference equation is (3.26). In continuous time, this difference equation is replaced by the following differential equation:

$$\frac{d}{dt}\lambda(t) = \omega \left(\Lambda \bar{F}(G(\lambda(t)) + \delta(t)) - \lambda(t) \right) . \qquad (3.42)$$

(We are assuming here and henceforth in this section that the utility function is generated by potential customers arriving at rate Λ with random rewards with distribution function $F(\cdot)$.) As in the discrete-time case, the quantity in parentheses on the r.h.s. of the differential equation is the difference between the actual and the predicted arrival rates at time t. If that difference is positive (negative) then rate of increase in the predicted arrival rate at time t is positive (negative).

To apply this algorithm to the case of individual optimization, we take $\delta(t) = \delta$, where δ is the given fixed toll. For the case of social optimization, we take $\delta(t) = \lambda(t)G'(\lambda(t))$ – the external effect at time t.

Does this variant of the continuous-time dynamic algorithm converge, and if so, does it converge to the appropriate value (e.g., the unique Nash equilibrium in the case of individual optimality, the unique global maximum of $\mathcal{U}(\lambda)$ in the case of social optimality)? The answer to both questions is "yes," as may be seen by the following argument.

Using the identity, $\lambda = \Lambda \bar{F}(U'(\lambda))$, we can rewrite the differential equation (3.42) in equivalent form as

$$\frac{d}{dt}\lambda(t) = \omega \left(\Lambda \bar{F}(G(\lambda(t)) + \delta(t)) - \Lambda \bar{F}(U'(\lambda(t))) \right) . \qquad (3.43)$$

Since $\bar{F}(\cdot)$ is strictly decreasing and continuous, it follows that

$$\frac{d}{dt}\lambda(t) \begin{Bmatrix} > \\ = \\ < \end{Bmatrix} 0 \Leftrightarrow U'(\lambda(t)) - [G(\lambda(t)) + \delta(t)] \begin{Bmatrix} > \\ = \\ < \end{Bmatrix} 0 .$$

Now consider the case of social optimality, in which $\delta(t) = \lambda(t)G'(\lambda(t))$. (The argument in the case of individual optimality is similar.) Recall the net utility function,

$$\mathcal{U}(\lambda) = U(\lambda) - \lambda G(\lambda) ,$$

whose first derivative is given by

$$\mathcal{U}'(\lambda) = U'(\lambda) - G(\lambda) - \lambda G'(\lambda) \ .$$

Consider $\mathcal{U}(\lambda(t))$, where $\lambda(t)$ satisfies the differential equation (3.43). We have

$$
\begin{aligned}
\frac{d}{dt}\mathcal{U}(\lambda(t)) &= \mathcal{U}'(\lambda(t)) \cdot \frac{d}{dt}\lambda(t) \\
&= (U'(\lambda(t)) - [G(\lambda(t)) + \lambda(t)G'(\lambda(t))]) \cdot \left(\frac{d}{dt}\lambda(t)\right) \\
&= (U'(\lambda(t)) - [G(\lambda(t)) + \delta(t)]) \cdot \left(\frac{d}{dt}\lambda(t)\right) \ .
\end{aligned}
$$

But the above argument showed that the quantity in the first parentheses on the r.h.s. of the last line of this equation is $(> 0, = 0, < 0)$, respectively, if and only if the quantity in the second parentheses is $(> 0, = 0, < 0)$. Hence, the r.h.s. of the equation is strictly positive for all t (i.e., $\mathcal{U}(\lambda(t))$ is strictly increasing over time), unless it equals zero, in which case $\lambda(t)$ is a solution to the first-order conditions. It follows that the algorithm converges (with strictly increasing net utility) to the unique solution to the first-order conditions for social optimality.

The argument in the case of a fixed toll δ (individual optimality) uses the transformed net utility function, $\hat{\mathcal{U}}(\cdot)$ (cf. Section 3.4.2) and proceeds along similar lines. We leave it to the reader to fill in the details.

3.6 Endnotes

Section 3.2

As mentioned already, the material in this section is taken from Rump and Stidham [170], in which one may find proofs of the results.

For a background on exponential smoothing, see Bails and Peppers [10]. Adaptive expectations are discussed more fully in Carter and Maddock [31]. This model of customer behavior is more general and less myopic than the model in Chapter 2 and in most flow-control studies in the literature on communication networks (cf. [97, 22, 30, 24, 25, 57, 56, 26, 58, 200, 119, 187, 70, 71]) which, using $\omega = 1$, assume customers have static expectations.

For a discussion of the classical supply-driven model in economics, see, e.g., Chiarella [39].

Section 3.2.1

Following Friedman and Landsberg [70], if we assume linear waiting costs and bounded rewards ($a < \infty$) it is not hard to show that an equilibrium will be globally stable under adaptive expectations provided there is sufficient capacity. However, these assumptions make this result rather weak, as we have just established in Theorem 3.1 that a capacity $\mu > \underline{\mu}$ will suffice in general.

For this special case, Stidham [187] and Friedman and Landsberg [70] studied the dynamics of the arrival rate $\lambda^{(n)} = \Lambda \bar{F}(\hat{\pi}^{(n)})$ under static expectations: $\hat{\pi}_n = \pi^{(n-1)} = \pi + hW_\mu(\lambda^{(n-1)})$. In this case, $\lambda^{(n)} = T_\mu(\lambda^{(n-1)})$, where $T_\mu(\lambda) := \Lambda\bar{F}(\pi + hW_\mu(\lambda))$. Friedman and Landsberg [70] showed that there exists a μ_g such that, for $\mu > \mu_g$ and for all $\lambda \in (0, \Lambda\bar{F}(0))$, $|(\partial/\partial\lambda)T_\mu(\lambda)| < 1$. This implies that $T_\mu(\lambda)$ is a contraction mapping and hence the equilibrium arrival rate is globally stable, whenever there is sufficient capacity μ. (We assume that, for all $\lambda, \varepsilon > 0$, there exists a μ such that $h(\partial/\partial\lambda)W_\mu(\lambda) \le \varepsilon$.)

See Sandefur [172] for a proof that the equilibrium is stable (unstable) if $|\frac{\partial}{\partial\hat{\pi}}\hat{\Pi}_\mu(\pi_\mu^e)| < 1$ $(|\frac{\partial}{\partial\hat{\pi}}\hat{\Pi}_\mu(\pi_\mu^e)| > 1)$. For a proof of Theorem 3.4, see Sandefur [172], Theorem 4.6. For a proof of Theorem 3.5, see Lorenz [133].

For a formal definition of chaos we refer the reader to Devaney [53], [133] and [172], all excellent introductions to this topic.

Chaotic behavior in queueing systems is a topic that has been addressed in the literature only recently. Chase, Serrano, and Ramadge [33] analyzed multiclass, deterministic fluid models operating at full capacity. They examined both a switched-arrival and a switched-server model. For the former model, operating under a threshold-type policy (switch the incoming flow to a queue whenever its level drops below a threshold), they showed that chaos can occur, by analyzing the process embedded at the switching times as a discrete-time dynamic system. Whitt [198] studied a deterministic multiclass network of queues and showed that the evolution of queue lengths at the various nodes exhibits behavior that is sensitive to initial conditions – a characteristic of chaotic systems. The model in this section differs from both these papers in that it allows a stochastic queueing process, but we are interested in stability or chaos of the dynamic input-pricing mechanism rather than the queueing process itself. Since we assume that the customers' adaptive response to congestion is based on average waiting, however, we are able to formulate the multi-period pricing problem as a deterministic dynamic system.

Masuda and Whang [141] studied a Jackson network of queues and analyzed the stability region for a discrete-time dynamic pricing algorithm with adaptive expectations (exponential smoothing), among other adaptive algorithms. Their paper may be regarded as an extension of the *M/M/1* model of this section to a network of queues. They did not, however, consider the possibility of chaotic behavior in the region of instability.

Section 3.4.1

In the language of dynamical systems theory (cf., e.g., Perko [156]), the dynamical system defined by (3.32)-(3.33) is a *gradient system* (cf. Definition 3 on p. 176 of [156]), and $\mathcal{U}$ is the associated Lyapunov function. Kelly et al. [109] introduced a generalization of this algorithm for a model of the Internet, which may be interpreted as a multiclass queueing network. The algorithm of [109] may be viewed as a simplified abstraction of the *TCP/IP* protocol in use in the Internet (see Jacobsen [96]). See Chapters 7 and 8 for extensions to single-class and multiclass networks of queues.

CHAPTER 4

Optimal Arrival Rates in a Multiclass Queue

Choosing the arrival rates for multiclass queues was one of the topics introduced in Chapter 1. Using a simple example with two classes, we saw that the objective function for social optimization could have multiple local maxima, which might complicate the search for the optimal arrival rates and tolls (entrance fees) that will implement these rates. In this chapter we return to this topic and examine it in more detail. We begin by introducing a general model for multiclass queues, and then consider several special cases, depending on the structure of the waiting costs and the queue discipline used to choose which class to serve. Specifically, we consider the *FIFO* discipline and static priority disciplines.

We consider the familiar optimality criteria – individual, social, and facility optimality – as well as a new criterion: class optimality. We show how the equilibrium solution generated by individually optimizing customers can be transformed into a social, class, or facility optimal solution by charging an appropriate entrance fee or toll. We also study the stability of the equilibrium solution under various dynamic algorithms for adaptive customer behavior, both in discrete and continuous time.

Our results demonstrate that the answers to the questions – (1) is there a unique solution to the first-order optimality conditions? and (2) are dynamic algorithms stable? – depend rather intricately on the model assumptions.

4.1 General Multiclass Model: Formulation

Our model is a multiclass generalization of the models presented in Chapter 2 in which the arrival rate was the decision variable. Here one must choose the arrival rate for each of several classes of customers. A special case in which there were just two classes was introduced in Section 1.4 of Chapter 1.

As in our development of the single-class model in Chapter 2, we begin with an abstract formulation of the model, defined by arrival rates, entrance fees, and utility functions and waiting-cost functions with specified properties. Then we present specific stochastic models (e.g., the $M/GI/1$ queue in steady state) which instantiate these properties.

We consider a facility which provides service to customers of m distinct classes, labeled $i \in M$, where $M := \{1, 2, \ldots, m\}$. Class-i jobs arrive with mean arrival rate λ_i. The class arrival rates are decision variables. The set of feasible values for λ_i is denoted A_i. Our default assumption is that $A_i = [0, \infty)$, $i \in M$. As with the single-class model, alterations to our model and results to allow for more general feasible sets are usually straightforward.

We assume that class i gains a gross utility per unit time which is a function, $U_i(\lambda_i)$, of λ_i. Our default assumption is that $U_i(\lambda_i)$ is nondecreasing, differentiable, and concave in $\lambda_i \in A_i$. In Chapter 2 we showed how the theory needs to be altered when these assumptions (in particular, differentiability) are relaxed. We also showed how a value function of this form can arise when there is a Poisson process of potential arriving customers and a probabilistic joining rule is followed. We do not need to repeat this analysis here, since it applies to each class considered separately.

Balanced against the benefit of throughput is the cost to customers caused by the time they spend in the system. For a given vector of arrival rates, $\boldsymbol{\lambda} = (\lambda_i, i \in M)$, let $G_i(\boldsymbol{\lambda})$ denote the average waiting cost of a class-i customer, averaged over all class-i customers who arrive during the period in question. Our default assumption is that $G_i(\boldsymbol{\lambda})$ takes values in $[0, \infty]$, is differentiable and strictly increasing in $\lambda_i \in A_i$, and is differentiable and nondecreasing in $\lambda_k \geq 0$, $k \neq i$. A special case that we shall consider in some detail is the case of a linear waiting cost, in which $G_i(\boldsymbol{\lambda}) = h_i W_i(\boldsymbol{\lambda})$, where $W_i(\boldsymbol{\lambda})$ is the average waiting time in the system of a class-i customer when the vector of arrival rates is $\boldsymbol{\lambda}$, and $h_i > 0$ is the cost per time unit spent by a class-i customer in the system.

For complete generality, we have allowed the class-i waiting cost to depend on all the arrival rates. In special cases treated in subsequent sections, this dependence is sometimes only through the sum of the arrival rates in all or a subset of the classes.

Finally, we assume that each class-i customer who joins the system is charged an entrance fee or toll δ_i. The sum of the toll and the waiting cost constitutes the *full price* of admission for class i, which we denote in general by π_i, or $\pi_i(\boldsymbol{\lambda})$, when we want to emphasize its dependence on $\boldsymbol{\lambda}$ for a fixed δ_i. Thus we have

$$\pi_i(\boldsymbol{\lambda}) = \delta_i + G_i(\boldsymbol{\lambda}) \ .$$

Before analyzing the various optimality criteria, we consider some examples of stochastic models that satisfy our basic assumptions. Detailed development and analysis of these models will be done in subsequent sections.

4.1.1 Example: M/GI/1 Queue with Linear Waiting Costs

To fix ideas let us consider a stochastic queueing model whose steady-state behavior instantiates the properties of our general model. The model is a multiclass *M/GI/1* queue.

Suppose customers of class i $(i \in M)$ arrive according to a Poisson process with mean arrival rate λ_i, where the parameter λ_i is a decision variable, chosen from the set $A_i = [0, \infty)$. The class interarrival times and service times are independent of one another. Specifically, the nth arriving customer of class

i brings a random amount of work $S_i^{(n)}$, where $S_i^{(1)}, S_i^{(2)}, \ldots$ are mutually independent and identically distributed as a generic random variable, S_i. The single server performs work deterministically at unit rate. The service time of a class-i job is thus distributed as S_i. Let the mean and second moment of S_i be given by $\mathrm{E}[S_i] = 1/\mu_i$ and $\mathrm{E}[S_i^2] = 2\beta_i/\mu_i^2$, $\beta_i \geq 1/2$.

4.1.1.1 FIFO *Queue Discipline*

Suppose that customers are served one at a time, without interruption, in order of arrival (a *first-in, first-out* or *FIFO* queue discipline). Let $D(\lambda)$ denote the steady-state expected delay (waiting time in the queue) for this system. Since the queue discipline does not discriminate among jobs according to class, the steady-state expected delay is the same for all classes.

The Pollaczek-Khintchine formula gives

$$D(\lambda) = \frac{\sum_{k \in M} \lambda_k \beta_k / \mu_k^2}{1 - \sum_{k \in M} \lambda_k / \mu_k} , \qquad (4.1)$$

subject to the stability condition,

$$\sum_{k \in M} \lambda_k / \mu_k < 1 . \qquad (4.2)$$

When this condition does not hold, $D(\lambda) = \infty$.

Let $W_i(\lambda)$ denote the expected waiting time in the system of a class-i job in steady state. Then (again subject to (4.2))

$$\begin{aligned} W_i(\lambda) &= D(\lambda) + \frac{1}{\mu_i} \\ &= \frac{\sum_{k \in M} \lambda_k \beta_k / \mu_k^2}{1 - \sum_{k \in M} \lambda_k / \mu_k} + \frac{1}{\mu_i} . \end{aligned}$$

It follows that the waiting-cost function, $G_i(\lambda)$ for class-i is given by

$$\begin{aligned} G_i(\lambda) &= h_i W_i(\lambda) \\ &= h_i \left(\frac{1}{\mu_i} + \frac{\sum_{k \in M} \lambda_k \beta_k / \mu_k^2}{1 - \sum_{k \in M} \lambda_k / \mu_k} \right) , \end{aligned} \qquad (4.3)$$

with $G_i(\lambda) = \infty$ if (4.2) does not hold.

It is easily verified that $G_i(\lambda)$ defined in this way satisfies our basic assumptions; that is, $G_i(\lambda)$ is differentiable and strictly increasing in $\lambda_i \geq 0$, and differentiable and nondecreasing (in fact, strictly increasing) in $\lambda_k \geq 0$, $k \neq i$.

In the special case where the service times are class independent, with mean $1/\mu$ and second moment $2\beta/\mu^2$, the steady-state expected delay depends only on the total arrival rate, $\lambda = \sum_{k \in M} \lambda_k$ and is given by

$$D(\lambda) = \frac{\lambda \beta}{\mu(\mu - \lambda)} , \qquad (4.4)$$

which (of course) coincides with the steady-state expected delay in a single-class $M/GI/1$ queueing system. The steady-state expected waiting time in the system also depends only on the total arrival rate λ and is given by

$$W(\lambda) = \frac{\lambda\beta}{\mu(\mu - \lambda)} + \frac{1}{\mu} .$$

It follows that the waiting-cost function for class-i depends only on λ and is given by

$$\begin{aligned} G_i(\lambda) &= h_i W(\lambda) \\ &= h_i \left(\frac{\lambda\beta}{\mu(\mu - \lambda)} + \frac{1}{\mu} \right) . \end{aligned} \tag{4.5}$$

4.1.1.2 Nonpreemptive Priority Queue Discipline

Suppose customers are served in order of class, with class 1 receiving the highest priority, class 2 the next highest, and so forth. When a customer arrives to an empty system, that customer is placed immediately into service. At the completion of a service, a customer from the highest-priority class among those in the queue is placed immediately into service. Service of a customer, once begun, is never interrupted. (There is no preemption.)

In this case the steady-state expected delay (waiting time in queue) for class i, denoted $D_i(\boldsymbol{\lambda})$, is given by

$$D_i(\boldsymbol{\lambda}) = \frac{\sum_{k=1}^{m} \lambda_k \beta_k / \mu_k^2}{(1 - \sum_{k=1}^{i-1} \lambda_k / \mu_k)(1 - \sum_{k=1}^{i} \lambda_k / \mu_k)} , \tag{4.6}$$

subject to the stability condition,

$$\sum_{k=1}^{i} \lambda_k / \mu_k < 1 . \tag{4.7}$$

When this condition does not hold, $D_i(\boldsymbol{\lambda}) = \infty$. The steady-state expected waiting time in the system for class i is given by

$$W_i(\boldsymbol{\lambda}) = D_i(\boldsymbol{\lambda}) + \frac{1}{\mu_i} . \tag{4.8}$$

It follows that the waiting-cost function, $G_i(\boldsymbol{\lambda})$ for class i is given by

$$\begin{aligned} G_i(\boldsymbol{\lambda}) &= h_i W_i(\boldsymbol{\lambda}) \\ &= h_i \left(\frac{\sum_{k=1}^{m} \lambda_k \beta_k / \mu_k^2}{(1 - \sum_{k=1}^{i-1} \lambda_k / \mu_k)(1 - \sum_{k=1}^{i} \lambda_k / \mu_k)} + \frac{1}{\mu_i} \right) , \end{aligned} \tag{4.9}$$

when (4.7) holds, with $G_i(\boldsymbol{\lambda}) = \infty$ when (4.7) does not hold.

It is easily verified that $G_i(\boldsymbol{\lambda})$ defined in this way satisfies our basic assumptions; that is, $G_i(\boldsymbol{\lambda})$ is differentiable and strictly increasing in $\lambda_i \geq 0$, and differentiable and nondecreasing in $\lambda_k \geq 0$, $k \neq i$.

4.1.1.3 Preemptive-Resume Priority Queue Discipline

Suppose customers are served in order of class, with class 1 receiving the highest priority, class 2 the next highest, and so forth. When a customer arrives to an empty system, that customer is placed immediately into service. At the completion of a service, a customer from the highest-priority class among those in the queue is placed immediately into service. If a higher-priority customer arrives while a lower-priority customer is in service, that service is interrupted and the higher-priority customer is placed in service immediately. At the next instant when there are no higher-priority customers in the system, service of the lower-priority customer is resumed where it left off.

In this case the steady-state expected delay (waiting time in the queue), $D_i(\boldsymbol{\lambda})$, depends on $\boldsymbol{\lambda}$ only through its first i components and is given by

$$D_i(\boldsymbol{\lambda}) = \frac{\sum_{k=1}^{i} \lambda_k \beta_k / \mu_k^2}{(1 - \sum_{k=1}^{i-1} \lambda_k / \mu_k)(1 - \sum_{k=1}^{i} \lambda_k / \mu_k)} \,, \tag{4.10}$$

when (4.7) holds, with $D_i(\boldsymbol{\lambda}) = \infty$ when (4.7) does not hold. In this context delay is defined as the waiting time in the queue before entering service for the first time. The steady-state expected waiting time in the system for class i is given by

$$W_i(\boldsymbol{\lambda}) = D_i(\boldsymbol{\lambda}) + \frac{(\sum_{k=1}^{i-1} \rho_k)(1/\mu_i)}{1 - \sum_{k=1}^{i-1} \rho_i} + \frac{1}{\mu_i} \,. \tag{4.11}$$

The second term in this expression is the expected waiting time in the queue after entering service for the first time, caused by preemption. The third term is the expected time in service. The sum of the second and third terms, namely,

$$\frac{1/\mu_i}{1 - \sum_{k=1}^{i-1} \rho_i} \,,$$

constitutes the expected *completion time*, where the completion time is defined as the time from the first entrance into service until service is completed.

The waiting-cost function, $G_i(\boldsymbol{\lambda})$ for class i is given by

$$G_i(\boldsymbol{\lambda}) = h_i W_i(\boldsymbol{\lambda}) \tag{4.12}$$

when (4.7) holds, with $G_i(\boldsymbol{\lambda}) = \infty$ when (4.7) does not hold.

Again, it is easily verified that $G_i(\boldsymbol{\lambda})$ defined in this way satisfies our basic assumptions; that is, $G_i(\boldsymbol{\lambda})$ is differentiable and strictly increasing in $\lambda_i \geq 0$, and differentiable and nondecreasing in $\lambda_k \geq 0$, $k \neq i$.

4.2 General Multiclass Model: Optimal Solutions

As in our previous models, the solution to the decision problem depends on who is making the decision. In the case of multiple classes, the decision may be made by the individual customers, each concerned only with its own net utility (*individual optimality*), by an agent for the class as a whole (*class optimality*), or by a system operator, who might be interested in maximizing the aggregate net utility to all customers (*social optimality*) or in maximizing profit (*facility optimality*). Note that class optimality is a new concept which is relevant in a multiclass system. Class optimality is intermediate between individual and social optimality in that the class agent takes into account the interactions (including external effects) within its class, but not those with other classes.

As usual, we shall be interested in whether and how the facility operator can choose toll values that will induce individually optimizing customers to behave in a way that is consistent with social or facility optimality. In general, the tolls are allowed (and need) to be class dependent. Class-dependent tolls presume that the facility operator who is charging the tolls can reliably discern which class an arriving customer belongs to.

4.2.1 Individually Optimal Arrival Rates

Returning now to the general model, let us consider the decision problem from the point of view of an arriving customer of class i concerned only with its own net utility, which it wishes to maximize (individual optimality). Suppose we are given the full price of admission, $\pi_i(\boldsymbol{\lambda})$, as a function of $\boldsymbol{\lambda} \geq 0$. We assume that $\pi_i(\cdot)$ takes values in $[0, \infty]$ and is differentiable and strictly increasing in $\lambda_i \geq 0$, for each set of fixed values of $\lambda_k \geq 0$, $k \neq i$. (In the single-facility model of this chapter, $\pi_i(\boldsymbol{\lambda}) = \delta_i + G_i(\boldsymbol{\lambda})$, the sum of the toll and the waiting cost for class i. But, in keeping with our usual approach, we shall first develop the theory in the more general setting in which the full price $\pi_i(\boldsymbol{\lambda})$ is left unspecified. This generality will be useful when we deal with multiclass networks of queues in Chapter 8.)

Our analysis of individual optimality for class-i customers parallels the analysis of the single-class model in Chapter 2. For a particular value π_i of the full price of admission, an arriving customer of class i concerned only with maximizing its own net utility will join if the value it receives from joining exceeds π_i, balk if it is lower, and be indifferent between joining and balking if it equals π_i. The marginal utility, $U_i'(\lambda_i)$, represents the value received by the marginal class-i user when the arrival rate for class i equals λ_i. At an individually optimal arrival rate, the marginal user will be indifferent between joining and balking, so that

$$U_i'(\lambda_i) = \pi_i \,, \tag{4.13}$$

if this equation has a solution in $A_i = [0, \infty)$. If $U_i'(0) < \pi_i$,then there is no solution to (4.13) in $A_i = [0, \infty)$; in this case no user has any incentive to join and we set $\lambda_i = 0$. If $U_i'(0) \geq \pi_i$, then since $U_i'(\lambda_i)$ is continuous and nonincreasing in λ_i, there is a solution to (4.13) in $A_i = [0, \infty)$. (To avoid trivialities we shall assume that $\lim_{\lambda_i \to \infty} U_i'(\lambda_i) < \pi$.) Thus for a fixed price π_i an individually optimal arrival rate for class i is characterized by the following conditions:

$$\begin{aligned} U_i'(\lambda_i) &\leq \pi_i \,, \\ U_i'(\lambda_i) &= \pi_i \,, \text{ if } \lambda_i > 0 \,. \end{aligned}$$

Now suppose $\pi_i = \pi_i(\boldsymbol{\lambda})$ for $\boldsymbol{\lambda} = (\lambda_1, \ldots, \lambda_i, \ldots, \lambda_m)$, where λ_i satisfies these conditions for each class $i \in M$. Then the system is in equilibrium: no individually optimizing customer acting unilaterally will have any incentive to deviate from its current action. In this case $\boldsymbol{\lambda} = (\lambda_i, i \in M)$ satisfies the following conditions for each class $i \in M$:

$$U_i'(\lambda_i) \leq \pi_i(\boldsymbol{\lambda}) \tag{4.14}$$
$$U_i'(\lambda_i) = \pi_i(\boldsymbol{\lambda}) \text{ , if } \lambda_i > 0 \tag{4.15}$$

These are the equilibrium conditions which must be satisfied by an individually optimal vector of arrival rates, which we shall denote by

$$\boldsymbol{\lambda}^e = (\lambda_i^e, i \in M) \ .$$

Equivalent equilibrium conditions are the following:

$$\pi_i(\boldsymbol{\lambda}) - U_i'(\lambda_i) \geq 0 \text{ , for all } i \in M \ ,$$
$$\lambda_i \geq 0 \text{ , for all } i \in M \ ,$$
$$\lambda_i(\pi_i(\boldsymbol{\lambda}) - U_i'(\lambda_i)) = 0 \text{ , for all } i \in M \ .$$

Note that the equality constraints take the form of complementary-slackness conditions. This form of the equilibrium conditions will facilitate comparison of individual optimization with social and facility optimization.

4.2.1.1 Existence and Uniqueness of Individually Optimal Solution

Now let us focus on the case where $\pi_i(\boldsymbol{\lambda}) = G_i(\boldsymbol{\lambda}) + \delta_i$, $i \in M$. The equilibrium conditions satisfied by the vector $\boldsymbol{\lambda}^e$ of individually optimal arrival rates now become

$$U_i'(\lambda_i) \leq G_i(\boldsymbol{\lambda}) + \delta_i \tag{4.16}$$
$$U_i'(\lambda_i) = G_i(\boldsymbol{\lambda}) + \delta_i \text{ , if } \lambda_i > 0 \tag{4.17}$$

for all $i \in M$. Under a mild technical assumption, we can show that there exists a unique solution to these equilibrium conditions.

Assumption 1   There exists a compact polyhedral convex set, $\tilde{A} \subseteq A := \prod_{i \in M} A_i$, such that

1. $G_i(\boldsymbol{\lambda}) < \infty$ for all $i \in M$ and $\boldsymbol{\lambda} \in \tilde{A}$;
2. for all $i \in M$, $U_i'(\lambda_i) < G_i(\boldsymbol{\lambda}) + \delta_i$ for all $\boldsymbol{\lambda} \in A - \tilde{A}$.

Theorem 4.1 *Under Assumption 1 there exists a unique solution to the equilibrium conditions (4.16), (4.17), which characterize the vector $\boldsymbol{\lambda}^e$ of individually optimal arrival rates.*

The proof of this theorem is deferred until the next section, since it depends on constructing a class optimization problem that is equivalent to the individual optimization problem.

4.2.2 Class-Optimal Arrival Rates

We now consider the problem from the point of view of the manager of each class. The class manager for class i is concerned with maximizing the net benefit, $B_i(\boldsymbol{\lambda})$, received per unit time by jobs of class i, where

$$B_i(\boldsymbol{\lambda}) := U_i(\lambda_i) - \lambda_i(G_i(\boldsymbol{\lambda}) + \delta_i) . \tag{4.18}$$

Given fixed strategies of the other players (that is, fixed values of the arrival rates, λ_k, $k \neq i$) an optimal strategy for the manager of class i is therefore to choose a value for λ_i such that

$$\lambda_i = \arg \max_{\{\lambda_i \geq 0\}} B_i(\lambda_1, \ldots, \lambda_i, \ldots, \lambda_m) =: \Gamma_i(\boldsymbol{\lambda}) .$$

(Note that $\Gamma_i(\boldsymbol{\lambda})$ depends on $\boldsymbol{\lambda} = (\lambda_1, \ldots, \lambda_i, \ldots, \lambda_m)$ only through λ_k, $k \neq i$.) Let $\boldsymbol{\lambda}^c = (\lambda_1^c, \ldots, \lambda_m^c)$ denote a Nash equilibrium for this m-player game, that is, a vector $\boldsymbol{\lambda}$ at which no player has an incentive to deviate unilaterally from its current strategy. Thus,

$$\lambda_i^c = \Gamma_i(\boldsymbol{\lambda}^c) , \ i \in M .$$

A Nash equilibrium $\boldsymbol{\lambda}^c$ is therefore a solution to the fixed-point equation,

$$\boldsymbol{\lambda} = \boldsymbol{\Gamma}(\boldsymbol{\lambda}) ,$$

where $\boldsymbol{\Gamma}(\boldsymbol{\lambda}) := (\Gamma_1(\boldsymbol{\lambda}), \ldots, \Gamma_m(\boldsymbol{\lambda}))$.

For each $i \in M$, the following *KKT* conditions are necessary for λ_i to maximize $B_i(\boldsymbol{\lambda})$, with λ_k fixed for $k \neq i$:

$$U_i'(\lambda_i) \ \leq \ G_i(\boldsymbol{\lambda}) + \lambda_i \frac{\partial}{\partial \lambda_i} G_i(\boldsymbol{\lambda}) + \delta_i , \tag{4.19}$$

$$U_i'(\lambda_i) \ = \ G_i(\boldsymbol{\lambda}) + \lambda_i \frac{\partial}{\partial \lambda_i} G_i(\boldsymbol{\lambda}) + \delta_i , \text{ if } \lambda_i > 0 . \tag{4.20}$$

Therefore, (4.19) and (4.20) must hold simultaneously for all $i \in M$ in order for an allocation $\boldsymbol{\lambda} = (\lambda_i, i \in M)$ to be class optimal.

4.2.2.1 Existence and Uniqueness of Class-Optimal Solution

Under mild technical assumptions, we can show that there exists a unique solution to the *KKT* conditions (4.19) and (4.19), hence a unique fixed point of the operator $\Gamma(\cdot)$ – a unique class-optimal solution.

Assumption 2 There exists a compact convex set, $\tilde{A} \subseteq A = \prod_{i \in M} A_i$, such that

1. $G_i(\boldsymbol{\lambda}) < \infty$ for all $i \in M$ and $\boldsymbol{\lambda} \in \tilde{A}$;

2. for all $\boldsymbol{\lambda} \in A$, there exists a $\boldsymbol{\lambda}' = (\lambda_k', k \in M) \in \tilde{A}$ such that $B_i(\boldsymbol{\lambda}') \geq B_i(\boldsymbol{\lambda})$ for all $i \in M$.

For all $i \in M$, and all $\boldsymbol{\lambda}$, let $H_i(\boldsymbol{\lambda}) := \lambda_i G_i(\boldsymbol{\lambda})$.

Assumption 3   For all $i \in M$, $H_i(\boldsymbol{\lambda})$ is a convex function of λ_i, for each set of fixed values of λ_k, $k \neq i$, such that $\boldsymbol{\lambda} = (\lambda_k, k \in M) \subset \tilde{A}$.

Assumption 2 ensures that the search for a class-optimal solution can be restricted, without loss of optimality, to the compact, convex set $\tilde{A}$ (in which the net-benefit payoff functions, $B_i(\boldsymbol{\lambda})$ are all finite valued). Therefore an equilibrium solution to the class-optimization problem is a point $\boldsymbol{\lambda}^c \in \tilde{A}$ such that

$$B_i(\boldsymbol{\lambda}^c) = \max_{\{\lambda_i\}}\{B_i(\lambda_1^c, \ldots, \lambda_i, \ldots, \lambda_m^c) | (\lambda_1^c, \ldots, \lambda_i, \ldots, \lambda_m^c) \in \tilde{A}\} \, , \, i \in M \, .$$

(4.21)

Assumption 3 ensures that $B_i(\boldsymbol{\lambda})$ is continuous in $\boldsymbol{\lambda}$ and concave in λ_i, for each set of fixed values of λ_j, $j \neq i$, such that $\boldsymbol{\lambda} = (\lambda_1, \ldots, \lambda_i, \ldots, \lambda_m) \in \hat{A}$.

4.2.2.2 Example: M/GI/1 *Queue with Linear Waiting Costs and* FIFO *Queue Discipline.*

In this example, we have

$$
\begin{aligned}
G_i(\boldsymbol{\lambda}) &= h_i W_i(\boldsymbol{\lambda}) \\
&= h_i \left(\frac{1}{\mu_i} + \frac{\sum_{k \in M} \lambda_k \beta_k / \mu_k^2}{1 - \sum_{k \in M} \lambda_k / \mu_k} \right)
\end{aligned}
$$

(cf. equation (4.3) in Section 4.1.1.1).

Exercise 1   Show that in this case $G_i(\boldsymbol{\lambda})$ is convex and nondecreasing in λ_i, for each set of fixed values of λ_k, $k \neq i$, and that as a result Assumption 3 holds.

Exercise 2   Suppose $U_i(\lambda_i) \to 0$ as $\lambda_i \to \infty$, for all $i \in M$. Show that Assumption 2 holds.

Under Assumptions 2 and 3, the set of constrained net-benefit optimization problems (4.21) satisfies the general assumptions for a concave m-player non-cooperative game (cf. Rosen [161]). Theorem 1 of [161] then establishes the existence of an equilibrium solution to the class-optimization problem.

It remains to show that the equilibrium point is unique. Since the strategy space of each player of the game may now depend on the strategies of the other players through the constraint, $\boldsymbol{\lambda} \in \tilde{A}$, the simultaneous strategy space $\tilde{A}$ for all players is what Rosen calls a *coupled constraint set*. In the general case of a m-player concave game with a coupled constraint set the nonnegative multipliers which satisfy the *KKT* conditions are usually unrelated. The multipliers can be normalized, giving rise to the more restricted concept of a *normalized equilibrium*, which depends on the normalizing weights (cf. Rosen [161]). Uniqueness can only be proved in general for the normalized equilibrium point corresponding to each given set of weights.

In order to avoid this limitation on the uniqueness of the equilibrium point, we need to strengthen Assumption 2 slightly.

Assumption 4   There exists a compact polyhedral convex set, $\tilde{A} \subseteq A := \prod_{i \in M} A_i$, such that

1. $G_i(\boldsymbol{\lambda}) < \infty$ for all $i \in M$ and $\boldsymbol{\lambda} \in \tilde{A}$;
2. for each $\boldsymbol{\lambda} \in A$, there exists a $\boldsymbol{\lambda}' = (\lambda'_k, k \in M) \in \tilde{A}$ for which the only active constraints among those defining $\tilde{A}$ are (possibly) some of the nonnegativity constraints, and for which $B_i(\boldsymbol{\lambda}') \geq B_i(\boldsymbol{\lambda})$ for all $i \in M$.

The effect of the strengthened requirement in Assumption 4 (2) is to ensure, not only that there exists an equilibrium solution $\boldsymbol{\lambda}^c$ to the class-optimization problem that lies in the compact convex set $\tilde{A}$, but also the restriction to normalized equilibrium points referred to above is not necessary. Assumption 4 is satisfied, for example, in queueing models with a single server working at rate μ in which $G_i(\boldsymbol{\lambda}) \uparrow \infty$ as $\lambda \uparrow \mu$. Since the (stability) constraint, $\sum_{k \in M} \lambda_k = \lambda < \mu$, will never be binding at equilibrium, the m multipliers that ensure complementary slackness at this constraint for each of the m players must all equal zero at optimality. This removes the coupling concern and any dependence of the equilibrium on the normalizing weights.

Following Rosen [161], we define a weighted sum of the payoff functions, $\sum_{i \in M} \omega_i B_i(\boldsymbol{\lambda})$, for each vector $\boldsymbol{\omega}$ of nonnegative weights, $\omega_i \geq 0$, $i \in M$. We then define the *pseudogradient*, $\mathbf{g}(\boldsymbol{\lambda}, \boldsymbol{\omega})$, to have components $g_i(\boldsymbol{\lambda}, \boldsymbol{\omega}) = \omega_i \frac{\partial}{\partial \lambda_i} B_i(\boldsymbol{\lambda})$, $i \in M$. Therefore, the Jacobian of the pseudogradient, $\mathbf{g}'(\boldsymbol{\lambda}, \boldsymbol{\omega})$, has ij-th component

$$g'_{ij}(\boldsymbol{\lambda}, \boldsymbol{\omega}) = \begin{cases} -\omega_i \psi_i & , i \neq j \\ -\omega_i(\psi_i + \eta_i) & , i = j \end{cases} ,$$

where

$$\psi_i := \frac{\partial}{\partial \lambda_i} G_i(\boldsymbol{\lambda}) + \lambda_i \frac{\partial^2}{\partial \lambda_i^2} G_i(\boldsymbol{\lambda}) > 0 ,$$

and

$$\eta_i := \frac{\partial}{\partial \lambda_i} G_i(\boldsymbol{\lambda}) - U_i''(\lambda_i) > 0 .$$

In order to establish uniqueness of the equilibrium it suffices via Theorems 2 and 6 of Rosen [161] to show that $\mathbf{g}'(\boldsymbol{\lambda}, \boldsymbol{\omega}) + (\mathbf{g}'(\boldsymbol{\lambda}, \boldsymbol{\omega}))^T$ is negative definite for some $\boldsymbol{\omega} > 0$. It therefore suffices to show that $\mathbf{g}'(\boldsymbol{\lambda}, \boldsymbol{\omega})$ is negative definite for some $\boldsymbol{\omega} > 0$. Letting $\omega_i = \psi_i^{-1}$ gives

$$-\mathbf{g}'(\boldsymbol{\lambda}, \boldsymbol{\omega}) = \begin{pmatrix} \frac{\psi_1 + \eta_1}{\psi_1} & 1 & \cdots & 1 \\ 1 & \frac{\psi_2 + \eta_2}{\psi_2} & \cdots & 1 \\ \vdots & \vdots & \ddots & \vdots \\ 1 & 1 & \cdots & \frac{\psi_m + \eta_m}{\psi_m} \end{pmatrix} ,$$

which is clearly symmetric and positive definite. Therefore $\mathbf{g}'(\boldsymbol{\lambda}, \boldsymbol{\omega})$ is negative definite.

These arguments provide the following general existence and uniqueness result.

Theorem 4.2 *Under Assumptions 3 and 4 the solution to the class-optimal m-player net benefit optimization problem (4.21) exists and defines a unique Nash equilibrium.*

4.2.2.3 Proof of Theorem 4.1

As promised in the Section 4.2.1.1, we can also use this result to establish the existence and uniqueness of the individually optimal solution under Assumption 1, by constructing an equivalent class optimization problem.

Suppose Assumption 1 holds. For $\boldsymbol{\lambda} = (\lambda_i, i \in M)$ and $i \in M$, define

$$\hat{H}_i(\boldsymbol{\lambda}) \quad := \quad \int_0^{\lambda_i} G_i(\lambda_1, \ldots, \nu_i, \ldots, \lambda_m) d\nu_i \,,$$

$$\hat{B}_i(\boldsymbol{\lambda}) \quad := \quad U_i(\lambda_i) - (\hat{H}_i(\boldsymbol{\lambda}) + \lambda_i \delta_i) \,.$$

Thus, $\hat{B}_i(\boldsymbol{\lambda})$ can be interpreted as the net benefit per unit time to class i when the class-i waiting cost per unit time is $\hat{H}_i(\boldsymbol{\lambda})$ rather than $H_i(\boldsymbol{\lambda}) = \lambda_i G_i(\boldsymbol{\lambda})$. Since $G_i(\boldsymbol{\lambda})$ is increasing in λ_i, it follows that $\hat{H}_i(\boldsymbol{\lambda})$ is convex in λ_i, for each set of fixed values of λ_j, $j \neq i$. Therefore, Assumption 3 holds with H_i replaced by $\hat{H}_i$. It follows that $\hat{B}_i(\boldsymbol{\lambda})$ is concave in λ_i for each set of fixed values of λ_j, $j \neq i$. But Assumption 1 (2) implies that

$$\frac{\partial}{\partial \lambda_i} \hat{B}_i(\boldsymbol{\lambda}) = U_i'(\lambda_i) - (G_i(\boldsymbol{\lambda}) + \delta_i) < 0 \,,$$

for all $\boldsymbol{\lambda} \in A - \tilde{A}$. This in turn implies that Assumption 4 (2) holds with B_i replaced by $\hat{B}_i$. Since Assumption 1 (1) coincides with Assumption 4 (1), it follows from Theorem 4.2 that there exists a unique solution to the equilibrium conditions for the class-optimal problem with H_i replaced by $\hat{H}_i$. But the equilibrium conditions for this class-optimal problem coincide with the equilibrium conditions for the individually optimal problem. Therefore, there exists a unique solution to the equilibrium conditions for the individually optimal problem. This completes the proof of Theorem 4.1. ∎

4.2.2.4 Comparison of Class Optimal and Individually Optimal Solutions

From observing the optimality conditions for individual and class optimization, it follows that an individually optimal solution will also be optimal for the class-optimization problem with tolls δ_i, $i \in M$, if the toll charged to entering class-i customers in the individual-optimization problem equals

$$\delta_i^e := \lambda_i^c \frac{\partial}{\partial \lambda_i} G_i(\boldsymbol{\lambda}^c) + \delta_i \,.$$

The additional toll for class i is just the external effect on other class-i customers – the marginal increase in the total waiting cost incurred by all existing flows of class i – of a marginal increase in the flow of class i (evaluated at the class-optimal flow allocation, $\boldsymbol{\lambda}^c$). In this way the individually optimizing customers are forced to take account of the effect of their decision on other members of their own class (but not, of course, on members of other classes). Since $G_i(\boldsymbol{\lambda})$ is increasing in λ_i, the external effect is nonnegative (positive if $\lambda_i^c > 0$) and hence $\delta_i^e \geq \delta_i$ ($\delta_i^e > \delta_i$ if $\lambda_i^c > 0$).

4.2.3 Socially Optimal Arrival Rates

Consider a flow allocation, $\boldsymbol{\lambda} = (\lambda_i, i \in M)$. The net benefit per unit time for class i is

$$B_i(\boldsymbol{\lambda}) = U_i(\lambda_i) - \lambda_i G_i(\boldsymbol{\lambda}) . \tag{4.22}$$

A *socially optimal* allocation of arrival rates $\boldsymbol{\lambda}^s = (\lambda_i^s, i \in M)$ to the classes is defined as one that maximizes the aggregate net benefit to all classes. It may thus be found by solving the following optimization problem:

$$\textbf{(P)} \quad \max_{\{\lambda_i \geq 0, i \in M\}} \quad \mathcal{U}(\boldsymbol{\lambda}) := \sum_{i \in M} U_i(\lambda_i) - \sum_{i \in M} \lambda_i G_i(\boldsymbol{\lambda})$$

$$\text{s.t.} \quad \lambda_i \geq 0 , \; i \in M .$$

(Since tolls are simply transfer fees, they do not appear in the objective function for social optimality.)

A socially optimal allocation of arrival rates $\boldsymbol{\lambda}^s = (\lambda_i^s, i \in M)$ must satisfy the following *KKT* conditions, which are necessary for an optimal solution to Problem (**P**): for all $i \in M$,

$$U_i'(\lambda_i) \;\leq\; G_i(\boldsymbol{\lambda}) + \sum_{k \in M} \lambda_k \frac{\partial}{\partial \lambda_i} G_k(\boldsymbol{\lambda}) ; \tag{4.23}$$

$$U_i'(\lambda_i) \;=\; G_i(\boldsymbol{\lambda}) + \sum_{k \in M} \lambda_k \frac{\partial}{\partial \lambda_i} G_k(\boldsymbol{\lambda}) , \text{ if } \lambda_i > 0 . \tag{4.24}$$

4.2.3.1 Comparison of Socially Optimal and Toll-Free Individually Optimal Solutions

Comparing these equations to the equilibrium conditions for an individually optimal solution, we see that the individually optimal arrival rates will also be socially optimal if entering customers of class i are charged a toll

$$\delta_i^s := \sum_{k \in M} \lambda_k^s \frac{\partial}{\partial \lambda_i} G_k(\boldsymbol{\lambda}^s) .$$

As expected, the socially optimal toll for class i is just the *external effect* – the marginal increase in the total waiting cost incurred by all existing flows – of a marginal increase in the flow of class i (evaluated at the socially optimal flow allocation, $\boldsymbol{\lambda}^s$). By charging this socially optimal toll at each resource, the facility operator can induce individually optimizing classes to behave in a socially optimal way, thereby making the Nash-equilibrium flow allocation for individual optimality coincide with a socially optimal solution.

Remark 1   For this general model in which the waiting-cost functions, $G_k(\boldsymbol{\lambda})$, and their partial derivatives may depend on the entire vector, $\boldsymbol{\lambda} =$

$(\lambda_i, i \in M)$, the socially optimal tolls will in general be class dependent. Later (see Section 4.4) we shall consider a special case in which each $G_k(\boldsymbol{\lambda})$ depends on $\boldsymbol{\lambda}$ only through the total arrival rate, $\lambda = \sum_{i \in M} \lambda_i$. In this case the socially optimal toll does not depend on the class.

Since for all $k \neq i$, $G_k(\boldsymbol{\lambda})$ is nondecreasing in λ_i, and $G_i(\boldsymbol{\lambda})$ is strictly increasing in λ_i, the external effect is nonnegative (positive if $\lambda_i^s > 0$) and hence $\delta_i \geq 0$ ($\delta_i > 0$ if $\lambda_i^s > 0$).

These observations facilitate a comparison between the socially optimal solution and the individually optimal allocation when no tolls are charged. Since $U_i'(\cdot)$ is nonincreasing (by the concavity of $U_i(\cdot)$), we have the following theorem.

Theorem 4.3 *Suppose $\delta_i = 0$, $i \in M$, and let λ_i^e, $i \in M$, be the corresponding individually optimal arrival rates. Then for each class i, $\lambda_i^s \leq \lambda_i^e$. The inequality is strict if $U_i'(\cdot)$ is strictly decreasing and $\lambda_i^s > 0$.*

4.2.3.2 Comparison of Socially Optimal and Class-Optimal Solutions

From comparing the optimality conditions for social and class optimization, it follows that a class-optimal solution will also be optimal for the social-optimization problem if the toll charged to entering class-i customers in the class-optimization problem is

$$\delta_i = \sum_{k \in M} \lambda_k^s \frac{\partial}{\partial \lambda_i} G_k(\boldsymbol{\lambda}^s) - \lambda_i^s \frac{\partial}{\partial \lambda_i} G_i(\boldsymbol{\lambda}^c) \,.$$

Since entering class-i customers are already charged the class-i external effect,

$$\lambda_i^s \frac{\partial}{\partial \lambda_i} G_i(\boldsymbol{\lambda}^c) \,,$$

at the class-optimal allocation, we must first replace this term by the class-i external effect at the socially optimal allocation, then add the external effects on the other classes at the socially optimal allocation.

4.2.3.3 Existence and Uniqueness of Socially Optimal Solution

Are the *KKT* conditions sufficient as well as necessary for a vector of arrival rates to be socially optimal? If not, there may be multiple solutions to these conditions, only one of which will (typically) be the socially optimal allocation. From the general theory we know that the *KKT* conditions are sufficient if the objective function is concave in $\boldsymbol{\lambda} = (\lambda_i, i \in M)$. Since the utility term, $\sum_{i \in M} U_i(\lambda_i)$, is concave in $\boldsymbol{\lambda}$, the objective function will be concave if the waiting cost term, $\sum_{i \in M} \lambda_i G_i(\boldsymbol{\lambda})$, is convex in $\boldsymbol{\lambda}$.

We shall find, however, that our model is susceptible to anomalies which result from the fact that the class-i waiting-cost per unit time, $H_i(\boldsymbol{\lambda}) = \lambda_i G_i(\boldsymbol{\lambda})$, may not be convex in $\boldsymbol{\lambda}$, even if $G_i(\boldsymbol{\lambda})$ is convex. In this case, the objective function may not be jointly concave in $\boldsymbol{\lambda} = (\lambda_i, i \in M)$. (By contrast, in the model with a single class considered in Chapter 2, the objective function is always jointly concave.) We shall give examples of this phenomenon below when we consider special cases of our general model.

4.2.4 Facility-Optimal Arrival Rates

Now consider the system from the point of view of a facility operator whose goal is to choose class-dependent tolls, δ_i, $i \in M$, that will maximize revenue. Assuming that the arriving customers are individual optimizers, choosing a toll δ_i for each class i will result in vector $\boldsymbol{\lambda} = (\lambda_i, i \in M)$ of arrival rates that satisfy the equilibrium conditions ($i \in M$):

$$\lambda_i(\delta_i + G_i(\boldsymbol{\lambda}) - U_i'(\lambda_i)) = 0 \,,$$
$$\delta_i + G_i(\boldsymbol{\lambda}) - U_i'(\lambda_i) \geq 0 \,,$$
$$\lambda_i \geq 0 \,.$$

The revenue-maximizing facility operator thus seeks values of δ_i, $i \in M$, that solve the following optimization problem:

$$\max_{\{\delta_i, \lambda_i, i \in M\}} \sum_{i \in M} \lambda_i \delta_i \,,$$
$$\text{s.t.} \quad \lambda_i(\delta_i + G_i(\boldsymbol{\lambda}) - U_i'(\lambda_i)) = 0 \,, \ i \in M \,,$$
$$\delta_i + G_i(\boldsymbol{\lambda}) - U_i'(\lambda_i) \geq 0 \,, \ i \in M \,,$$
$$\lambda_i \geq 0 \,, \ i \in M \,.$$

Subtracting $\sum_{i \in M} \lambda_i(\delta_i + G_i(\boldsymbol{\lambda}) - U_i'(\lambda_i))$ (which equals zero by the first set of constraints) from the objective function and simplifying leads to the following equivalent formulation:

$$\max_{\{\delta_i, \lambda_i, i \in M\}} \sum_{i \in M} \lambda_i(U_i'(\lambda_i) - G_i(\boldsymbol{\lambda}))$$
$$\text{s.t.} \quad \delta_i \geq U_i'(\lambda_i) - G_i(\boldsymbol{\lambda}) \,, \ i \in M$$
$$\delta_i = U_i'(\lambda_i) - G_i(\boldsymbol{\lambda}) \,, \text{ if } \lambda_i > 0 \,, \ i \in M \,,$$
$$\lambda_i \geq 0 \,, \ i \in M \,.$$

As in the single-class model (cf. Chapter 2), we shall assume for now that $\lim_{\lambda_i \to 0} \lambda_i U_i'(\lambda_i) = 0$, $i \in M$. (See Section 4.2.4.1 below for further discussion of this point.)

The first two sets of constraints now serve simply to define δ_i (nonuniquely), given λ_i, $i \in M$. Therefore, it suffices to solve the following problem, with λ_i, $i \in M$, as the only decision variables,

$$\max_{\{\lambda_i \geq 0, i \in M\}} \sum_{i \in M} \lambda_i(U_i'(\lambda_i) - G_i(\boldsymbol{\lambda})) \,,$$

and then choose δ_i, $i \in M$, to satisfy

$$\delta_i \geq U_i'(\lambda_i) - G_i(\boldsymbol{\lambda}) \,, \ i \in M \,,$$
$$\delta_i = U_i'(\lambda_i) - G_i(\boldsymbol{\lambda}) \,, \text{ if } \lambda_i > 0 \,, \ i \in M \,.$$

Let $\boldsymbol{\lambda}^f = (\lambda_1^f, \ldots, \lambda_m^f)$ denote a facility optimal allocation of arrival rates. The vector $\boldsymbol{\lambda}$ satisfies the following necessary *KKT* conditions for optimality:

$$U_i'(\lambda_i) + \lambda_i U_i''(\lambda_i) \leq G_i(\boldsymbol{\lambda}) + \sum_{j \in M} \lambda_j \frac{\partial}{\partial \lambda_i} G_j(\boldsymbol{\lambda}) ; \qquad (4.25)$$

$$U_i'(\lambda_i) + \lambda_i U_i''(\lambda_i) = G_i(\boldsymbol{\lambda}) + \sum_{j \in M} \lambda_j \frac{\partial}{\partial \lambda_i} G_j(\boldsymbol{\lambda}) \text{ if } \lambda_i > 0 , \quad (4.26)$$

for all $i = 1, \ldots, m$.

It follows that a facility optimal toll for class i is given by

$$\delta_i^f = \sum_{j \in M} \lambda_j^f \frac{\partial}{\partial \lambda_i} G_j(\boldsymbol{\lambda}^f) - \lambda_i^f U_i''(\lambda_i^f) , \ i \in M .$$

As in the single-facility model (see Chapter 2), note that the facility optimization problem takes the same form as the social optimization problem, but with modified utility functions, $\tilde{U}_i(\lambda_i) := \lambda_i U_i'(\lambda_i)$, for each class $i \in M$. But, as remarked in Chapter 2, the modified utility functions, $\tilde{U}_i(\lambda_i)$, may not be concave or even nondecreasing. Thus in this case, in contrast to the socially optimal problem, concavity of the objective function may fail even if the class-i waiting-cost per unit time, $H_i(\boldsymbol{\lambda}) = \lambda_i G_i(\boldsymbol{\lambda})$, is convex in $\boldsymbol{\lambda}$.

Since the objective function for facility optimization may fail to be concave, the necessary conditions, (4.25) and (4.26), may not be sufficient for an arrival-rate vector, $\boldsymbol{\lambda} = (\lambda_i, i \in M)$, to be globally facility optimal. Conditions (4.25) and (4.26) may have multiple solutions, including local maxima, local minima, or saddle-points.

4.2.4.1 Heavy-Tailed Rewards

Now we consider what happens if we relax our technical assumption that $\lim_{\lambda \to 0} \lambda U'(\lambda) = 0$. Basically the analysis in Sections 2.1.3 and 2.3.3 of Chapter 2 carries over to the multiclass model with very little modification.

In particular, suppose $\lim_{\lambda_i \to 0} \lambda_i U_i'(\lambda_i) = \kappa_i$, where $0 < \kappa_i \leq \infty$, for some class $i \in M$. (As in the single-class model, this is the case if $\lim_{x \to \infty} x \bar{F}_i(x) = \kappa$, where $\bar{F}_i(x) := P\{R_i > x\}$ and R_i is the random reward received by a class-i customer who joins the system. Recall that this implies that $\bar{F}(\cdot)$ has a heavy tail: cf. Section 2.3.3 in Chapter 2.)

Consider the case, $\kappa_i = \infty$. In this case, for any fixed values of λ_k, $k \neq i$, the profit-maximizing facility operator can earn an arbitrarily large profit, $\tilde{U}(\boldsymbol{\lambda})$, by choosing an arbitrarily small value for λ_i, which corresponds to choosing an arbitrarily large toll, δ_i, for class i. To see this, recall that

$$\tilde{U}(\boldsymbol{\lambda}) = \sum_{k \in M} \lambda_k U_k'(\lambda_k) - \sum_{k \in M} \lambda_k G_k(\boldsymbol{\lambda}) .$$

Now $G_i(\boldsymbol{\lambda})$ is strictly increasing in λ_i and $G_k(\boldsymbol{\lambda})$ is nondecreasing in λ_i for all

$k \neq i$. Hence, the total waiting cost per unit time, $\sum_{k \in M} \lambda_k G_k(\boldsymbol{\lambda})$, strictly decreases as $\lambda_i \to 0$. Meanwhile, the sum, $\sum_{k \neq i} \lambda_k U'_k(\lambda_k)$ remains constant, but $\lambda_i U'_i(\lambda_i) \to \infty$ as $\lambda_i \to 0$. Therefore, $\tilde{\mathcal{U}}(\boldsymbol{\lambda}) \to \infty$ as $\lambda_i \to 0$ while λ_k, $k \neq i$, remain fixed.

Note that, if there are several classes i such that $\lim_{\lambda_i \to 0} \lambda_i U'_i(\lambda_i) = \infty$, then the facility operator may pursue this strategy for each such class.

4.3 General Multiclass Model: Dynamic Algorithms

We now turn our attention to the application of dynamic adaptive algorithms to multiclass queueing facilities. We introduced dynamic adaptive algorithms in the context of a single-class queueing facility in Chapter 3. The extension of these algorithms to multiclass facilities involves a number of intricacies, but the basic concepts carry over with few changes. We shall consider continuous-time algorithms only, leaving to the reader the extension of discrete-time algorithms to the multiclass model.

4.3.1 Continuous-Time Dynamic Algorithms

As usual, our starting point is the basic model with individually optimizing customers. Class-optimal, socially optimal, or facility optimal solutions may then be implemented by charging entering customers an appropriate toll.

Recall the equilibrium conditions for an individually optimal solution ($i \in M$):

$$U'_i(\lambda_i) \quad \leq \quad G_i(\boldsymbol{\lambda}) + \delta_i \; ; \tag{4.27}$$

$$U'_i(\lambda_i) \quad = \quad G_i(\boldsymbol{\lambda}) + \delta_i \; , \text{ if } \lambda_i > 0 \; . \tag{4.28}$$

Now suppose the arrival rates and tolls evolve in continuous time. For each $i \in M$, let $\lambda_i(t)$ denote the arrival rate for class i and $\delta_i(t)$ the toll charged each entering customer of class i at time $t \geq 0$. Let ω_i be a positive constant. The vector of arrival rates at time t is $\boldsymbol{\lambda}(t) = (\lambda_i(t), i \in M)$. Consider the system of differential equations

$$\frac{d}{dt} \lambda_i(t) = \omega_i \left(U'_i(\lambda_i(t)) - [G_i(\boldsymbol{\lambda}(t)) + \delta_i(t)] \right) \; , \; i \in M \; . \tag{4.29}$$

We interpret (4.29) as follows. The facility operator charges a toll $\delta_i(t)$ per entering customer of class i at time t. Equation (4.29) corresponds to a rate-control algorithm in which the arrival rate for each class i is adjusted in proportion to the current difference between the marginal utility and the full price (the waiting cost plus the toll) of entering the system for class i. If this difference is positive (that is, if the marginal utility exceeds the full price), then the flow is increased; if negative then the flow is decreased. (As usual, if the current value of $\lambda_i(t)$ is zero and the r.h.s. of the differential equation is negative, it is understood that the rate-control algorithm keeps this flow at zero rather than reducing it further and making it negative.)

It is therefore reasonable to conjecture that this algorithm will converge to a value of $\boldsymbol{\lambda} = (\lambda_i, i \in M)$ satisfying the Nash equilibrium conditions, (4.27), (4.28), provided that $\delta_i(t) \to \delta_i$, for each $i \in M$.

The specific behavior of the algorithm depends on how we choose the tolls, $\delta_i(t)$, $i \in M$. If $\delta_i(t) = \delta_i$, $i \in M$ (fixed tolls), then the algorithm converges to the individually optimal arrival rate vector, $\boldsymbol{\lambda}^e$, associated with the fixed tolls, δ_i, $i \in M$. On the other hand, if we choose the $\delta_i(t)$ equal to the appropriate tolls for class optimality, social optimality, or facility optimality, then the algorithm converges (not surprisingly) to a class-optimal, socially optimal, or facility optimal vector of arrival rates, respectively. These assertions are proved in the following four sections, using the properties of optimal solutions established in Section 4.2.

4.3.1.1 Algorithm with Fixed Toll: Individual Optimality

Consider the dynamical system governed by the differential equation (4.29), but now suppose that the toll charged each entering customer of class i is a fixed quantity, δ_i, $i \in M$. The system of differential equations is now

$$\frac{d}{dt} \lambda_i(t) = \omega_i \left(U_i'(\lambda_i(t)) - [G_i(\boldsymbol{\lambda}(t)) + \delta_i] \right) , \ i \in M . \quad (4.30)$$

A stationary point $\boldsymbol{\lambda}$ of this system, at which the r.h.s. of each of the differential equations (4.30) equals zero (or is less than or equal to zero, if $\lambda_i = 0$), satisfies (4.27), (4.28), and is therefore a Nash equilibrium for individually optimizing customers when entering customers of class i are charged the toll δ_i, $i \in M$. By Theorem 4.1, this equilibrium solution is unique under Assumption 1.

Recall that we proved Theorem 4.1 by constructing a class-optimization problem with the same optimality conditions as the individual optimization problem. The same construction allows us to conclude that the dynamic algorithm (4.30) converges globally to the unique equilibrium solution for the individual optimization problem. (In the next section we shall prove global convergence in the class optimization problem.) Thus we have the following theorem.

Theorem 4.4 *Under Assumption 1, the dynamic algorithm (4.30) converges globally to the unique solution to the equilibrium conditions (4.14), (4.15), which characterize the vector $\boldsymbol{\lambda}^e$ of individually optimal arrival rates.*

4.3.1.2 Algorithm with Time-Varying Toll: Class Optimality

Consider the dynamical system governed by the differential equation (4.29), and suppose that each entering customer of class i is charged a time-varying toll

$$\delta_i(t) = \frac{\partial}{\partial \lambda_i} G_i(\boldsymbol{\lambda}(t)) + \delta_i ,$$

$i \in M$, where the δ_i, $i \in M$, are given, fixed tolls. The system of differential equations is now

$$\frac{d}{dt}\lambda_i(t) = \omega_i \left(U_i'(\lambda_i(t)) - [G_i(\boldsymbol{\lambda}(t)) + \frac{\partial}{\partial \lambda_i}G_i(\boldsymbol{\lambda}(t)) + \delta_i] \right) , \; i \in M \; . \quad (4.31)$$

Recall that by Theorem 4.2 there is a unique class-optimal solution under Assumptions 3 and 4, which satisfies the equilibrium conditions, (4.19) and (4.20). Using equation (4.18) we can rewrite these conditions equivalently as

$$\frac{\partial}{\partial \lambda_i}B_i(\boldsymbol{\lambda}) \;\; \leq \;\; 0 \, , \quad\quad\quad\quad (4.32)$$

$$\frac{\partial}{\partial \lambda_i}B_i(\boldsymbol{\lambda}) \;\; = \;\; 0 \, . \, , \text{ if } \lambda_i > 0 \quad\quad\quad (4.33)$$

We can similarly rewrite the dynamical system (4.31) equivalently as

$$\frac{d}{dt}\lambda_i(t) = \omega_i \cdot \frac{\partial}{\partial \lambda_i}B_i(\boldsymbol{\lambda}(t)) \, , \; i \in M \; . \quad\quad (4.34)$$

From Theorem 9 of Rosen [161] we obtain the following result.

Theorem 4.5 *Under Assumptions 3 and 4 the dynamic algorithm (4.34) converges globally to the unique solution to the equilibrium conditions (4.32), (4.33), which characterize the vector* $\boldsymbol{\lambda}^c$ *of class optimal arrival rates.*

4.3.1.3 Algorithm with Time-Varying Toll: Social Optimality

Consider the dynamical system governed by the differential equation (4.29), and suppose now that each entering customer of class i is charged a time-varying toll

$$\delta_i(t) = \sum_{k \in M} \frac{\partial}{\partial \lambda_i}G_k(\boldsymbol{\lambda}(t)) \, , \quad\quad\quad\quad (4.35)$$

$i \in M$. The system of differential equations is now

$$\frac{d}{dt}\lambda_i(t) = \omega_i \left(U_i'(\lambda_i(t)) - [G_i(\boldsymbol{\lambda}(t)) + \sum_{k \in M} \frac{\partial}{\partial \lambda_i}G_k(\boldsymbol{\lambda}(t))] \right) , \; i \in M \; . \quad (4.36)$$

In this version of the dynamic algorithm the toll $\delta_i(t)$ (given by (4.35)) equals the overall external effect of class-i flow: the marginal increase in the total waiting cost of all existing flows in all classes as a result of a marginal increase in λ_i.

The stability analysis of the algorithm in this case parallels that for social optimality in the single-class model (cf. Section 3.4.1 of Chapter 3), with the important difference that there may now be multiple points to which the algorithm can converge. Recall the objective function for social optimality in the multiclass model:

$$\mathcal{U}(\boldsymbol{\lambda}) = \sum_{i \in M} U_i(\lambda_i) - \sum_{i \in M} \lambda_i G_i(\boldsymbol{\lambda}) \; .$$

We have ($i \in M$)

$$\frac{\partial}{\partial \lambda_i} \mathcal{U}(\boldsymbol{\lambda}) = U_i'(\lambda_i) - [G_i(\boldsymbol{\lambda}) + \sum_{k \in M} \frac{\partial}{\partial \lambda_i} G_k(\boldsymbol{\lambda})] . \tag{4.37}$$

Now consider $\mathcal{U}(\boldsymbol{\lambda}(t))$, where $\boldsymbol{\lambda}(t) := (\lambda_i(t), i \in M)$ and $\lambda_i(t)$ satisfies the set of differential equations (4.36). It follows from (4.36) and (4.37) that

$$\begin{aligned}
\frac{d}{dt} \mathcal{U}(\boldsymbol{\lambda}(t)) &= \sum_{i \in M} \frac{\partial}{\partial \lambda_i} \mathcal{U}(\boldsymbol{\lambda}(t)) \cdot \frac{d}{dt} \lambda_i(t) \\
&= \sum_{i \in M} \omega_i \left(U_i'(\lambda_i(t) - [G_i(\boldsymbol{\lambda}(t)) + \sum_{k \in M} \frac{\partial}{\partial \lambda_i} G_k(\boldsymbol{\lambda}(t))] \right)^2 ,
\end{aligned}$$

from which we see that $\mathcal{U}(\boldsymbol{\lambda}(t))$ is strictly increasing over time unless the quantity in parentheses on the last line equals zero for all $i \in M$, that is, unless $\boldsymbol{\lambda}(t)$ is a solution to the *KKT* conditions. It follows that the algorithm converges (with strictly increasing aggregate utility) to a solution to the *KKT* conditions. We have observed that there may in general be several such solutions, each of which is a Nash equilibrium for individually optimizing customers when charged the appropriate tolls. Which equilibrium is approached depends on the starting point. We shall see several examples of this phenomenon in subsequent sections.

4.3.1.4 Algorithm with Time-Varying Toll: Facility Optimality

To keep the exposition simple, we now assume that $\lambda_i U_i'(\lambda_i) \to 0$ as $\lambda_i \to 0$, for all $i \in M$. The modifications to the following analysis that are required when this condition is not satisfied are straightforward.

Consider the dynamical system governed by the differential equation (4.29), and suppose now that each entering customer of class i is charged a time-varying toll

$$\delta_i(t) = \sum_{k \in M} \lambda_k(t) \frac{\partial}{\partial \lambda_i} G_k(\boldsymbol{\lambda}(t)) - \lambda_i U_i''(t) , \tag{4.38}$$

$i \in M$. The system of differential equations is now

$$\begin{aligned}
\frac{d}{dt} \lambda_i(t) &= \omega_i (U_i'(\lambda_i(t)) + \lambda_i(t) U_i''(t) \\
&\quad - [G_i(\boldsymbol{\lambda}(t)) + \sum_{k \in M} \lambda_k(t) \frac{\partial}{\partial \lambda_i} G_k(\boldsymbol{\lambda}(t))]) , \ i \in M . \tag{4.39}
\end{aligned}$$

At a stationary point $\boldsymbol{\lambda}$ of this system, the r.h.s. of each of the differential equations (4.39) equals zero (or is less than or equal to zero, if $\lambda_i = 0$). That is,

$$U_i'(\lambda_i) + \lambda_i U_i''(\lambda_i) - [G_i(\boldsymbol{\lambda}) + \sum_{k \in M} \lambda_k \frac{\partial}{\partial \lambda_i} G_k(\boldsymbol{\lambda})] \ \leq \ 0 ,$$

$$U_i'(\lambda_i) + \lambda_i U_i''(\lambda_i) - [G_i(\boldsymbol{\lambda}) + \sum_{k \in M} \lambda_k \frac{\partial}{\partial \lambda_i} G_k(\boldsymbol{\lambda})] \quad = \quad 0 \text{ , if } \lambda_i > 0 \text{ ,}$$

$i \in M$. These are the necessary *KKT* conditions (cf. (4.25), (4.26)) for a facility optimal solution, $\boldsymbol{\lambda}^f$.

To show that the dynamic algorithm defined by the differential equation (4.39) converges to such a stationary point, we shall once again make use of the equivalence, noted in Remark 3 in Chapter 2, between the facility optimization problem and a social optimization problem with modified utility functions, $\tilde{U}_i(\lambda_i) := \lambda_i U_i'(\lambda_i)$, $i \in M$. For this transformed social optimization problem the associated dynamic algorithm is governed by the system of differential equations,

$$\frac{d}{dt}\lambda_i(t) = \omega_i \left(\tilde{U}_i'(\lambda_i(t)) - [G_i(\boldsymbol{\lambda}(t)) + \sum_{k \in M} \lambda_k(t) \frac{\partial}{\partial \lambda_i} G_k(\boldsymbol{\lambda}(t))] \right) , \quad (4.40)$$

$i \in M$, which is equivalent to the system (4.39). Define a new net utility function, $\hat{\mathcal{U}}$, as follows:

$$\tilde{\mathcal{U}}(\boldsymbol{\lambda}) := \sum_{k \in M} (\tilde{U}_k(\lambda_k) - \lambda_k G_k(\boldsymbol{\lambda})) .$$

Note that

$$\frac{\partial}{\partial \lambda_i}\tilde{\mathcal{U}}(\boldsymbol{\lambda}) \quad = \quad \tilde{U}_i'(\lambda_i) - [G_i(\boldsymbol{\lambda}) + \sum_{k \in M} \lambda_k \frac{\partial}{\partial \lambda_i} G_k(\boldsymbol{\lambda})]$$

$$= \quad U_i'(\lambda_i) + \lambda_i U_i''(\lambda_i) - [G_i(\boldsymbol{\lambda}) + \sum_{k \in M} \lambda_k \frac{\partial}{\partial \lambda_i} G_k(\boldsymbol{\lambda})] \text{ , } (4.41)$$

$i \in M$. The function $\tilde{\mathcal{U}}(\cdot)$ is a Lyapunov function for the dynamical system governed by the differential equation (4.40), as can be seen by the following argument. Consider $\tilde{\mathcal{U}}(\boldsymbol{\lambda}(t))$, where $\boldsymbol{\lambda}(t)$ satisfies the differential equation (4.40). We have

$$\frac{d}{dt}\tilde{\mathcal{U}}(\boldsymbol{\lambda}(t)) \quad = \quad \sum_{i \in M} \frac{\partial}{\partial \lambda_i}\tilde{\mathcal{U}}(\boldsymbol{\lambda}(t)) \cdot \frac{d}{dt}\lambda_i(t)$$

$$= \quad \sum_{i \in M} \omega_i \left(\tilde{U}_i'(\lambda_i(t)) - [G_i(\boldsymbol{\lambda}(t)) + \sum_{k \in M} \frac{\partial}{\partial \lambda_i} G_k(\boldsymbol{\lambda}(t))] \right)^2 \text{ ,}$$

from which we see that $\tilde{\mathcal{U}}(\boldsymbol{\lambda}(t))$ is strictly increasing over time unless the quantity in parentheses on the last line equals zero (or is less than or equal to zero if $\lambda_i(t) = 0$) for all $i \in M$, that is, unless $\boldsymbol{\lambda}(t)$ is a solution to the *KKT* conditions. It follows that the algorithm converges (with strictly increasing aggregate utility) to *a* solution to the *KKT* conditions. As in the social optimization problem, there may in general be several such solutions, each of which is a Nash equilibrium for individually optimizing customers when charged the appropriate tolls. Which equilibrium is approached depends on the starting point.

4.4 Waiting Costs Dependent on Total Arrival Rate

Now we consider the special case of our general model in which the waiting cost for each class depends on the vector $\boldsymbol{\lambda} = (\lambda_i, i \in M)$ only through the total arrival rate, $\lambda = \sum_{i \in M} \lambda_i$. In this case we can say quite a bit more about the structure of the allocations under the various optimality criteria and the relations between them.

Suppose $G_i(\boldsymbol{\lambda}) = G_i(\lambda)$, where $\lambda = \sum_{k \in M} \lambda_k$ and $G_i(\lambda)$ is strictly increasing and differentiable in the scalar variable λ. (Here we are abusing notation slightly by using the same symbol "G_i" for both functions.)

Consider the example of an $M/GI/1$ queue with linear waiting cost, introduced in Section 4.1.1. When the queue discipline is *FIFO* and the service times are class independent (with mean $1/\mu$), we noted that the waiting-cost function for class-i customers depends only on the total arrival rate λ and is given by (4.5), which we rewrite here for convenience ($i \in M$):

$$G_i(\lambda) = h_i \left(\frac{1}{\mu} + \frac{\lambda \beta}{\mu(\mu - \lambda)} \right) . \tag{4.42}$$

In the special case of an exponential service-time distribution (an $M/M/1$ queue), we have $\beta = 1$, so that

$$G_i(\lambda) = h_i \left(\frac{1}{\mu - \lambda} \right) , \; i \in M . \tag{4.43}$$

This expression is also valid for an $M/GI/1$ queue with processor-sharing distribution.

We shall use these examples frequently in our discussion of optimal flow allocations and dynamic algorithms below.

4.4.1 Individual Optimality

The equilibrium conditions satisfied by the vector $\boldsymbol{\lambda}^e$ of individually optimal arrival rates now take the form

$$U_i'(\lambda_i) \; \leq \; G_i(\lambda) + \delta_i \tag{4.44}$$
$$U_i'(\lambda_i) \; = \; G_i(\lambda) + \delta_i , \; \text{if } \lambda_i > 0 \tag{4.45}$$

for all $i \in M$, where $\lambda = \sum_{i \in M} \lambda_i$.

Since the model under consideration is a special case of our general model (cf. Section 4.2.1), we know that there exists a unique individually optimal allocation under Assumption 1.

Assumption 5 For all $i \in M$, $G_i(\lambda) \to \infty$ as $\lambda \to \infty$.

Exercise 3   Prove that Assumption 1 holds for the model with waiting costs dependent only on the total arrival rate, if Assumption 5 holds.

Note that Assumption 5 holds in the example of an $M/GI/1$ queue with linear waiting costs and $FIFO$ discipline. (See Section 4.1.1.1.)

4.4.2 Class Optimality

Recall that in the case of class optimization the manager for class i is concerned with maximizing the net benefit, $B_i(\boldsymbol{\lambda})$, received per unit time by jobs of class i, which now takes the form

$$B_i(\boldsymbol{\lambda}) = U_i(\lambda_i) - \lambda_i(G_i(\lambda) + \delta_i) , \qquad (4.46)$$

where $\lambda = \sum_{k \in M} \lambda_k$. Given fixed values of the arrival rates, λ_k, $k \neq i$, the optimal strategy for the manager of class i is therefore to choose a value for λ_i such that

$$\lambda_i = \arg \max_{\{\lambda_i \geq 0\}} \{U_i(\lambda_i) - \lambda_i(G_i(\lambda_i + \bar{\lambda}_i) + \delta_i)\} , \qquad (4.47)$$

where

$$\bar{\lambda}_i := \sum_{k \neq i} \lambda_k = \lambda - \lambda_i .$$

For each $i \in M$ the following KKT conditions are necessary for λ_i to maximize $B_i(\boldsymbol{\lambda})$, with λ_k fixed for $k \neq i$:

$$U_i'(\lambda_i) \leq G_i(\lambda_i + \bar{\lambda}_i) + \delta_i + \lambda_i G_i'(\lambda_i + \bar{\lambda}_i) \qquad (4.48)$$
$$U_i'(\lambda_i) = G_i(\lambda_i + \bar{\lambda}_i) + \delta_i + \lambda_i G_i'(\lambda_i + \bar{\lambda}_i) , \text{ if } \lambda_i > 0 \qquad (4.49)$$

Therefore, (4.48) and (4.49) must hold simultaneously for all $i \in M$ in order for an allocation $\boldsymbol{\lambda} = (\lambda_i, i \in M)$ to be class optimal.

We shall make the following assumption.

Assumption 6

1. For all $i \in M$, $G_i(\lambda) \to \infty$ as $\lambda \to \infty$.
2. For all $i \in M$,
$$2G_i'(\lambda) + \lambda_i G_i''(\lambda) \geq 0 ,$$
   for all $\boldsymbol{\lambda} \in A = \prod_{i \in M} A_i$.

Assumption 6 (1) implies Assumption 4. Assumption 6 (2) implies that $H_i(\boldsymbol{\lambda}) = \lambda_i G_i(\lambda)$ is a convex function of λ_i, for each set of fixed values of λ_k, $k \neq i$, where $\lambda = \sum_{k \in M} \lambda_k$. Hence Assumption 3 holds. (Note that, in order for Assumption 6 (2) to hold, it suffices that $G_i(\cdot)$ be convex, which is the case in the example of an $M/GI/1$ queue with linear waiting costs and $FIFO$ discipline.)

Thus, as a corollary of Theorem 4.2, we have the following theorem.

Theorem 4.6 *Under Assumption 6 there exists a unique class-optimal allocation $\boldsymbol{\lambda}^c$ for the model in which the waiting costs depend on the total arrival rate.*

In the case where $\delta_i = 0$, note that the optimization problem for class i takes the same form as the social optimization problem for a single-class problem with (variable) arrival rate λ_i with utility function $U_i(\lambda_i)$ and waiting-cost function $\tilde{G}_i(\lambda_i)$, where

$$\tilde{G}_i(\lambda_i) := G_i(\lambda_i + \bar{\lambda}_i) .$$

4.4.3 Social Optimality

The social optimization problem now takes the form:

$$\max_{\{\lambda; \lambda_i, i \in M\}} \quad \mathcal{U}(\boldsymbol{\lambda}) := \sum_{i \in M} U_i(\lambda_i) - \sum_{i \in M} \lambda_i G_i(\lambda)$$

$$\text{s.t.} \quad \sum_{i \in M} \lambda_i = \lambda \; ; \; \lambda_i \geq 0 \, , \, i \in M .$$

A socially optimal allocation of arrival rates $\boldsymbol{\lambda}^s = (\lambda_i^s, i \in M)$ must satisfy the following *KKT* conditions, which are necessary for an optimal solution to this optimization problem:

$$U_i'(\lambda_i) \quad \leq \quad G_i(\lambda) + \sum_{k \in M} \lambda_k G_k'(\lambda) \, , \tag{4.50}$$

$$U_i'(\lambda_i) \quad = \quad G_i(\lambda) + \sum_{k \in M} \lambda_k G_k'(\lambda) \, , \text{ if } \lambda_i > 0 \, , \tag{4.51}$$

$i \in M$, where

$$\lambda = \sum_{k \in M} \lambda_k .$$

Comparing these conditions to those for an individually optimal solution, we see that individually optimizing customers may be induced to behave in a socially optimal way if we charge each entering customer – *regardless of class* – the socially optimal toll,

$$\delta^s := \sum_{k \in M} \lambda_k^s G_k'(\lambda^s) \, ,$$

$i \in M$ where $\lambda^s = \sum_{k \in M} \lambda_k^s$. As usual, the socially optimal toll for class i equals the external effect of a marginal increase in λ_i on the total waiting cost per unit time associated with the current flows in all the classes. But now, since the waiting costs depend only on the total arrival rate, this external effect – the socially optimal toll – is independent of class.

4.4.3.1 Nonconcavity of Objective Function

In our discussion of the general model, we noted that the objective function for the social optimization problem may not be jointly concave in $\boldsymbol{\lambda} = (\lambda_i, i \in M)$, in spite of the convexity/concavity properties of its components. As a consequence, the *KKT* conditions may have multiple solutions, some of which are local maxima, local minima, or saddle-points. In the special case we are considering, in which the waiting costs depend only on the total arrival rate, one might hope that this problem would not occur. As we shall see, however, even in this much simpler model the total waiting-cost per unit time is not in general jointly convex in $\boldsymbol{\lambda}$, even when each class-i waiting-cost function, $G_i(\lambda)$, is convex in λ. As a result the objective function for social optimization may fail to be jointly concave.

To see what can happen, consider a system with two classes. (The extension to more than two classes is straightforward.) In this case, the objective function for social optimization takes the form,

$$\mathcal{U}(\lambda_1, \lambda_2) = U_1(\lambda_1) + U_2(\lambda_2) - f(\lambda_1, \lambda_2) \, ,$$

where $f(\lambda_1, \lambda_2) := \lambda_1 G_1(\lambda_1 + \lambda_2) + \lambda_2 G_2(\lambda_1 + \lambda_2)$. That is, $f(\lambda_1, \lambda_2)$ is the total waiting cost per unit time expressed as a function of the two flows, λ_1 and λ_2.

Let us assume that $G_i(\lambda)$ is convex in λ, and $G_i(\lambda) \to \infty$ as $\lambda \to \infty$, $i = 1, 2$ (as is the case, for example, in the $M/GI/1$ queue with linear waiting costs and $FIFO$ queue discipline.) Then Assumptions 5 and 6 both hold. We have shown that both the individual optimization and the class optimization problem have unique solutions in this case, but, as we shall see presently, the social optimization problem may not.

As a consequence of the assumed convexity of $G_i(\cdot)$, $i = 1, 2$, the combined waiting-cost function, $f(\lambda_1, \lambda_2)$ is convex in λ_1 and convex in λ_2. To check for joint convexity, we evaluate

$$\Delta := \left(\frac{\partial^2 f}{\partial \lambda_1^2} \right) \left(\frac{\partial^2 f}{\partial \lambda_2^2} \right) - \left(\frac{\partial^2 f}{\partial \lambda_1 \lambda_2} \right)^2$$

and check to see if Δ is nonnegative. Now

$$\frac{\partial^2 f}{\partial \lambda_1^2} = 2A + C \, , \quad \frac{\partial^2 f}{\partial \lambda_2^2} = 2B + C \, , \quad \frac{\partial^2 f}{\partial \lambda_1 \lambda_2} = A + B + C \, ,$$

where

$$A := G_1'(\lambda_1 + \lambda_2) \, ; \quad B := G_2'(\lambda_1 + \lambda_2) \, ; \quad C := \lambda_1 G_1''(\lambda_1 + \lambda_2) + \lambda_2 G_2''(\lambda_1 + \lambda_2) \, .$$

It follows that

$$\begin{aligned}
\Delta &= (2A + C)(2B + C) - (A + B + C)^2 \\
&= 4AB + 2(A + B)C + C^2 - (A + B)^2 - 2(A + B)C - C^2 \\
&= 4AB - A^2 - 2AB - B^2 \\
&= -(A - B)^2 \, ,
\end{aligned}$$

which is strictly negative unless $A = B$, that is, unless $G_1'(\lambda_1 + \lambda_2) = G_2'(\lambda_1 + \lambda_2)$. Thus $f(\lambda_1, \lambda_2)$ is *not* in general a jointly convex function of λ_1 and λ_2.

In the special case of a system with linear waiting costs, we have $G_i(\cdot) = h_i \cdot W(\cdot)$, where $W(\lambda)$ is the expected waiting time in the system for a customer (regardless of class). In this case we see that $\Delta = -((h_1 - h_2)W'(\lambda_1 + \lambda_2))^2$, which is strictly negative unless $h_1 = h_2$, that is, unless the classes are homogeneous with respect to their sensitivity to delay. In this case the conditions for joint convexity fail at *every* point in the feasible region if the classes are heterogeneous, that is, if $h_1 \neq h_2$ (cf. Section 1.4.2 of Chapter 1).

Note that the only properties that we have assumed are that the waiting cost per customer for each class is an increasing, convex, and differentiable function of the sum of the flows, and that the waiting cost per unit time for each class is the product of the flow (number of customers arriving per unit time) and the waiting cost per customer. All these properties are weak and common in the literature. The assumption that each $G_i(\cdot)$ is convex is simply an assumption that each class's marginal cost of waiting does not decrease as the total flow increases. As we have observed, this property holds in our canonical example: an $M/GI/1$ queue with linear waiting costs and *FIFO* queue discipline. In queueing terms, the relation – waiting cost per unit time = (arrival rate) × (waiting cost per customer) – is a special case of $H = \lambda G$, the generalization of $L = \lambda W$, which holds under weak assumptions. (See El-Taha and Stidham [60], Chapter 6. In the case of linear waiting cost, it just follows from $L = \lambda W$ itself.)

As we noted, nonconvexity of the waiting-cost function may result in the objective function for the social optimization problem failing to be jointly concave. In the case of linear utility functions, the failure is immediate and universal: the objective function for social optimality is always nonconcave. Because of the attention given them in the literature and because they lead to class dominance, we shall study linear utility functions in detail in Section 4.5 below.

In general, when the objective function for social optimality is not jointly concave, the *KKT* (necessary) conditions may not be sufficient for a socially optimal allocation. There may be multiple solutions to these conditions, some of which may be saddle-points or local minima, rather than local, let alone global, maxima. In general joint concavity of the objective function depends on "how strictly concave" the $U_i(\lambda_i)$ functions are. Section 4.6 exhibits examples of value functions for which joint concavity holds and examples for which it does not hold, all in the context of an $M/M/1$ queue with a linear waiting-cost function.

4.4.4 Facility Optimality

Now consider the system from the point of view of a facility operator whose goal is to choose class-dependent tolls, δ_i, $i \in M$, that will maximize revenue.

Proceeding as in the general model, we can find facility optimal arrival rates by solving the following problem, with λ_i, $i \in M$, as the only decision variables,

$$\max_{\{\lambda_i \geq 0, i \in M\}} \sum_{i \in M} \lambda_i \left(U_i'(\lambda_i) - G_i \left(\sum_{k \in M} \lambda_k \right) \right),$$

and then choose δ_i, $i \in M$, to satisfy

$$\delta_i \geq U_i'(\lambda_i) - G_i(\sum_{k \in M} \lambda_k) , \ i \in M ,$$

$$\delta_i = U_i'(\lambda_i) - G_i(\sum_{k \in M} \lambda_k) , \text{ if } \lambda_i > 0 , \ i \in M .$$

Let $\boldsymbol{\lambda}^f = (\lambda_1^f, \ldots, \lambda_m^f)$ denote a facility optimal allocation of arrival rates. The vector $\boldsymbol{\lambda}$ satisfies the following necessary KKT conditions for optimality:

$$U_i'(\lambda_i) + \lambda_i U_i''(\lambda_i) \leq G_i(\lambda) + \sum_{j \in M} \lambda_j G_j'(\lambda) ; \tag{4.52}$$

$$U_i'(\lambda_i) + \lambda_i U_i''(\lambda_i) = G_i(\lambda) + \sum_{j \in M} \lambda_j G_j'(\lambda) \text{ if } \lambda_i > 0 ; \tag{4.53}$$

$$\sum_{k \in M} \lambda_k = \lambda , \tag{4.54}$$

for all $i \in M$.

It follows that the facility optimal toll for class i is given by

$$\delta_i^f = \sum_{k \in M} \lambda_k^f G_k'(\sum_{k \in M} \lambda_k^f) - \lambda_i^f U_i''(\lambda_i^f) ,$$

for all $i \in M$.

As in the general model of this chapter, the objective function for this problem may not be concave in $\boldsymbol{\lambda}$, even under the (operative) assumptions that $U_i(\lambda_i)$ is concave, because $\lambda_i U_i'(\lambda_i)$ may not be concave. Thus in this case, in contrast to the socially optimal problem, concavity of the objective function may fail even if the class-i waiting-cost per unit time, $H_i(\boldsymbol{\lambda}) = \lambda_i G_i(\boldsymbol{\lambda})$, is jointly convex in $\boldsymbol{\lambda}$.

Since the objective function for facility optimization may fail to be concave, the necessary conditions, (4.52), (4.52), and (4.54), may not be sufficient for an arrival-rate vector, $\boldsymbol{\lambda} = (\lambda_i, i \in M)$, to be globally optimal. These conditions may have multiple solutions, including local maxima, local minima, or saddle-points.

4.5 Linear Utility Functions: Class Dominance

In this section we shall spend some time discussing the "extreme" case of linear utility functions, mainly in the context of the model with waiting costs dependent on total arrival rate discussed in the previous section. We do this not only because linear utility functions have been extensively considered in the literature, but also because the consequences of the lack of joint concavity of the objective function are most dramatic in this case. Specifically, it turns out that optimal allocations exhibit *class dominance* – that is, a single class has positive arrival rate and all other classes have zero arrival rate – under individual and social optimality. (Class-optimal allocations are not class dominant in general.)

Suppose then that the utility function for each class is linear:

$$U_i(\lambda_i) = r_i \lambda_i \,, \ i \in M \,. \tag{4.55}$$

We assume that $r_1 > r_2 > \ldots > r_m$. Entering customers of class i are charged a toll δ_i, which may be class dependent.

With respect to waiting-cost functions, we mainly restrict attention to the model of the previous section. That is, we assume a general (differentiable, nondecreasing, and convex) waiting-cost function, $G_i(\lambda)$, for each class $i \in M$, which depends on the total arrival rate, $\lambda = \sum_{i \in M} \lambda_i$. Recall that an example in which these assumptions are instantiated is a steady-state $M/GI/1$ queue with linear waiting costs and class-independent service times, operating under *FIFO* queue discipline.

We also consider a variant of this model, with class-dependent service times, in which the waiting costs do not depend just on the total arrival rate, but the individually optimal, socially optimal, and facility optimal allocations still exhibit class dominance. (See Section 4.5.5.)

4.5.1 Individually Optimal Arrival Rates

When the utility functions take the form (4.55), an allocation, $\boldsymbol{\lambda} = (\lambda_i, i \in M)$, will be a Nash equilibrium for individually optimizing customers if and only if

$$
\begin{aligned}
r_i &\leq G_i(\lambda) + \delta_i \\
r_i &= G_i(\lambda) + \delta_i \,, \text{ if } \lambda_i > 0 \,,
\end{aligned}
$$

for all $i \in M$, where $\lambda = \sum_{i \in M} \lambda_i$. Thus, in order to have a positive arrival rate in two classes, say $i = 1$ and $i = 2$, we must have

$$
\begin{aligned}
r_1 &= G_1(\lambda) + \delta_1 \,, \tag{4.56} \\
r_2 &= G_2(\lambda) + \delta_2 \,. \tag{4.57}
\end{aligned}
$$

Note that, if both (4.56) and (4.57) hold for a particular value of λ, then *any* allocation (λ_1, λ_2) such that $\lambda_1 + \lambda_2 = \lambda$, $\lambda_1 \geq 0$, $\lambda_2 \geq 0$, will be individually optimal.

Under what circumstances can we expect there to be a positive value of λ satisfying both (4.56) and (4.57)? Note first that, for an arbitrary value of λ, if we are free to choose the tolls, δ_1 and δ_2, and they are permitted to be class dependent, then we can choose them such that both (4.56) and (4.57) hold for that value of λ. (Set $\delta_1 = r_1 - G_1(\lambda)$ and $\delta_2 = r_2 - G_2(\lambda)$.) On the other hand, if δ_1 and δ_2 are given, then in order for both (4.56) and (4.57) to hold, we must be able to find a value of λ such that

$$
\begin{aligned}
G_1(\lambda) &= r_1 - \delta_1 \,, \\
G_2(\lambda) &= r_2 - \delta_2 \,,
\end{aligned}
$$

which in general will not be possible. For example, in the case of linear waiting-cost functions – $G_i(\lambda) = h_i \cdot W(\lambda)$ – in order for this to be possible we must have

$$(r_1 - \delta_1)/h_1 = (r_2 - \delta_2)/h_2 = W(\lambda) .$$

Thus we see that having a positive arrival rate for more than one class in an individually optimal allocation is a "rare event" when the utility functions are linear and the tolls are given (in particular, when both tolls are zero). The prevailing scenario is one in which there is *class dominance*, that is, a single class has a positive arrival rate and all other classes have zero arrival rate. In fact, even if both (4.56) and (4.57) do hold, then there still exists a class-dominant solution, e.g., $\lambda_1 = \lambda$, $\lambda_2 = 0$. (Recall that any nonnegative allocation on the line $\lambda_1 + \lambda_2 = \lambda$, including both endpoints, is individually optimal in this case.)

To summarize: there always exists a class-dominant solution to the individual optimization problem in the case of linear utility functions. A dominant class i and the associated arrival rate λ are characterized by the following conditions:

$$\begin{aligned} r_i &= G_i(\lambda) + \delta_i \\ r_k &\leq G_k(\lambda) + \delta_k \text{ , for all } j \neq i . \end{aligned}$$

4.5.1.1 *Linear Waiting Costs*

In the case of linear waiting costs – $G_i(\lambda) = h_i \cdot W(\lambda)$, $i \in M$ – we have

$$(r_i - \delta_i)/h_i = \max_{\{k \in M\}} \{(r_k - \delta_k)/h_k\} ,$$

and λ is the solution to the equation

$$W(\lambda) = (r_i - \delta_i)/h_i .$$

When all the tolls equal zero, we have

$$r_i/h_i = \max_{\{k \in M\}} \{r_k/h_k\} ,$$

with

$$W(\lambda) = r_i/h_i .$$

That is, a dominant class is one that maximizes the reward gained per unit of waiting cost rate.

4.5.2 *Class-Optimal Arrival Rates*

Recall that in the case of class optimization the manager for class i is concerned with maximizing the net benefit, $B_i(\boldsymbol{\lambda})$, received per unit time by jobs of class i, which now takes the form

$$B_i(\boldsymbol{\lambda}) = \lambda_i(r_i - G_i(\lambda) - \delta_i) , \tag{4.58}$$

where $\lambda = \sum_{k \in M} \lambda_k$. Without loss of generality, we shall assume here that $\delta_i = 0$, for all $i \in M$. (To accommodate nonzero tolls, simply replace r_i by $r_i - \delta_i$ throughout the analysis.)

Given fixed values of the arrival rates, λ_k, $k \neq i$, the optimal strategy for the manager of class i is therefore to choose a value for λ_i such that

$$\lambda_i = \arg \max_{\{\lambda_i \geq 0\}} \{\lambda_i(r_i - G_i(\lambda_i + \bar{\lambda}_i))\} , \tag{4.59}$$

where

$$\bar{\lambda}_i := \sum_{k \neq i} \lambda_k = \lambda - \lambda_i .$$

For each $i \in M$ the following *KKT* conditions are necessary for λ_i to maximize $B_i(\boldsymbol{\lambda})$, with λ_k fixed for $k \neq i$:

$$r_i \leq G_i(\lambda_i + \bar{\lambda}_i) + \lambda_i G_i'(\lambda_i + \bar{\lambda}_i) , \tag{4.60}$$

$$r_i = G_i(\lambda_i + \bar{\lambda}_i) + \lambda_i G_i'(\lambda_i + \bar{\lambda}_i) , \text{ if } \lambda_i > 0 . \tag{4.61}$$

Therefore, (4.48) and (4.49) must hold simultaneously for all $i \in M$ in order for an allocation $\boldsymbol{\lambda} = (\lambda_i, i \in M)$ to be class optimal.

We shall focus our attention in this section on the special case of an *M/GI/1* queue with linear waiting costs and *FIFO* discipline. We shall also assume that the service times are class independent. To summarize, then, we make the following assumption:

Assumption 7

1. the utility function for each class is linear: $U_i(\lambda_i) = r_i \cdot \lambda_i$, $i \in M$, where $r_1 > r_2 > \cdots > r_m$;

2. the queue discipline is *FIFO*;

3. the facility operates as an *M/GI/1* queue in steady state, with a service-time distribution that is class independent with mean $\mathrm{E}[\boldsymbol{S}] = 1/\mu$ and second moment $\mathrm{E}[\boldsymbol{S}^2] = 2\beta/\mu^2$;

4. the waiting-cost function for each class i is linear in the expected (steady-state) waiting time in the system with h_i = waiting cost per unit time, $i \in M$.

We already know (cf. Section 4.4.2) that in this case Assumption 6 is satisfied and therefore there exists a unique class-optimal allocation, $\boldsymbol{\lambda}^c = (\lambda_1^c, \ldots, \lambda_m^c)$. Under Assumption 7 we shall be able to derive closed-form expressions for the class-optimal arrival rates.

Let $W(\lambda)$ denote the expected steady-state waiting time in the system when the $\lambda = \sum_{k \in M} \lambda_i$, the total arrival rate. (Assumption 7 (2) and (3) imply that the expected steady-state waiting time in the system is independent of class and depends only on the total arrival rate λ.) From the Pollatczek-Khintchine formula we have

$$W(\lambda) = \frac{1}{\mu} + \frac{\lambda\beta}{\mu(\mu - \lambda)} ,$$

$$= \frac{\mu - \beta'\lambda}{\mu(\mu - \lambda)} \,, \tag{4.62}$$

for $0 \le \lambda < \mu$, where $\beta' := 1 - \beta$. It follows from Assumption 7 (4) that $G_i(\boldsymbol{\lambda}) = h_i W(\lambda)$, $i \in M$. Together with Assumption 7 (1), this implies that the net benefit per unit time for class i is given by

$$B_i(\boldsymbol{\lambda}) = r_i \lambda_i - \lambda_i h_i W(\lambda) \,, \tag{4.63}$$

where $W(\lambda)$ is given by (4.62).

As we have observed, under Assumption 7 there exists a unique class-optimal solution, $\boldsymbol{\lambda}^c = (\lambda_1^c, \dots, \lambda_m^c)$, which is therefore the unique solution to the necessary conditions, (4.19), (4.20), $i \in M$. In this case, these conditions take the following form ($i \in M$):

$$r_i \ \le \ h_i(W(\lambda) + \lambda_i W'(\lambda)) \tag{4.64}$$

$$r_i \ = \ h_i(W(\lambda) + \lambda_i W'(\lambda)) \,, \text{ if } \lambda_i > 0 \,, \tag{4.65}$$

where $\lambda = \sum_{i \in M} \lambda_i$.

We shall focus on the case of an interior solution, $\lambda_i^c > 0$, $i \in M$. For each $i \in M$, let $\mu_i := \mu - \sum_{k \ne i} \lambda_k$. Note that μ_i can be interpreted as the service capacity remaining for class i after allocating flow λ_k to each class k, $k \ne i$. Note also that $\mu_i - \lambda_i = \mu - \lambda$, for all $i \in M$. It then follows, for each $i \in M$, that

$$
\begin{aligned}
W(\lambda) + \lambda_i W'(\lambda) \ &= \ \frac{\mu - \beta'\lambda}{\mu(\mu - \lambda)} + \frac{(1 - \beta')\lambda_i}{(\mu - \lambda)^2} \\
&= \ \frac{(\mu - \beta'\lambda)(\mu - \lambda) + (1 - \beta')\mu\lambda_i}{\mu(\mu - \lambda)^2} \\
&= \ \frac{((1 - \beta')\mu + \beta'\mu_i - \beta'\lambda_i)(\mu_i - \lambda_i) + (1 - \beta')\mu\lambda_i}{\mu(\mu_i - \lambda_i)^2} \\
&= \ \frac{\beta'\lambda_i^2 - 2\beta'\mu_i\lambda_i + ((1 - \beta')\mu + \beta'\mu_i)\lambda_i}{\mu(\mu_i - \lambda_i)^2} \,.
\end{aligned}
$$

Substituting this expression into (4.65) yields the equilibrium conditions

$$h_i(\beta'\lambda_i^2 - 2\beta'\mu_i\lambda_i + ((1 - \beta')\mu + \beta'\mu_i)\lambda_i) = r_i\mu(\mu_i - \lambda_i)^2 \,, \ i \in M \,. \tag{4.66}$$

Combining terms, (4.66) reduces to

$$h_i(1 - \beta')\mu\mu_i = (r_i\mu - h_i\beta')(\mu_i - \lambda_i)^2 \,, \ i \in M \,.$$

Since $r_i\mu - h_i\beta' > 0$ follows from Assumption 7 that $r_i\mu > h_i$ and the fact that $\beta' \le 1/2$, we divide through by this positive quantity, yielding the following nonlinear system of m equations in m unknowns:

$$\nu_i\mu_i = (\mu_i - \lambda_i)^2 \,, \ i \in M \,, \tag{4.67}$$

where $\nu_i := (h_i(1 - \beta')\mu)/(r_i\mu - h_i\beta') > 0$, $i \in M$. Solving (4.67) for λ_i yields

$$\lambda_i = \mu_i - \sqrt{\nu_i\mu_i} \,, \ i \in M \,. \tag{4.68}$$

Note that this expression for λ_i has the same form as the expression for the socially optimal arrival rate in the single-class case (cf.),

$$\lambda^s = \mu - \sqrt{\nu\mu} \,,$$

and reduces to it when $m = 1$. The crucial difference in the multiclass case is that (4.68) does not give an explicit expression for λ_i, inasmuch as the parameter μ_i depends on λ_j, $j \neq i$.

To solve the nonlinear system (4.67) for λ_i, $i \in M$, first note that, since $\mu_i - \lambda_i = \mu - \lambda$ for all $i \in M$, we can rewrite (4.67) as

$$\nu_i\mu_i = \nu^2 \,, \ i \in M \,, \tag{4.69}$$

where the unknown $\nu = \mu - \lambda$ can be interpreted as the "excess" capacity of the system as a whole. Rewriting these equations and combining them with the equation defining ν yields the following system of $m + 1$ equations in the $m + 1$ unknowns, $\lambda_1, \ldots, \lambda_m, \nu$:

$$\mu - \sum_{k \neq i} \lambda_k \ = \ \nu^2 \nu_i^{-1} \,, \ i \in M \tag{4.70}$$

$$\mu - \sum_{k \in M} \lambda_k \ = \ \nu \tag{4.71}$$

The i-th of the equations (4.70) and equation (4.71) together give

$$\nu^2 \nu_i^{-1} = \mu - \sum_{k \neq i} \lambda_k = \nu + \lambda_i \,,$$

from which we obtain the following expression for λ_i in terms of ν:

$$\lambda_i = \nu^2 \nu_i^{-1} - \nu \,, \ i \in M \,. \tag{4.72}$$

Summing the m equations (4.70) yields the quadratic equation,

$$\nu^2 \sum_{k \in M} \nu_k^{-1} - (m - 1)\nu - \mu = 0$$

which when solved gives the following explicit expression for ν:

$$\nu = \left(2 \sum_{k \in M} \nu_k^{-1} \right)^{-1} \left(m - 1 + \left[(m - 1)^2 + 4\mu \sum_{k \in M} \nu_k^{-1} \right]^{1/2} \right) . \tag{4.73}$$

Recall that we were assuming the existence of an interior solution when we set out to solve the equations (4.65). To ensure that (4.72) yields $\lambda_i > 0$, $i \in M$, we require that $\nu_i < \nu$ for all $i \in M$.

4.5.2.1 Continuous-Time Dynamic Algorithm

Consider the continuous-time dynamic algorithm discussed in Section 4.3.1.2. For the model under consideration in this section, the differential equations governing this algorithm take the form:,

$$\frac{d}{dt}\lambda_i(t) = \omega_i\left(r_i - h_i\left[W(\lambda(t)) + \lambda_i(t)W'(\lambda(t))\right]\right) , \ i \in M .$$

where $\lambda(t) = \sum_{k \in M} \lambda_k(t)$. Since this model satisfies Assumption 6, we know that the dynamic algorithm converges globally to the unique class-optimal allocation, $\boldsymbol{\lambda}^c$, whose explicit form we have just derived.

4.5.2.2 Discrete-Time Dynamic Algorithm

We now consider a discrete-time dynamic algorithm. Specifically, we shall study a relaxed sequential algorithm (cf. Section 3.2 of Chapter 3) operating at discrete stages, $n = 0, 1, \ldots$. At each stage n, the class managers take turns (in sequence) updating their flow-control strategies. The ordering is not unique and any sequence will suffice provided that it is the same at each step. After a particular class adjusts its strategy, the value of its new flow rate is broadcast to all the other classes. The next class in the sequence then proceeds to update its strategy.

At step $n + 1 \geq 1$ of the sequential algorithm, the best-reply strategy for class i is given by

$$\nu_i^{(n)} := \arg \max_{\{\lambda_i \geq 0\}} B_i(\lambda_1^{(n+1)}, \ldots, \lambda_{i-1}^{(n+1)}, \lambda_i, \lambda_{i+1}^{(n)}, \ldots, \lambda_m^{(n)}) , \ i \in M ,$$

with arbitrary (admissible) initial strategy vector $\boldsymbol{\lambda}^{(0)} = (\lambda_1^{(0)}, \ldots, \lambda_m^{(0)})$. We shall assume, however, that each class manager uses a *relaxed* update, which is a linear combination of the previous strategy, $\lambda_i^{(n)}$, and the best-reply strategy, $\nu_i^{(n)}$, viz.,

$$\lambda_i^{(n+1)} = (1 - \omega)\lambda_i^{(n)} + \omega\nu_i^{(n)} , \tag{4.74}$$

where ω is an arbitrary (nonnegative) relaxation parameter.

Note that when $\omega = 1$ there is no relaxation and the procedure reduces to best reply. The case $\omega < 1$ ($\omega > 1$) is called *under-relaxation* (*over-relaxation*). Note also that when $m = 1$ the procedure reduces to one of the variants of the single-class discrete-time algorithm introduced in Chapter 3 (cf. equation (3.24) in Section 3.3).

In order to study the asymptotic behavior of the sequential discrete-time algorithm, we begin by re-examining the equilibrium-seeking behavior of class i. Recall that the class-i manager, given the arrival rates, λ_k, $k \neq i$, of the other classes, will seek to maximize the class-i net benefit. The resulting arrival rate, λ_i, is given by

$$\lambda_i = \mu_i - \sqrt{\nu_i \mu_i} ,$$

where $\mu_i = \mu - \sum_{k \neq i} \lambda_k$ (cf. equation (4.68)). That is, knowing the strategies, λ_k, $k \neq i$, of the other classes, the best-reply for class i is given by the nonlinear mapping $\Gamma_i(\boldsymbol{\lambda}) = \mu_i - (\nu_i \mu_i)^{1/2}$.

To analyze the nonlinear system, $\boldsymbol{\Gamma}(\boldsymbol{\lambda}) = (\Gamma_1(\boldsymbol{\lambda}), \ldots, \Gamma_m(\boldsymbol{\lambda}))$, we shall use Lyapunov's first method in which we examine a linearized approximation of the system near the equilibrium fixed point $\boldsymbol{\lambda}^c$ of the system of difference equations, $\boldsymbol{\lambda}^{(n+1)} = \boldsymbol{\Gamma}(\boldsymbol{\lambda}^{(n)})$ The $m \times m$ Jacobian matrix of $\boldsymbol{\Gamma}(\boldsymbol{\lambda})$ has elements

$$\frac{\partial}{\partial \lambda_j} \Gamma_i(\boldsymbol{\lambda}) = \begin{cases} -1 + (1/2)(\nu_i/\mu_i)^{1/2} & , \ i \neq j \\ 0 & , \ i = j \end{cases}.$$

At equilibrium, equation (4.69) holds and hence the Jacobian matrix is given by

$$\boldsymbol{J} = \begin{pmatrix} 0 & -\eta_1 & \cdots & -\eta_1 \\ -\eta_2 & 0 & \cdots & -\eta_2 \\ \vdots & \vdots & \ddots & \vdots \\ -\eta_m & -\eta_m & \cdots & 0 \end{pmatrix},$$

where $\eta_i := 1 - (1/2)(\nu_i/\nu) \in (\frac{1}{2}, 1)$, $i \in M$.

Defining a vector of perturbations, $\boldsymbol{y}^{(n)} := \boldsymbol{\lambda}^{(n)} - \boldsymbol{\lambda}^c$, from the equilibrium point provides a linearized fixed-point equation, $\boldsymbol{y} = \boldsymbol{J}\boldsymbol{y}$. This matrix equation can be rewritten as the equivalent linear system of equations, $\boldsymbol{B}\boldsymbol{y} = \boldsymbol{0}$, where $\boldsymbol{B} := \boldsymbol{I} - \boldsymbol{J}$. It will be convenient to consider also the equivalent transformed system, $\overline{\boldsymbol{B}}\boldsymbol{y} = \boldsymbol{0}$, where

$$\overline{\boldsymbol{B}} := \boldsymbol{\Delta}^{-1}\boldsymbol{B} = \begin{pmatrix} \eta_1^{-1} & 1 & \cdots & 1 \\ 1 & \eta_2^{-1} & \cdots & 1 \\ \vdots & \vdots & \ddots & \vdots \\ 1 & 1 & \cdots & \eta_m^{-1} \end{pmatrix},$$

with $\boldsymbol{\Delta} := \text{diag}(\eta_1, \ldots, \eta_m)$. Clearly, $\overline{\boldsymbol{B}}$ is symmetric and positive definite.

This linearized system will govern the behavior of the perturbation $\boldsymbol{y}$ when it is small, and so determine the behavior of the original nonlinear system within a local neighborhood of the equilibrium. The behavior of the relaxed sequential algorithm in a local neighborhood of the equilibrium can therefore be approximated in terms of our linearized system of perturbations by

$$\begin{aligned} \boldsymbol{y}^{(n+1)} &= (1 - \omega)\boldsymbol{y}^{(n)} + \omega(\boldsymbol{J}^l \boldsymbol{y}^{(n+1)} + \boldsymbol{J}^u \boldsymbol{y}^{(n)}) \\ &= \omega \boldsymbol{J}^l \boldsymbol{y}^{(n+1)} + [(1 - \omega)\boldsymbol{I} + \omega \boldsymbol{J}^u]\boldsymbol{y}^{(n)}, \end{aligned}$$

where $\boldsymbol{J}^l$ and $\boldsymbol{J}^u$ are respectively the lower and upper triangle matrices of $\boldsymbol{J}$. This approximating algorithm corresponds to the Gauss-Seidel matrix-iteration procedure, $\boldsymbol{y}^{(n+1)} = \boldsymbol{G}_\omega \boldsymbol{y}^{(n)}$ for solving a linear system of equations (cf., e.g., Varga [194]). Here the Gauss-Seidel matrix is given by

$$\boldsymbol{G}_\omega = (\boldsymbol{I} - \omega \boldsymbol{J}^l)^{-1}[(1 - \omega)\boldsymbol{I} + \omega \boldsymbol{J}^u],$$

and arises from the splitting of the system matrix, $\boldsymbol{B} = \boldsymbol{I} - \boldsymbol{J}$, given by

$$\boldsymbol{B} = \boldsymbol{M}_\omega - \boldsymbol{N}_\omega,$$

with nonsingular $\boldsymbol{M}_\omega := \frac{1}{\omega}\boldsymbol{I} - \boldsymbol{J}^l$ and $\boldsymbol{N}_\omega := \frac{1-\omega}{\omega}\boldsymbol{I} + \boldsymbol{J}^u$.

4.5.2.3 Discrete-Time Dynamic Algorithm: Equivalence to Power Maximization

Here we remark that, with respect to local stability of the class-optimal equilibrium, the net-benefit-maximization model has an equivalent counterpart under the power criterion (cf. Section 2.5 of Chapter 2), which has been extensively studied in the telecommunications literature. For the multiclass model, the (generalized) *power* for class i takes the form

$$P_i(\boldsymbol{\lambda}) := \frac{\lambda_i^{\alpha_i}}{W(\lambda)} \, ,$$

$i \in M$, with (as usual) $\lambda = \sum_{k \in M} \lambda_k$. Under the power criterion, the objective of the manager of class i is to choose $\lambda_i \geq 0$ to maximize $P_i(\boldsymbol{\lambda})$, for given values of λ_k, $k \neq i$. Here, the parameter $\alpha_i > 0$ is a weighting factor that may be used by the manager of class i to achieve a desired trade-off between throughput and average waiting time.

For an *M/M/1* queueing system with service rate μ, we have

$$\begin{aligned} P_i(\boldsymbol{\lambda}) &= \lambda_i^{\alpha_i}(\mu - \lambda) \\ &= \lambda_i^{\alpha_i}(\mu_i - \lambda_i) \, , \end{aligned}$$

where $\lambda = \sum_{k \in M} \lambda_k$ and $\mu_i = \mu - \sum_{k \neq i} \lambda_k$. The manager for class i chooses λ_i to maximize $P_i(\boldsymbol{\lambda})$ for given λ_k, $k \neq i$ – equivalently, for given μ_i. This problem takes the same form as the single-class power maximization problem considered in Section 2.5.1 of Chapter 2. The necessary condition for λ_i to maximize $P_i(\boldsymbol{\lambda})$ for given μ_i is

$$\frac{\partial}{\partial \lambda_i} P(\boldsymbol{\lambda}) = \lambda_i^{\alpha_i - 1}(\alpha_i(\mu_i - \lambda_i) - \lambda_i) = 0 \, ,$$

which has the unique solution,

$$\lambda_i = \left(\frac{\alpha_i}{\alpha_i + 1}\right)\mu_i \, . \tag{4.75}$$

In a class-optimal solution under the power criterion, this relation must hold simultaneously for all $i \in M$. Since $\mu_i = \mu - \sum_{k \neq i} \lambda_k$, we have the following system of linear equations:

$$\lambda_i + \left(\frac{\alpha_i}{\alpha_i + 1}\right) \sum_{k \neq i} \lambda_k = \left(\frac{\alpha_i}{\alpha_i + 1}\right) \mu \, , \ i \in M \, .$$

In matrix-vector terms, this system may be written as $\boldsymbol{P\lambda} = \boldsymbol{b}$, where

$$\boldsymbol{P} := \begin{pmatrix} 1 & \frac{\alpha_1}{\alpha_1+1} & \cdots & \frac{\alpha_1}{\alpha_1+1} \\ \frac{\alpha_2}{\alpha_2+1} & 1 & \cdots & \frac{\alpha_2}{\alpha_2+1} \\ \vdots & \vdots & \ddots & \vdots \\ \frac{\alpha_m}{\alpha_m+1} & \frac{\alpha_m}{\alpha_m+1} & \cdots & 1 \end{pmatrix} \quad \text{and} \quad \boldsymbol{b} := \begin{pmatrix} \frac{\alpha_1}{\alpha_1+1} \\ \frac{\alpha_2}{\alpha_2+1} \\ \vdots \\ \frac{\alpha_m}{\alpha_m+1} \end{pmatrix} \mu \, .$$

Since the system is linear, no linearization in a local neighborhood of the equilibrium need take place. This affords the opportunity to establish global convergence results.

With the substitution, $\eta_i = \alpha_i/(1+\alpha_i)$, the functional form of the linearized system matrix B for the net-benefit criterion is the same as the system matrix P under the power criterion. Using the fact that $\eta_i = 1 - (1/2)(\nu_i/\nu) \in (\frac{1}{2}, 1)$ for the net-benefit criterion, we see that the above substitution reduces the linearized version of the net-benefit problem to an instance of the power problem with $\alpha_i = \eta_i/(1-\eta_i) = (\eta_i^{-1}-1)^{-1} > 1$. Since matrix-iterative algorithms (in particular, the standard and relaxed Gauss-Seidel and Jacobi techniques) which solve these systems are simply splittings of the system matrix, the behavior of each system is the same. Thus the global behavior of an $M/M/1$ system operating under the power criterion with $\alpha_i > 1$ for all $i \in M$ is equivalent to the local behavior of an $M/GI/1$ system operating under the net-benefit criterion with linear waiting costs and fixed rewards. We note here that additional results derived for our net-benefit problem, particularly those developed for the relaxed algorithms, have equivalent counterparts under the power criterion.

4.5.2.4 Discrete-Time Dynamic Algorithm: Convergence of the Relaxed Sequential Version

Now we examine the asymptotic behavior of the relaxed sequential algorithm developed in Section 4.5.2.2. Since convergence will depend on the spectral radius of the relaxed Gauss-Seidel matrix G_ω, we appeal to Theorem 3.5 in [194], which establishes that $\rho(G_\omega) \geq |1 - \omega|$. Thus the relaxed sequential algorithm diverges when $|1 - \omega| \geq 1$, allowing us to restrict attention to the case $0 < \omega < 2$.

Theorem 4.7 *The relaxed sequential algorithm with $0 < \omega < 2$ converges locally to the interior Nash equilibrium λ^c for all $m \geq 1$.*

Proof   We extend the proof in Zhang and Douligeris [200] to the relaxed algorithm. The relaxed Gauss-Seidel splitting $\overline{B} = \overline{M}_\omega - \overline{N}_\omega$ with nonsingular $\overline{M}_\omega = \Delta^{-1}M_\omega$ and $\overline{N}_\omega = \Delta^{-1}N_\omega$ defines a corresponding relaxed Gauss-Seidel matrix

$$
\begin{aligned}
\overline{G}_\omega &= (\overline{M}_\omega)^{-1}\overline{N}_\omega = (\overline{M}_\omega)^{-1}(\overline{M}_\omega - \overline{B}) \\
&= I - (\overline{M}_\omega)^{-1}\overline{B} = I - (\Delta^{-1}M_\omega)^{-1}\Delta^{-1}B = I - (M_\omega)^{-1}B \\
&= (M_\omega)^{-1}(M_\omega - B) = (M_\omega)^{-1}N_\omega = G_\omega \, .
\end{aligned}
$$

Since $\omega\overline{M}_\omega = \Delta^{-1}(I - \omega J^l)$ is nonsingular and $\overline{B}$ is symmetric, positive definite for all ω, $\rho(G_\omega) = \rho(\overline{G}_\omega) < 1$ follows from Theorem 3.6 in Varga [194]. Thus we have convergence for all $m \geq 1$.   ∎

4.5.3 Socially Optimal Arrival Rates

The social optimization problem now takes the form:

$$\max_{\{\lambda;\lambda_i,i\in M\}} \mathcal{U}(\boldsymbol{\lambda}) = \sum_{i\in M} \lambda_i \left(r_i - h_i G_i(\lambda)\right)$$

$$\text{s.t.} \quad \sum_{i\in M} \lambda_i = \lambda$$

$$\lambda_i \geq 0 , \ i \in M$$

Without loss of generality, we shall again consider the case of two classes: $M = \{1,2\}$. In this case the social optimization problem can be written as follows:

$$\max_{\{\lambda;\lambda_1,\lambda_2\}} \mathcal{U}(\lambda_1,\lambda_2) = r_1\lambda_1 - \lambda_1 h_1 G_1(\lambda) + r_2\lambda_2 - \lambda_2 h_2 G_2(\lambda)$$

$$\text{s.t.} \quad \lambda_1 + \lambda_2 = \lambda ,$$

$$\lambda_1 \geq 0 , \ \lambda_2 \geq 0 .$$

Suppose that the total flow equals λ under an optimal flow allocation. The locus of all points (λ_1,λ_2) with this value for the total flow is just the straight line,

$$\lambda_1 + \lambda_2 = \lambda .$$

An optimal flow allocation must therefore solve the following "knapsack" problem:

$$\max_{\{\lambda;\lambda_1,\lambda_2\}} (r_1 - G_1(\lambda))\lambda_1 + (r_2 - G_2(\lambda))\lambda_2$$

$$\text{s.t.} \quad \lambda_1 + \lambda_2 = \lambda$$

$$\lambda_1 \geq 0 , \ \lambda_2 \geq 0 .$$

Since both the objective function and the constraint are linear (recall that λ is fixed), an optimal solution to this problem will lie at an extreme point with either λ_1 or λ_2 equal to λ and the other variable equal to zero. Therefore, a socially optimal flow allocation will exhibit class dominance.

A further observation is the following. Consider the straight line, $\lambda_1+\lambda_2 = \lambda$ where λ is chosen such that

$$r_1 - G_1(\lambda) = r_2 - G_2(\lambda) . \tag{4.76}$$

Along this line both the total flow λ and the aggregate net utility, $\mathcal{U}(\lambda_1,\lambda_2)$, are constant: $\mathcal{U}(\lambda_1,\lambda_2) = u$, say. In particular, the two extreme points on this line, $(\lambda,0)$, and $(0,\lambda)$, share this net utility; that is,

$$\mathcal{U}(\lambda,0) = \mathcal{U}(0,\lambda) = u .$$

But

$$\mathcal{U}(\lambda,0) \ \leq \ \mathcal{U}(\lambda_1^*,0) ,$$
$$\mathcal{U}(0,\lambda) \ \leq \ \mathcal{U}(0,\lambda_2^*) ,$$

where λ_i^* is the optimal flow allocation to class i when only class i receives positive flow. Thus we see that all flow allocations along the line, $\lambda_1 + \lambda_2 = \lambda$, when λ satisfies (4.76), are dominated by *both* the optimal single-class allocations. We shall use this result below.

Remark 2   Of course, our example is special in that the class utility functions, $U_i(\lambda_i)$, are linear. As we observed before, if the $U_i(\lambda_i)$ are "sufficiently strictly concave," this may override the nonconvexity of the waiting costs and render an interior point optimal. But even then one must be wary of multiple solutions to the first-order conditions, some of which will be stationary points or local maxima of the objective function. We give several examples with nonlinear utility functions in Section 4.6 below.

A question remains: what is the significance of an interior-point solution to the first-order conditions, if one exists? Since the functions involved are all differentiable, we know that the first-order conditions are *necessary* for an interior point to be socially optimal. But the non-concavity of the objective function implies that they may not be sufficient.

For our example, it turns out that, not only are the first-order conditions not sufficient, but also an interior-point solution to these conditions always has a net utility that is no greater than that of *each* of the optimal single-class allocations, $(\lambda_1^*, 0)$ and $(0, \lambda_2^*)$. To see this, write the first-order necessary conditions for an interior-point to be optimal:

$$\frac{\partial}{\partial \lambda_1} \mathcal{U}(\lambda_1, \lambda_2) = r_1 - G_1(\lambda_1 + \lambda_2) - [\lambda_1 G_1'(\lambda_1 + \lambda_2) + \lambda_2 G_2'(\lambda_1 + \lambda_2)]$$
$$= 0 \, ;$$

$$\frac{\partial}{\partial \lambda_2} \mathcal{U}(\lambda_1, \lambda_2) = r_2 - G_2(\lambda_1 + \lambda_2) - [\lambda_1 G_1'(\lambda_1 + \lambda_2) + \lambda_2 G_2'(\lambda_1 + \lambda_2)]$$
$$= 0 \, .$$

Let $(\tilde{\lambda}_1, \tilde{\lambda}_2)$ be a solution to these conditions and suppose $\tilde{\lambda}_1 > 0$, $\tilde{\lambda}_2 > 0$, and $\tilde{\lambda}_1 + \tilde{\lambda}_2 < 1$. It follows that

$$r_1 - G_1(\tilde{\lambda}_1 + \tilde{\lambda}_2) = r_2 - G_2(\tilde{\lambda}_1 + \tilde{\lambda}_2) \, ,$$

so that an interior solution to the first-order conditions must lie on the line $\tilde{\lambda}_1 + \tilde{\lambda}_2 = \tilde{\lambda}$, where $\tilde{\lambda}$ satisfies the equation

$$r_1 - G_1(\tilde{\lambda}) = r_2 - G_2(\tilde{\lambda}) \, .$$

That is, an interior solution to the first-order conditions must lie on the line defined by (4.76) with $\lambda = \tilde{\lambda}$. Thus (cf. the discussion above)

$$\mathcal{U}(\tilde{\lambda}_1, \tilde{\lambda}_2) \leq \mathcal{U}(\lambda_1^*, 0) \, ,$$
$$\mathcal{U}(\tilde{\lambda}_1, \tilde{\lambda}_2) \leq \mathcal{U}(0, \lambda_2^*) \, .$$

That is, any interior solution to the first-order conditions is dominated by *both* the optimal single-class allocations. In other words, the system achieves a greater net utility by allocating all flow to a single class, *regardless of which class*, than by using an interior allocation satisfying the first-order conditions!

A final observation has to do with external effects, congestion tolls, and equilibrium solutions. Note that charging each entering customer (regardless of class) a toll δ, where

$$\delta = \tilde{\lambda}_1 G_1'(\tilde{\lambda}_1 + \tilde{\lambda}_2) + \tilde{\lambda}_2 G_2'(\tilde{\lambda}_1 + \tilde{\lambda}_2) \, ,$$

makes $(\tilde{\lambda}_1, \tilde{\lambda}_2)$ a Nash equilibrium for individually optimizing customers: no customer has an incentive to deviate from its choice to enter or not, assuming that no other customer makes a change. (This is true because each flow, λ_i, satisfies the corresponding first-order condition, which is sufficient for λ_i to attain the maximum of the objective function with the other variables held fixed, by the concavity of the objective function in λ_i.) Thus, we see that, even by charging the "correct" toll (namely, a toll equal to the external effect), the system cannot be certain that individually optimizing customers will be directed to a socially optimal flow allocation. Rather, the resulting allocation, even though it is a Nash equilibrium, may be dominated by both of the optimal single-class allocations.

Thus we have a dramatic example of the pitfalls of marginal-cost allocation when the classes are heterogeneous in their sensitivities to congestion.

4.5.4 Facility Optimal Arrival Rates

The problem for a facility operator interested in maximizing its revenue from charging tolls now takes the form (after the usual transformation: cf. Section 4.4.4):

$$\max_{\{\lambda; \lambda_i, i \in M\}} \quad \sum_{i \in M} \lambda_i (r_i - G_i(\lambda))$$

$$\text{s.t.} \quad \sum_{k \in M} \lambda_k = \lambda$$

$$\lambda_i \geq 0 \, , \, i \in M \, .$$

This optimization problem is identical to the social optimization problem. Hence the facility optimal and socially optimal arrival rates and tolls coincide in the case of linear class utility functions.

We could also have deduced this equivalence from our observation in Section 4.2.4 that the facility optimization problem with class utility functions $U_i(\lambda_i)$ is equivalent to the social optimization problem with class utility functions $\tilde{U}_i(\lambda_i) := \lambda_i U_i'(\lambda_i)$. When $U_i(\lambda_i) = r_i \cdot \lambda_i$, we have $\tilde{U}_i(\lambda_i) = \lambda_i \cdot r_i = U_i(\lambda_i)$.

4.5.5 Generalization: Class-Dependent Service Rates

We now study a generalization of the $M/GI/1$ model which we have been using to illustrate our results. We consider an $M/GI/1$ queue with *FIFO* queue discipline, linear waiting costs, and class-dependent service rates. Although in this model the average waiting time no longer depends only on the total arrival rate, the form of the waiting-cost functions still leads to class dominance when the class utility functions are linear. We shall show this by a variant of the argument used above for the case of equal service rates (see Section 4.5.3).

Recall that the waiting-cost function for class i for this model takes the following form (cf. equation (4.3) in Section 4.1.1):

$$G_i(\boldsymbol{\lambda}) \;=\; h_i W_i(\boldsymbol{\lambda})$$

$$\;=\; h_i \left(\frac{1}{\mu_i} + D(\boldsymbol{\lambda}) \right) \tag{4.77}$$

where

$$D(\boldsymbol{\lambda}) = \frac{\sum_{k \in M} \lambda_k \beta_k / \mu_k^2}{1 - \sum_{k \in M} \lambda_k / \mu_k} \, . \tag{4.78}$$

4.5.5.1 Individually Optimal Arrival Rates

We shall show that there always exists a class-dominant solution to the individual optimization problem for this model. In fact, our argument works provided only that $G_i(\boldsymbol{\lambda})$ is given by the formula (4.77) and does not depend on the particular form of $D(\boldsymbol{\lambda})$. The crucial property is that $D(\boldsymbol{\lambda})$ is class independent (a consequence of our assumption that the queue discipline is *FIFO*.)

When the utility functions take the linear form (4.55) and the waiting-cost functions are given by (4.77), an allocation, $\boldsymbol{\lambda} = (\lambda_i, i \in M)$, will be a Nash equilibrium for individually optimizing customers if and only if

$$r_i \;\leq\; h_i/\mu_i + h_i D(\boldsymbol{\lambda}) + \delta_i$$
$$r_i \;=\; h_i/\mu_i + h_i D(\boldsymbol{\lambda}) + \delta_i \, , \text{ if } \lambda_i > 0 \, ,$$

for all $i \in M$. Thus, in order to have a positive arrival rate in two classes, say $i = 1$ and $i = 2$, we must have

$$r_1 \;=\; h_1/\mu_1 + h_1 D(\boldsymbol{\lambda}) + \delta_1 \, , \tag{4.79}$$
$$r_2 \;=\; h_2/\mu_2 + h_2 D(\boldsymbol{\lambda}) + \delta_2 \, . \tag{4.80}$$

Under what circumstances can we expect there to be an arrival vector $\boldsymbol{\lambda}$ such that both (4.79) and (4.80) hold? Note first that, for an arbitrary $\boldsymbol{\lambda}$ such that

$$h_i D(\boldsymbol{\lambda}) \leq r_i - h_i/\mu_i \, , \; i = 1, 2 \, ,$$

if we are free to choose the tolls, δ_1 and δ_2, and they are permitted to be class dependent, then we can choose them such that both (4.79) and (4.80) hold for that value of $\boldsymbol{\lambda}$. (Set $\delta_1 = r_1 - h_1/\mu_1 - h_1 D(\boldsymbol{\lambda})$ and $\delta_2 = r_2 - h_2/\mu_2 - h_2 D(\boldsymbol{\lambda})$.)

On the other hand, if δ_1 and δ_2 are given, then in order for both (4.56) and (4.57) to hold, we must be able to find a value of $\boldsymbol{\lambda}$ such that

$$D(\boldsymbol{\lambda}) \;=\; (r_1 - h_1/\mu_1 - \delta_1)/h_1 \, ,$$
$$D(\boldsymbol{\lambda}) \;=\; (r_2 - h_2/\mu_2 - \delta_2)/h_2 \, ,$$

which can happen only if

$$(r_1 - h_1/\mu_1 - \delta_1)/h_1 = (r_2 - h_2/\mu_2 - \delta_2)/h_2 . \qquad (4.81)$$

Thus we see that having a positive arrival rate for more than one class in an individually optimal allocation is a "rare event" when the utility functions are linear and the tolls are given (in particular, when both tolls are zero). The prevailing scenario is one in which there is class dominance: a single class has a positive arrival rate and all other classes have zero arrival rate. In fact, even if both (4.79) and (4.80) do hold, then there still exists a class-dominant solution, provided that there exists a $\boldsymbol{\lambda}$ with only one positive λ_i for which (4.81) holds. (This will always be the case when $D(\boldsymbol{\lambda})$ is given by (4.78), for example.)

To summarize: for the *M/GI/1* model with class-dependent service rates, *FIFO* discipline, linear utility functions, and linear waiting-cost functions, there always exists a class-dominant solution to the individual optimization problem. A dominant class i and the associated arrival rate vector $\boldsymbol{\lambda}$ are characterized by the following conditions:

$$\begin{aligned} D(\boldsymbol{\lambda}) &= (r_i - h_i/\mu_i - \delta_i)/h_i \\ &= \min_{k \in M}(r_k - h_k/\mu_k - \delta_k)/h_k \end{aligned}$$

where $\boldsymbol{\lambda} = (0, \ldots, \lambda_i^e, \ldots, 0)$ and λ_i^e is the individually optimal arrival rate for the single-class system consisting of only class-i customers. (We shall call this the *i-system*.)

In the *M/GI/1* model we are considering (in which $D(\boldsymbol{\lambda})$ is given by (4.78)), the expected waiting-time in the queue for the i-system with arrival rate λ is given by the Pollaczek-Khintchine formula:

$$D_i(\lambda) = \frac{\lambda \beta_i/\mu_i^2}{1 - \lambda/\mu_i} .$$

In this case $D(\boldsymbol{\lambda}) = D_i(\lambda_i^e)$ and λ_i^e is the unique solution to the equation

$$r_i = h_i \left(\frac{1}{\mu_i} + \frac{\lambda 2\beta_i/\mu_i^2}{2(1 - \lambda/\mu_i)} \right) + \delta_i . \qquad (4.82)$$

4.5.5.2 Socially Optimal Arrival Rates

As usual in our analysis of social optimality we shall restrict attention to a two-class system (without loss of generality). For $m = 2$ the objective function for social optimization is:

$$\begin{aligned} \mathcal{U}(\boldsymbol{\lambda}) &= r_1\lambda_1 - \lambda_1 G_1(\boldsymbol{\lambda}) + r_2\lambda_2 - \lambda_2 G_2(\boldsymbol{\lambda}) \\ &= r_1\lambda_1 - \lambda_1 h_1/\mu_1 - \lambda_1 h_1 D(\boldsymbol{\lambda}) + \\ &\quad r_2\lambda_2 - \lambda_2 h_2/\mu_2 - \lambda_2 h_2 D(\boldsymbol{\lambda}) . \end{aligned}$$

Suppose that $D(\boldsymbol{\lambda}) = w$ under a socially optimal flow allocation. A socially optimal flow allocation, therefore, must solve the following problem:

$$\max_{\{\lambda_1, \lambda_2\}} \quad r_1\lambda_1 - \lambda_1 h_1/\mu_1 + r_2\lambda_2 - \lambda_2 h_2/\mu_2 - (\lambda_1 h_1 + \lambda_2 h_2)w$$

$$\text{s.t.} \quad \frac{\lambda_1 2\beta_1/\mu_1^2 + \lambda_2 2\beta_2/\mu_2^2}{2(1 - \lambda_1/\mu_1 - \lambda_2/\mu_2)} = w$$

$$\lambda_1 \geq 0, \ \lambda_2 \geq 0.$$

The constraint may be rewritten as a linear constraint by multiplying both sides of the equality by the denominator of the fractional expression and simplifying, yielding the following equivalent optimization problem (a "knapsack" problem):

$$\max_{\{\lambda_1, \lambda_2\}} \quad (r_1 - h_1/\mu_1 - h_1 w)\lambda_1 + (r_2 - h_2/\mu_2 - h_2 w)\lambda_2$$

$$\text{s.t.} \quad (2\beta_1/\mu_1^2 + 2w/\mu_1)\lambda_1 + (2\beta_2/\mu_2^2 + 2w/\mu_2)\lambda_2 = 2w$$

$$\lambda_1 \geq 0, \ \lambda_2 \geq 0.$$

Since both the objective function and the constraint are linear (recall that w is fixed), the solution will lie at an extreme point with either λ_1 or λ_2 equal to zero.

It follows that in a socially optimal solution one class will dominate and have positive flow, while all other classes will have zero flow.

Processor-Sharing Queue Discipline

Now suppose the queue discipline is processor sharing. That is, when there are n customers in the system, each customer receives $1/n$ of the processing rate of the server. In this case, a customer's expected waiting time in the system is proportional to its service time, with constant of proportionality equal to $(1 - \rho)^{-1}$, where ρ is the overall traffic intensity:

$$\rho = \lambda E[\boldsymbol{S}]$$

$$= \sum_{i \in M} \lambda_i/\mu_i.$$

Thus the expected waiting time in the system, $W_i(\boldsymbol{\lambda})$, of a class-$i$ customer is given by

$$W_i(\boldsymbol{\lambda}) = \frac{1/\mu_i}{1 - \sum_{i \in M} \lambda_i/\mu_i}$$

The objective function for social optimization now takes the form:

$$\mathcal{U}(\boldsymbol{\lambda}) = \sum_{i \in M} U_i(\lambda_i) - \sum_{i \in M} \lambda_i G_i(\boldsymbol{\lambda})$$

$$= \sum_{i \in M} U_i(\lambda_i) - \sum_{i \in M} \frac{\lambda_i h_i/\mu_i}{1 - \sum_{i \in M} \lambda_i/\mu_i}$$

Now let $\rho_i := \lambda_i/\mu_i$, the traffic intensity for class i. Substituting in the above expression for $\mathcal{U}(\boldsymbol{\lambda})$, we obtain

$$\mathcal{U}(\boldsymbol{\lambda}) \;\; = \;\; \sum_{i \in M} \hat{U}_i(\rho_i) - \sum_{i \in M} \frac{\rho_i h_i}{1 - \sum_{i \in M} \rho_i} \; ,$$

where $\hat{U}_i(\rho_i) := U_i(\rho_i \mu_i)$. Note that, considered as a function of $(\rho_1, \ldots, \rho_m)$, $\mathcal{U}(\cdot)$ assumes exactly the same form as the objective function (considered as a function of $(\lambda_i, i \in M)$) for the model in which the waiting time depends only on the total flow, $\lambda = \sum_{i \in M} \lambda_i$. Thus we can apply the analysis of Section 4.4, replacing λ_i by ρ_i to obtain similar results. In particular, in the case where the modified utility function for each class i is linear in ρ_i,

$$\hat{U}_i(\rho_i) = \alpha_i \rho_i \; , \tag{4.83}$$

say, we have the same extreme form of class dominance we observed there. Note that (4.83) holds if and only if

$$U_i(\lambda) = (\alpha_i / \mu_i) \lambda_i \; .$$

4.5.6 Finding the Optimal Single-Class Allocations

Once we know that class dominance holds, the next step is to find conditions that determine which class is dominant under each optimality criterion. We shall see that the order of domination under social optimality may differ from that under individual optimality. Moreover, the difference can be extreme: it is possible for the dominance order under social (and facility) optimality to be the reverse of the order under individual optimality.

Throughout this section we assume that no tolls are charged: $\delta_i = 0$ for all $i \in M$.

We shall use a special case of the $M/GI/1$ queue with $FIFO$ queue discipline, linear waiting costs, and class-dependent service times to illustrate these points. In this case (cf. equations (4.77) and (4.78) in Section 4.5.5) the waiting cost for a class-i customer is given by

$$\begin{aligned}
G_i(\boldsymbol{\lambda}) \;\; &= \;\; h_i W_i(\boldsymbol{\lambda}) \\
&= \;\; h_i \left(\frac{1}{\mu_i} + D(\boldsymbol{\lambda}) \right) \\
&= \;\; h_i \left(\frac{1}{\mu_i} + \frac{\sum_{k \in M} \lambda_k \beta_k / \mu_k^2}{1 - \sum_{k \in M} \lambda_k / \mu_k} \right) \; .
\end{aligned}$$

In Section 4.5.5 we showed that class dominance holds in this system for both individual and social optimization.

We shall consider the special case in which the service times in class k are exponentially distributed with first and second moments $1/\mu_k$ and $2/\mu_k^2$, respectively; that is, $\beta_k = 1$, $k \in M$. In this case, the unconditional service times have a hyper-exponential distribution with first and second moments

$$E[\boldsymbol{S}] = \sum_{k \in M} \frac{\lambda_k}{\mu_k} ,$$

$$E[\boldsymbol{S}^2] = \sum_{k \in M} \frac{2\lambda_k}{\mu_k^2} ,$$

respectively.

When class i is dominant with arrival rate λ, the system operates as a single-class $M/M/1$ queue with arrival rate λ and service rate μ_i (the i-system). Let $\mathcal{U}_i(\lambda)$ denote the net utility function for this system. Then

$$\mathcal{U}_i(\lambda) = r_i\lambda - \frac{h_i\lambda}{\mu_i - \lambda} .$$

Recall that, under both individual and social optimization, class i will have a positive arrival rate if and only if $r_i > h_i/\mu$, or, equivalently,

$$\frac{r_i}{h_i} > \frac{1}{\mu_i} . \tag{4.84}$$

So without loss of generality we assume that (4.84) holds for all $i \in M$.

The individually optimal single-class allocation for class i is

$$\lambda_i^e = \mu_i - \frac{h_i}{r_i} .$$

For the case of zero tolls, it follows from the results in Section 4.5.5 that class i dominates class k under individual optimality if and only if

$$\frac{r_i}{h_i} - 1/\mu_i \geq \frac{r_k}{h_k} - 1/\mu_k . \tag{4.85}$$

Now consider the single-class allocation under social optimality. This is found by comparing the objective-function values for social optimization under the various single-class allocations. The socially optimal single-class allocation for class i is

$$\lambda_i^s = \mu_i - \sqrt{\frac{\mu_i h_i}{r_i}} .$$

Substituting $\lambda = \lambda_i^s$ into the above expression for $\mathcal{U}_i(\lambda)$ yields

$$\mathcal{U}_i(\lambda_i^s) = r_i\mu_i - \sqrt{\mu_i h_i r_i} - \frac{h_i\left(\mu_i - \sqrt{\frac{\mu_i h_i}{r_i}}\right)}{\sqrt{\frac{\mu_i h_i}{r_i}}}$$

$$= r_i\mu_i - 2\sqrt{\mu_i h_i r_i} - h_i$$

$$= \mu\left(r_i - 2\sqrt{\frac{h_i r_i}{\mu}} - \frac{h_i}{\mu}\right)$$

$$= \left(\sqrt{\mu_i r_i} - \sqrt{h_i}\right)^2 .$$

We see that class i dominates class k under social optimality if and only if

$$\sqrt{\mu_i r_i} - \sqrt{h_i} > \sqrt{\mu_k r_k} - \sqrt{h_k} . \tag{4.86}$$

Compare this to the dominance criterion (4.85) for individual optimality. It is *ipso facto* clear that there can be situations in which class i dominates class k under individual optimization but the opposite is true under social optimization.

To illustrate this phenomenon concretely, let us consider the special case of equal service rates: $\mu_i = \mu$ for all $i \in M$. The individually optimal single-class allocation for class i is now

$$\lambda_i^e = \mu - \frac{h_i}{r_i} .$$

Class i dominates class k under individual optimality if and only if

$$\frac{r_i}{h_i} \geq \frac{r_k}{h_k} . \tag{4.87}$$

Note that this is the case if and only if $\lambda_i^e \geq \lambda_k^e$. That is, under individual optimality the class with the largest i.o. arrival rate dominates the facility.

Now consider the single-class allocation under social optimality. The socially optimal single-class allocation for class i is now

$$\lambda_i^s = \mu - \sqrt{\frac{\mu h_i}{r_i}} .$$

Substituting $\lambda = slam_i$ into the above expression for $\mathcal{U}_i(\lambda)$ yields

$$\mathcal{U}_i(\lambda) = \mu \left(\sqrt{r_i} - \sqrt{\frac{h_i}{\mu}} \right)^2 .$$

For notational convenience, let $\alpha_i := \sqrt{r_i}$, $\beta_i := \sqrt{h_i/\mu}$. We see that class i dominates class k under social optimality if and only if

$$\alpha_i - \beta_i > \alpha_k - \beta_k . \tag{4.88}$$

Compare this to the dominance criterion (4.87) for individual optimality, which can be rewritten as

$$\frac{\alpha_i}{\beta_i} > \frac{\alpha_k}{\beta_k} . \tag{4.89}$$

From these results it is evident that we can have a situation in which class i may dominate class k under individual optimization, but class k dominates class i under social optimization. Figure 4.1 graphically illustrates how this can occur.

Remark 3   The results in this section apply also to an $M/GI/1$ queue with processor-sharing service discipline, since the formula for the expected steady-state waiting time in that system coincides with that for the $M/M/1$ queue with *FIFO* discipline.

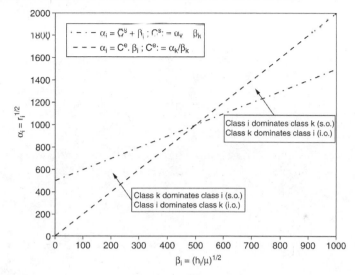

Figure 4.1 *Class Dominance Regions for Individual and Social Optimization*

4.6 Examples with Different Utility Functions

We now return to the case of general (concave, nondecreasing) class utility functions, in order to explore, via examples, the effect of linearity or nonlinearity of the class utility functions on the economic properties of our model. We focus on social optimality since (as we have seen) the objective function for social optimization can fail to be jointly concave and therefore can have multiple local optima, because the waiting-cost component of the objective function is not jointly convex. When the class utility functions are linear, the objective function is *never* jointly concave and in Section 4.5 we showed that this leads to class dominance. We also observed that, if the class utility functions are nonlinear and "sufficiently" strictly concave, the objective function may be jointly concave in spite of the failure of the waiting-cost component to be jointly convex. This observation is our motivation for considering various examples of nonlinear, strictly concave class utility functions in this section.

Our examples will show that multiple local optima and nonglobal convergence of the dynamic rate-control algorithm can occur with nonlinear as well as linear utility functions. We shall use the $M/M/1$ queue with *FIFO* queue discipline, equal service rates ($\mu_i = \mu$, $i \in M$), and linear waiting-cost functions as a vehicle for demonstrating these properties. As usual we shall restrict attention to the case $m = 2$. Without loss of generality, we assume that $\mu = 1$. (This amounts to a choice of time unit.)

Remark 4 The results in this section apply also to an $M/GI/1$ queue with processor-sharing service discipline, since the formula for the expected steady-state waiting time in that system coincides with that for the $M/M/1$ queue with *FIFO* discipline.

For the examples we consider, the problem of finding a socially optimal allocation of flows takes the form:

$$\max_{\{\lambda_1, \lambda_2\}} \quad \mathcal{U}(\lambda_1, \lambda_2) = U_1(\lambda_1) - \frac{h_1\lambda_1}{1 - \lambda_1 - \lambda_2} + U_2(\lambda_2) - \frac{h_2\lambda_2}{1 - \lambda_1 - \lambda_2}$$

$$\text{s.t.} \quad \lambda_1 + \lambda_2 < 1 \,, \; \lambda_1 \geq 0 \,, \; \lambda_2 \geq 0 \,.$$

In addition to linear utility functions, we shall consider several nonlinear concave utility functions which are well studied in the economics literature, namely:

1. Square-Root Utility Functions – $U_i(\lambda_i) = r_i\lambda_i + s_i\sqrt{\lambda_i}$, where $r_i \geq 0$, $s_i \geq 0$, $i = 1, 2$;

2. Logarithmic Utility Functions – $U_i(\lambda_i) = r_i\log(1 + \lambda_i)$, where $\alpha_i > 0$, $i = 1, 2$;

3. Quadratic Utility Functions – $U_i(\lambda_i) = r_i\lambda_i - s_i\lambda_i^2$, where $r_i \geq 0$, $s_i \geq 0$, $i = 1, 2$.

We give examples in which

1. $\mathcal{U}$ is nonconcave and there is a single interior solution to the *KKT* conditions which is a saddlepoint, dominated by both single-class optima, as in the linear case;

2. $\mathcal{U}$ is nonconcave, but one of the single-class optima is the unique solution to the *KKT* conditions;

3. $\mathcal{U}$ is concave and there is a unique solution to the *KKT* conditions, which is an interior global maximum of $\mathcal{U}$.

4.6.1 Linear Utility Functions

Suppose $U_i(\lambda_i) = r_i\lambda_i$, where $r_i \geq 0$, $i = 1, 2$. From the analysis in Section 4.5.3 we know that in this case the objective function, $\mathcal{U}(\lambda_1, \lambda_2)$, is *never* jointly concave. Moreover, we have shown (cf. Section 4.5) that, if $(\tilde{\lambda}_1, \tilde{\lambda}_2)$ is an interior-point solution to the *KKT* conditions, then $\mathcal{U}(\tilde{\lambda}_1, \tilde{\lambda}_2) \leq \mathcal{U}(\lambda_1^*, 0)$ and $\mathcal{U}(\tilde{\lambda}_1, \tilde{\lambda}_2) \leq \mathcal{U}(0, \lambda_2^*)$, where λ_i^* is the optimal flow allocation to user i when only that user receives positive flow. That is, an interior solution to the first-order conditions is dominated by *both* the optimal single-user allocations. We shall now show that there is at most one interior-point solution to the *KKT* conditions and it is a saddle-point of the objective function.

The social optimization problem takes the form:

$$\max_{\{\lambda_1, \lambda_2\}} \quad r_1\lambda_1 - \frac{h_1\lambda_1}{1 - \lambda_1 - \lambda_2} + r_2\lambda_2 - \frac{h_2\lambda_2}{1 - \lambda_1 - \lambda_2}$$

$$\text{s.t.} \quad \lambda_1 + \lambda_2 < 1$$

$$\lambda_1 \geq 0 \,, \; \lambda_2 \geq 0 \,.$$

Let $a := r_1/h_1$, $2\beta := r_2/h_2$, $c := h_1/h_2$. Then an equivalent form for the above problem is

$$\max_{\{\lambda_1,\lambda_2\}} \quad c\left(a\lambda_1 - \frac{\lambda_1}{1-\lambda_1-\lambda_2}\right) + 2\beta\lambda_2 - \frac{\lambda_2}{1-\lambda_1-\lambda_2} \qquad (4.90)$$

$$\text{s.t.} \quad \lambda_1 + \lambda_2 < 1$$

$$\lambda_1 \geq 0, \ \lambda_2 \geq 0.$$

The unique solution to the first-order conditions is given by:

$$\tilde{\lambda}_1 = \frac{b(c-1)}{(ca-b)^2} - \frac{1}{c-1};$$

$$\tilde{\lambda}_2 = \frac{c}{c-1} - \frac{ca(c-1)}{(ca-b)^2}.$$

It can be shown that the pair $(\tilde{\lambda}_1, \tilde{\lambda}_2)$ is an interior point ($\tilde{\lambda}_1 > 0$, $\tilde{\lambda}_2 > 0$, $\tilde{\lambda}_1 + \tilde{\lambda}_2 < 1$) if the parameters satisfy the following conditions:

$$b > a > 1;$$

$$c > \frac{b-1}{a-1};$$

$$a < \frac{(ca-b)^2}{(c-1)^2} < b.$$

As an example in which these conditions are satisfied, take $a = 4$, $b = 9$, and $c = 4$. In this case, the solution to the first-order conditions is

$$\tilde{\lambda}_1 = 0.218; \ \tilde{\lambda}_2 = 0.354.$$

The optimal single-user flow allocations are $\lambda_1^* = 0.500$ and $\lambda_2^* = 0.667$. The objective function values of these three flow allocations are:

$$\mathcal{U}(\tilde{\lambda}_1, \tilde{\lambda}_2) = 3.81;$$
$$\mathcal{U}(\lambda_1^*, 0) = 4.00;$$
$$\mathcal{U}(0, \lambda_2^*) = 4.00.$$

Thus we have an illustration of the general result derived in Section 4.5: the interior-point equilibrium flow allocation is dominated by both optimal single-user allocations. A trajectory of the dynamic algorithm converges either to $(\lambda_1^*, 0)$ or to $(0, \lambda_2^*)$, depending on the starting point.

For the example just presented, in which $r_1 = 16$, $r_2 = 9$, $h_1 = 4$, $h_2 = 1$, Figure 4.2 and Figure 4.3 show, respectively, a contour plot and graph of the response surface, $\{\mathcal{U}(\lambda_1, \lambda_2)\}$.

Figure 4.4 shows a graph of the response surface for another example with linear utility functions. In this example, class 2 has a much lower value per unit flow ($r_2 = 12$) than class 1 ($r_1 = 64$), but is completely insensitive to delay ($h_2 = 0$), whereas class 1 has a positive delay sensitivity ($h_1 = 9$). Note that there is a significant difference between the objective function values associated with the two local maxima: $\mathcal{U}(\lambda_1^*, 0) = 25$, $\mathcal{U}(0, \lambda_2^*) = 12$. Moreover,

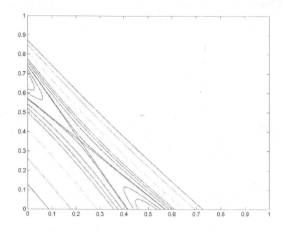

Figure 4.2 *Linear Utility Functions:* $\mathcal{U}(\lambda_1, \lambda_2) = 16\lambda_1 - 4\lambda_1/(1-\lambda_1-\lambda_2) + 9\lambda_2 - \lambda_2/(1-\lambda_1-\lambda_2)$

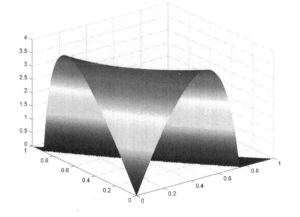

Figure 4.3 *Linear Utility Functions:* $\mathcal{U}(\lambda_1, \lambda_2) = 16\lambda_1 - 4\lambda_1/(1-\lambda_1-\lambda_2) + 9\lambda_2 - \lambda_2/(1-\lambda_1-\lambda_2)$

if the starting point is close to $(0, \lambda_2^*)$, the dynamic algorithm will converge to $(0, \lambda_2^*)$, with a loss of over 50% of the potential utility that could be gained at the global maximum, $(\lambda_1^*, 0)$.

As a possible application of this example, consider a communication network with two classes of traffic: real-time, high-value audio/video traffic that must be delivered rapidly (class 1) and file transfers, with a significantly lower intrinsic value but negligible urgency (class 2). Our results imply that, if the

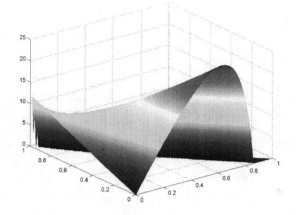

Figure 4.4 *Linear Utility Functions:* $\mathcal{U}(\lambda_1, \lambda_2) = 64\lambda_1 - 9\lambda_1/(1 - \lambda_1 - \lambda_2) + 12\lambda_2$

system starts with class-2 traffic predominating, then it will become "trapped" in such a state: the real-time traffic will never gain a foothold, even though the system as a whole would be much "better off" (in terms of total utility) if real-time traffic predominated. A similar example with nonlinear utilities is in the next section.

4.6.2 Square-Root Utility Functions

(**Note:** Figures for the examples in this section may be found in Section 4.9 at the end of the chapter.)

Now suppose $U_i(\lambda_i) = r_i\lambda_i + s_i\sqrt{\lambda_i}$, where $r_i \geq 0$, $s_i \geq 0$, $i = 1, 2$. When $s_i > 0$, $U_i(\lambda_i)$ is strictly concave with $U_i'(0) = +\infty$. But even under these conditions there can be multiple local maxima and nonglobal convergence of the dynamic algorithm. In the example illustrated in Figure 4.7, once again user 2 places a lower value on its flow ($r_2 = 15$, $s_2 = 0$) than user 1 ($r_1 = 64$, $s_1 = 8$), but is insensitive to delay ($h_2 = 0$), whereas user 1 has a positive delay sensitivity ($h_1 = 9$). Just as in the previous example with linear utility functions, we see that $\mathcal{U}(\lambda_1^*, 0)$ is significantly larger than $\mathcal{U}(0, \lambda_2^*)$, but if the starting point is close to $(0, \lambda_2^*)$, the dynamic algorithm will converge to $(0, \lambda_2^*)$ rather than the global maximum, $(\lambda_1^*, 0)$.

Figure 4.8 provides an example in which the opposite is true: $\mathcal{U}(0, \lambda_2^*)$ is significantly larger than $\mathcal{U}(\lambda_1^*, 0)$. In the example illustrated in Figure 4.9 the two values are more nearly equal, but the objective function $\mathcal{U}$ is still nonconcave, with a saddlepoint interior solution to the optimality conditions. Finally, Figure 4.10 illustrates an example in which the objective function is jointly concave, with an interior solution to the optimality conditions that is the unique global maximum.

4.6.3 Logarithmic Utility Functions

(**Note:** Figures for the examples in this section may be found in Section 4.9 at the end of the chapter.)

Now suppose $U_i(\lambda_i) = r_i \log(1 + \lambda_i)$, where $r_i > 0$, $i = 1, 2$. Figures 4.11 and 4.12 illustrate examples in which $\mathcal{U}$ is nonconcave with a saddlepoint interior solution to the optimality conditions. In the example illustrated in Figure 4.13, $\mathcal{U}$ is still nonconcave, but there is a unique solution to the optimality conditions, namely $(0, \lambda_2^*)$, which is the global maximum. Finally, Figure 4.14 illustrates an example in which the objective function is jointly concave, with an interior solution to the optimality conditions that is the unique global maximum.

4.6.4 Quadratic Utility Functions

(**Note:** Figures for the examples in this section may be found in Section 4.9 at the end of the chapter.)

Suppose $U_i(\lambda_i) = r_i \lambda_i - s_i \lambda_i^2$, where $r_i \geq 0$, $s_i \geq 0$, $i = 1, 2$. Figure 4.15 illustrates an example in which $\mathcal{U}$ is nonconcave, with an interior solution to the optimality equations that is a saddlepoint and two local maxima at $(\lambda_1^*, 0)$ and $(0, \lambda_2^*)$.

4.7 Multiclass Queue with Priorities

In this section we continue our study of the multiclass model, but now with classes served according to a (strict) priority discipline rather than in *FIFO* order. Again one must choose the arrival rate for each of several classes of customers, which may place different values on service and have different sensitivities to delay. Now, however, the waiting time – in the queue as well as in the system – depends on the class (and the priority discipline in effect).

Our starting point, as usual, is the general model introduced in Section 4.2. There are m classes of customers, labelled $i = 1, 2, \ldots, m$. Class i gains a gross utility per unit time $U_i(\lambda_i)$, where λ_i, the arrival rate of class i, is a decision variable, chosen from the feasible set $A_i = [0, \infty)$. We assume that $U_i(\lambda_i)$ is nondecreasing, differentiable, and concave in $\lambda_i \geq 0$. The average waiting cost of a class-i customer is $G_i(\boldsymbol{\lambda})$, where $\boldsymbol{\lambda} = (\lambda_i, i \in M)$. The average waiting cost incurred per unit time by class i is $H_i(\boldsymbol{\lambda}) = \lambda_i G_i(\boldsymbol{\lambda})$. In addition there is an entrance fee or toll, δ_i, for each class-i customer admitted to the system.

The decision may be made by the individual customers (*individual optimality*), by an agent for the class as a whole (*class optimality*), or by a system operator, who might be interested in maximizing profit (*facility optimality*) or in maximizing the aggregate net utility to all customers (*social optimality*).

In this section we shall confine our attention to the case where in each class the waiting cost for a job is a linear function of the average waiting time for a job of that class. That is,

$$G_i(\boldsymbol{\lambda}) = h_i W_i(\boldsymbol{\lambda}) , \ i \in M ,$$

where $W_i(\boldsymbol{\lambda})$ is the average waiting time in the system for a customer of class i. The net utility per unit time for class i is therefore

$$U_i(\lambda_i) - h_i \lambda_i W_i(\boldsymbol{\lambda}) . \tag{4.91}$$

The exact form for $W_i(\boldsymbol{\lambda})$ depends on the stochastic assumptions of the model and the queue discipline (e.g., nonpreemptive or preemptive-resume). We shall focus on an $M/GI/1$ model operating in steady state, in which class 1 receives the highest priority, class 2 the next highest, and so forth. We introduced this model in Section 4.1.1. The first and second moments of the service time for a class-i customer are $1/\mu_i$ and $2\beta_i/\mu_i^2$, respectively.

In the case of a nonpreemptive discipline, we have (cf. (4.8))

$$W_i(\boldsymbol{\lambda}) = \frac{\sum_{k=1}^{m} \lambda_k \beta_k / \mu_k^2}{(1 - \sum_{k=1}^{i-1} \lambda_k / \mu_k)(1 - \sum_{k=1}^{i} \lambda_k / \mu_k)} + \frac{1}{\mu_i} , \tag{4.92}$$

subject to the steady-state stability condition,

$$\sum_{k=1}^{i} \lambda_k / \mu_k < 1 . \tag{4.93}$$

When this condition does not hold, $W_i(\boldsymbol{\lambda}) = \infty$.

In the case of a preemptive-resume discipline, we have (cf. (4.11))

$$W_i(\boldsymbol{\lambda}) = \frac{\sum_{k=1}^{i} \lambda_k \beta_k / \mu_k^2}{(1 - \sum_{k=1}^{i-1} \lambda_k / \mu_k)(1 - \sum_{k=1}^{i} \lambda_k / \mu_k)} + \frac{1/\mu_i}{1 - \sum_{k=1}^{i-1} \rho_i} , \tag{4.94}$$

when (4.93) holds, with $W_i(\boldsymbol{\lambda}) = \infty$ when (4.93) does not hold.

4.7.1 Individually Optimal Arrival Rates

The equilibrium conditions satisfied by the vector $\boldsymbol{\lambda}^e$ of individually optimal arrival rates take the following form:

$$U_i'(\lambda_i) \ \leq \ h_i \left(W_i(\boldsymbol{\lambda}) + \delta_i \right) \tag{4.95}$$
$$U_i'(\lambda_i) \ = \ h_i \left(W_i(\boldsymbol{\lambda}) + \delta_i \right) , \ \text{if } \lambda_i > 0 \tag{4.96}$$

for all $i \in M$. Here $W_i(\boldsymbol{\lambda})$ is given by (4.92) in the case of a nonpreemptive queue discipline and by (4.94) if the discipline is preemptive resume, if the stability condition (4.93) holds, and $W_i(\boldsymbol{\lambda}) = \infty$ if not.

To show that there exists a unique solution to the equilibrium conditions (4.95) and (4.96), hence a unique individually optimal solution, it suffices to show that Assumption 1 holds. We leave this as an exercise for the reader.

Exercise 4   Show that Assumption 1 holds for the present model, for both nonpreemptive and preemptive-resume queue disciplines.

Thus, as a corollary to Theorem 4.1, we have the following theorem.

Theorem 4.8 *There exists a unique solution to (4.95), (4.96), hence a unique vector $\boldsymbol{\lambda}^e$ of individually optimal arrival rates for the M/GI/1 model with linear waiting costs, for both nonpreemptive and preemptive-resume queue disciplines.*

4.7.2 Class Optimal Arrival Rates

We now consider the problem from the point of view of the manager of each class. The class manager for class i is concerned with maximizing the net benefit, $B_i(\boldsymbol{\lambda})$, received per unit time by jobs of class i. In this case $B_i(\boldsymbol{\lambda})$ is given by

$$B_i(\boldsymbol{\lambda}) = U_i(\lambda_i) - \lambda_i(h_i W_i(\boldsymbol{\lambda}) + \delta_i) . \qquad (4.97)$$

Given fixed arrival rates, λ_k, $k \neq i$, the optimal strategy for the manager of class i is therefore to choose a value for λ_i such that

$$\lambda_i = \arg \max_{\{\lambda_i \geq 0\}} B_i(\lambda_1, \ldots, \lambda_i, \ldots, \lambda_m) . \qquad (4.98)$$

A Nash equilibrium $\boldsymbol{\lambda}^c$ for the class optimization problem is a vector $\boldsymbol{\lambda} = (\lambda_1, \ldots, \lambda_i, \ldots, \lambda_m)$ that simultaneously satisfies (4.98) for each $i \in M$.

For each $i \in M$, the following *KKT* conditions are necessary for λ_i to maximize $B_i(\boldsymbol{\lambda})$, with λ_k fixed for $k \neq i$:

$$U_i'(\lambda_i) \quad \leq \quad h_i \left(W_i(\boldsymbol{\lambda}) + \lambda_i \frac{\partial}{\partial \lambda_i} W_i(\boldsymbol{\lambda}) \right) + \delta_i ; \qquad (4.99)$$

$$U_i'(\lambda_i) \quad = \quad h_i \left(W_i(\boldsymbol{\lambda}) + \lambda_i \frac{\partial}{\partial \lambda_i} W_i(\boldsymbol{\lambda}) \right) + \delta_i , \text{ if } \lambda_i > 0 . \qquad (4.100)$$

Therefore, (4.99) and (4.100) must hold simultaneously for all $i \in M$ in order for an allocation $\boldsymbol{\lambda} = (\lambda_i, i \in M)$ to be class optimal.

4.7.2.1 Existence and Uniqueness of Class-Optimal Solution

To show that there exists a unique solution to the *KKT* conditions (4.19) and (4.19), hence a unique class-optimal solution, it suffices to show that Assumptions 3 and 4 hold. We leave this as an exercise for the reader.

Exercise 5   Show that in this case $G_i(\boldsymbol{\lambda})$ is convex and nondecreasing in λ_i, for each set of fixed values of λ_k, $k \neq i$, for both nonpreemptive and preemptive-resume queue disciplines, and that as a result Assumption 3 holds.

Exercise 6   Suppose $U_i(\lambda_i) \rightarrow 0$ as $\lambda_i \rightarrow \infty$, for all $i \in M$. Show that

Assumption 4 holds, for both nonpreemptive and preemptive-resume queue disciplines. (Hint. Take

$$\hat{A} = \{\boldsymbol{\lambda} : \lambda_k \geq 0, k \in M \; ; \; \sum_{k \in M} \lambda_k / \mu_k \leq 1\} \; .$$

Show by induction on $i = 1, 2, \ldots, m$, that any feasible $\boldsymbol{\lambda}$ can be replaced by $\boldsymbol{\lambda}' \in \hat{A}$ such that $\sum_{k=1}^{i} \lambda_k / \mu_k < 1$ and $B_i(\boldsymbol{\lambda}') \leq B_i(\boldsymbol{\lambda})$.)

Thus, as a corollary to Theorem 4.2, we have the following theorem.

Theorem 4.9 *There exists a unique solution to (4.99) and (4.100), hence a unique class-optimal allocation $\boldsymbol{\lambda}^c$ for the M/GI/1 model with linear waiting costs, for both nonpreemptive and preemptive-resume queue disciplines.*

4.7.2.2 Further Results for Preemptive-Resume Discipline

For simplicity in what follows we shall assume that $\delta_i = 0$ for all $i \in M$. (If not, define a new utility function, $\hat{U}_i(\lambda_i) := U_i(\lambda_i) - \lambda_i \delta_i$, and use $\hat{U}_i$ instead of U_i.) A salient characteristic of the *M/GI/1* model with preemptive-resume priorities is that the waiting time for class i depends only on the arrival rates of classes $k = 1, 2, \ldots, i$. It follows that a class-optimal solution may be found by solving a sequence of one-class optimization problems, in the order $k = 1, 2, \ldots, m$.

To see this, start with class 1 and note that the manager of class 1 seeks to choose λ_1 to maximize

$$
\begin{aligned}
B_1(\boldsymbol{\lambda}) &= U_1(\lambda_1) - \lambda_1 h_1 W_1(\boldsymbol{\lambda}) \\
&= U_1(\lambda_1) - \lambda_1 h_1 \left(\frac{\lambda_1 \beta_1 / \mu_1^2}{1 - \lambda_1 / \mu_1} + \frac{1}{\mu_1} \right) ,
\end{aligned}
$$

where $\boldsymbol{\lambda} = (\lambda_i, i \in M)$, with $\lambda_2, \ldots, \lambda_m$ fixed. But, as can be seen from the above expression, in this case $B_1(\boldsymbol{\lambda})$ is independent of $\lambda_2, \ldots, \lambda_m$ and takes the form of net benefit in the social optimization problem for a one-class *M/GI/1* queue with only customers of class 1. (See Chapter 2.) Thus, λ_1^c is the optimal value of the arrival rate for this problem.

Now with $\lambda_1 = \lambda_1^c$, the manager of class 2 seeks to maximize

$$
\begin{aligned}
B_2(\boldsymbol{\lambda}) &= U_2(\lambda_2) - \lambda_2 h_2 W_2(\boldsymbol{\lambda}) \\
&= U_2(\lambda_2) - \lambda_2 h_2 \left(\frac{\lambda_1 \beta_1 / \mu_1^2 + \lambda_2 \beta_2 / \mu_2^2}{(1 - \lambda_1 / \mu_1)(1 - \lambda_1 / \mu_1 - \lambda_2 / \mu_2)} \right. \\
&\quad \left. + \frac{1 / \mu_2}{1 - \lambda_1 / \mu_1} \right) \\
&= U_2(\lambda_2) - \lambda_2 \left(\frac{h_2}{1 - \lambda_1 / \mu_1} \right) \left(\frac{\lambda_1 \beta_1 / \mu_1^2 + \lambda_2 \beta_2 / \mu_2^2}{1 - \lambda_1 / \mu_1 - \lambda_2 / \mu_2} + \frac{1}{\mu_2} \right) ,
\end{aligned}
$$

where $\boldsymbol{\lambda} = (\lambda_1, \lambda_2 \ldots, \lambda_m)$, with $\lambda_3, \ldots, \lambda_m$ fixed. Here the expression for $B_2(\boldsymbol{\lambda})$ is independent of $\lambda_3, \ldots, \lambda_m$. Thus the manager for class 2 faces a single one-variable optimization problem, rather than a set of such problems, one for each set of fixed values of λ_k, $k \neq i$.

Continuing in this fashion, we see that the manager of each class i can take the values of λ_k, $k = 1, \ldots, i-1$, as given and equal to the class-optimal values for each of the associated classes, and ignore classes $i+1, \ldots, m$, choosing λ_i to maximize

$$
\begin{aligned}
B_i(\boldsymbol{\lambda}) &= U_i(\lambda_i) - \lambda_i h_i W_i(\boldsymbol{\lambda}) \\
&= U_i(\lambda_i) - \lambda_i h_i \left(\frac{\sum_{k=1}^{i} \lambda_k \beta_k / \mu_k^2}{(1 - \sum_{k=1}^{i-1} \lambda_k / \mu_k)(1 - \sum_{k=1}^{i} \lambda_k / \mu_k)} \right. \\
&\quad + \left. \frac{1/\mu_i}{1 - \sum_{k=1}^{i-1} \lambda_k / \mu_k} \right),
\end{aligned}
$$

4.7.3 Socially Optimal Arrival Rates

As usual, a *socially optimal* allocation of arrival rates $\boldsymbol{\lambda}^s = (\lambda_i^s, i \in M)$ to the classes is defined as one that maximizes the aggregate net benefit to all classes. It may thus be found by solving the following optimization problem:

$$
\max_{\{\lambda_i, i \in M\}} \quad \mathcal{U}(\boldsymbol{\lambda}) := \sum_{i \in M} U_i(\lambda_i) - \sum_{i \in M} \lambda_i h_i W_i(\boldsymbol{\lambda})
$$

$$
\text{s.t.} \quad \lambda_i \geq 0 , \ i \in M .
$$

Here $W_i(\boldsymbol{\lambda})$ is given by (4.92) in the nonpreemptive case and by (4.94) in the preemptive case, if the stability condition (4.93) holds, and $W_i(\boldsymbol{\lambda}) = \infty$ if not.

A socially optimal allocation of arrival rates $\boldsymbol{\lambda}^s = (\lambda_i^s, i \in M)$ satisfies the following *KKT* conditions, which are necessary for a solution to the above maximization problem:

$$
U_i'(\lambda_i) \leq h_i \left(W_i(\boldsymbol{\lambda}) + \sum_{k \in M} \lambda_k \frac{\partial}{\partial \lambda_i} W_k(\boldsymbol{\lambda}) \right) , \tag{4.101}
$$

$$
U_i'(\lambda_i) = h_i \left(W_i(\boldsymbol{\lambda}) + \sum_{k \in M} \lambda_k \frac{\partial}{\partial \lambda_i} W_k(\boldsymbol{\lambda}) \right) , \text{ if } \lambda_i > 0 . \tag{4.102}
$$

Our next task is to investigate the question of when the *KKT* conditions are sufficient as well as necessary for a global maximum of the socially optimal objective function. As in the model with *FIFO* discipline, a crucial issue here is whether the objective function, $\mathcal{U}(\boldsymbol{\lambda})$, for social optimality is jointly concave in the arrival rates, $\boldsymbol{\lambda} = (\lambda_i, i \in M)$. If so, then the vector of socially optimal arrival rates, $\boldsymbol{\lambda}^s = (\lambda_1^s, \ldots, \lambda_m^s)$, is the unique solution to the *KKT* first-order optimality conditions.

Since we are assuming that the utility functions, $U_i(\lambda_i)$, are all concave, a sufficient condition for $\mathcal{U}(\boldsymbol{\lambda})$ to be jointly concave in $\boldsymbol{\lambda}$ is that the total waiting cost, $\sum_{i \in M} \lambda_i h_i W_i(\boldsymbol{\lambda})$, is jointly convex in $\boldsymbol{\lambda}$. As we shall see, whether or not this is true depends on the specific structure of the priority discipline, in particular, on the relationship between the discipline and the waiting-cost and service-rate parameters, $h_i, \mu_i, \imath \in M$.

4.7.3.1 Existence and Uniqueness of Socially Optimal Solution under h-μ Scheduling Rule

In this section we shall analyze some special cases of our priority model and prove that the total waiting-cost function is jointly convex in $\boldsymbol{\lambda}$ if the scheduling rule (priority ordering) in use is the *h-μ rule*. Under this rule the classes are served in decreasing order of $h_i \cdot \mu_i$ – the product of the waiting-cost and service-rate parameters.* The special cases we consider are (1) the *M/GI/1* model with nonpreemptive discipline, and (2) the *M/M/1* model with preemptive-resume discipline.

Using numerical examples we also demonstrate that the total waiting-cost function may be nonconvex if a scheduling rule other than the *h-μ* rule is used.

Rather than show convexity of the total waiting-cost function under the *h-μ* rule directly, we shall instead use principles from the *achievable-region* approach to scheduling problems, based on the theory of *strong conservation laws*. In the problem under study the achievable region is the set of all waiting-time vectors, $(W_1, \ldots, W_m)$, that can be realized by admissible scheduling rules (queue disciplines). Using this approach we are able to consider a relaxed version of the social-optimization problem, in which both the scheduling rule and the arrival-rate vector $\boldsymbol{\lambda}$ are decision variables. We then show simultaneously that the *h-μ* rule is optimal for all $\boldsymbol{\lambda}$ and that the associated total waiting-cost function is jointly convex in $\boldsymbol{\lambda}$.

4.7.3.2 Relaxation of the Social Optimization Problem

In this section we introduce a relaxed version of the social optimization problem, in which the scheduling rule is a decision variable in addition to the vector of arrival rates, $\boldsymbol{\lambda} = (\lambda_i, i \in M)$. Let Φ denote the set of all admissible scheduling rules ϕ. An admissible scheduling rule is *nonanticipative*, *nonidling*, and *regenerative*.

Definition A scheduling rule is *nonanticipative* if the decision about which job to process at time $t \geq 0$ depends only the arrival times and work requirements of jobs that have arrive in $[0, t]$, and possibly on decisions taken before time t.

Definition A scheduling rule is *nonidling* if the server is always busy when there is at least one job in the system.

Definition A scheduling rule ϕ is *regenerative* if the decision about which job to process next does not use any information from previous busy cycles and coincides with the decision that would be made in the first busy cycle, given

---

* In the literature, this rule is commonly referred to as the *c-μ* rule because in the earliest references the class-*i* waiting-cost parameter was denoted by c_i rather than h_i.

the same information about previous arrival instants, work requirements, and decisions during the current busy cycle.

There may be additional restrictions on admissible scheduling rules (e.g., preemptive or nonpreemptive), depending on the problem context.

For the moment, let us consider an arbitrary fixed vector of arrival rates, $\boldsymbol{\lambda} = (\lambda_1, \ldots, \lambda_m)$. Let $W_i = W_i(\boldsymbol{\lambda})$, $i \in M$. Of course, W_i depends on the scheduling rule, $\phi \in \Phi$, and when appropriate we shall write W_i^ϕ (or $W_i^\phi(\boldsymbol{\lambda})$) to emphasize this dependence.

Suppose we can show that the waiting times, $(W_1^\phi, \ldots, W_m^\phi)$, $\phi \in \Phi$, satisfy the constraints,

$$\sum_{i \in M} \rho_i W_i = \alpha_M \tag{4.103}$$

$$\sum_{i \in S} \rho_i W_i \geq \alpha_S , \ S \subset M , \tag{4.104}$$

with $\rho_i := \lambda_i / \mu_i$, $i \in M$, where α_S, $S \subseteq M$ are given constants. Moreover, for each subset $S \subset M$, suppose the lower bound, α_S, is attained by any scheduling rule $\phi \in \Phi_S$, where Φ_S is the set of all scheduling rules that give strict priority to classes in $i \in S$ over classes $i \notin S$. That is,

$$\sum_{i \in S} \rho_i W_i^\phi = \alpha_S , \text{ for all } \phi \in \Phi_S , \ S \subset M . \tag{4.105}$$

(Presently we shall give two examples in which these conditions hold: an *M/GI/1* model with nonpreemptive discipline, and an *M/M/1* model with preemptive-resume discipline.)

Under these conditions, the system under study is said to satisfy *strong conservation laws* (see Section A.1 of Appendix A). The set of all waiting-time vectors, $(W_1, \ldots, W_m)$, satisfying (4.103) and (4.104) is called the *achievable region* for the problem under study. It follows that Theorem A.1 holds, with $x_i = \rho_i W_i$, $i \in S$. Letting $c_i = h_i \mu_i$ and substituting for c_i and x_i in Theorem A.1 then yields the following corollary.

Corollary 4.10 *For a fixed arrival-rate vector, $\boldsymbol{\lambda}$, suppose the waiting-time vector $(W_1^\phi(\boldsymbol{\lambda}), \ldots, W_m^\phi(\boldsymbol{\lambda}))$, $\phi \in \Phi$, satisfies strong conservation laws. Then the total waiting cost per unit time,*

$$\sum_{i \in M} \lambda_i h_i W_i^\phi(\boldsymbol{\lambda}) ,$$

is minimized among all scheduling rules $\phi \in \Phi$ by the h-μ rule: the queue discipline that gives strict priority to the classes in decreasing order of $h_i \mu_i$. That is, the classes are ordered so that

$$h_1 \mu_1 \geq h_2 \mu_2 \geq \cdots \geq h_m \mu_m .$$

Now let us return to our original problem, in which the arrival-rate vector $\boldsymbol{\lambda}$ is a decision variable and a fixed queue discipline $\phi \in \Phi$ is used. Suppose the queue discipline ϕ in use is the h-μ rule. Then, if strong conservation laws hold for every $\lambda \in A$, it follows from the above results that the total waiting cost per unit time,

$$\sum_{i \in M} \lambda_i h_i W_i^\phi(\lambda) \, ,$$

equals the optimal value of the objective function in the following optimization problem:

$$\min_{\{W_i\}} \quad \sum_{i \in M} \lambda_i h_i W_i$$

$$\text{s.t.} \quad \sum_{i \in M} \rho_i W_i = \alpha_M(\lambda)$$

$$\sum_{i \in S} \rho_i W_i \geq \alpha_S(\lambda) \, , \ S \subset M$$

Here $\alpha_S(\lambda) = \alpha_S$, $S \subseteq M$. (Since λ is once again a decision variable, we have written $\alpha_S(\lambda)$ to indicate the dependence of this parameter on λ.) Using the identities, $x_i = \rho_i W_i$, $i \in M$, we can write this optimization problem in equivalent form as

$$\min_{\{x_i\}} \quad \sum_{i \in M} (h_i \mu_i) x_i$$

$$\text{s.t.} \quad \sum_{i \in S} x_i \geq \alpha_S(\lambda) \, , \ S \subseteq M$$

(Note that we have replaced the equality constraint for $S = M$ with an inequality. This further relaxation is without loss of optimality, since all coefficients in the objective function and constraints are nonnegative and we are minimizing.)

Thus, to show that $\sum_{i \in M} \lambda_i h_i W_i^\phi(\lambda)$ is a jointly convex function of λ, it suffices to show that the optimal value of the objective function in the above linear-programming (LP) problem is a jointly convex function of the parameter λ. We establish this result in the following lemma, under the condition that $\alpha_S(\lambda)$ is convex in λ for each subset $S \subseteq M$.

First some notation. Let $x = (x_1, \ldots, x_m)$ and, for each $\lambda \in A = \prod_{i \in M} A_i$, let $x^*(\lambda) = (x_1^*(\lambda), \ldots, x_m^*(\lambda))$ denote a value of x that attains the minimum in the above LP. Let $f(\lambda) := \sum_{i \in M} (h_i \mu_i) x_i^*(\lambda)$. That is, $f(\lambda)$ is the minimum value of the objective function, expressed as a function of the vector of arrival rates, λ.

Lemma 4.11 *Suppose $\alpha_S(\lambda)$ is convex in λ for all $S \subseteq M$. Then $f(\lambda)$ is a jointly convex function of $\lambda \in A$.*

Proof Let $\lambda^1 \in A$, $\lambda^1 \in A$, and let $0 < \nu < 1$. First we show that the point $x = \nu x^*(\lambda^1) + (1-\nu) x^*(\lambda^2)$ is feasible, i.e., satisfies the constraints of the LP when $\lambda = \nu \lambda^1 + (1-\nu) \lambda^2$. Since $x^*(\lambda^k)$ is optimal for the LP with $\lambda = \lambda^k$, we have

$$\sum_{i \in S} x_i^*(\boldsymbol{\lambda}^k) \geq \alpha_S(\boldsymbol{\lambda}^k) , \ S \subseteq M ,$$

for $k = 1, 2$. Therefore, for all $S \subseteq M$,

$$
\begin{aligned}
\sum_{i \in M} (\nu x_i^*(\boldsymbol{\lambda}^1) + (1 - \nu) x_i^*(\boldsymbol{\lambda}^2)) &= \nu \sum_{i \in M} x_i^*(\boldsymbol{\lambda}^1) + (1 - \nu) \sum_{i \in M} x_i^*(\boldsymbol{\lambda}^2) \\
&\geq \nu \alpha_S(\boldsymbol{\lambda}^1) + (1 - \nu) \alpha_S(\boldsymbol{\lambda}^2) \\
&\geq \alpha_S(\nu \boldsymbol{\lambda}^1 + (1 - \nu) \boldsymbol{\lambda}^2) ,
\end{aligned}
$$

where the last inequality follows from the convexity of $\alpha_S(\boldsymbol{\lambda})$. Now, having established feasibility of the point $\boldsymbol{x} = \nu \boldsymbol{x}^*(\boldsymbol{\lambda}^1) + (1 - \nu) \boldsymbol{x}^*(\boldsymbol{\lambda}^2)$, we can assert that

$$
\begin{aligned}
f(\nu \boldsymbol{\lambda}^1 + (1 - \nu) \boldsymbol{\lambda}^2) &= \sum_{i \in M} (h_i \mu_i) x_i^*(\nu \boldsymbol{\lambda}^1 + (1 - \nu) \boldsymbol{\lambda}^2) \\
&\leq \sum_{i \in M} (h_i \mu_i)(\nu x_i^*(\boldsymbol{\lambda}^1) + (1 - \nu) x_i^*(\boldsymbol{\lambda}^2)) \\
&= \nu \sum_{i \in M} (h_i \mu_i) x_i^*(\boldsymbol{\lambda}^1) + (1 - \nu) \sum_{i \in M} (h_i \mu_i) x_i^*(\boldsymbol{\lambda}^2) \\
&= \nu f(\boldsymbol{\lambda}^1) + (1 - \nu) f(\boldsymbol{\lambda}^2) .
\end{aligned}
$$

Thus $f(\boldsymbol{\lambda})$ is convex in $\boldsymbol{\lambda}$.   ∎

The above results are summarized in the following theorem.

Theorem 4.12 *Let Φ be the set of admissible scheduling rules. Suppose that*

- *for all $\phi \in \Phi$, the waiting-time vector $(W_1^\phi(\boldsymbol{\lambda}), \dots, W_m^\phi(\boldsymbol{\lambda}))$ satisfies strong conservation laws, (4.103), (4.104), (4.105), for every $\boldsymbol{\lambda} \in A$, and*

- *$\alpha_S(\boldsymbol{\lambda})$ is convex in $\boldsymbol{\lambda} \in A$, for all $S \subseteq M$.*

Now let $\phi \in \Phi$ denote the h-μ rule. Then the total-waiting-cost function,

$$\sum_{i \in M} \lambda_i h_i W_i^\phi(\boldsymbol{\lambda}) ,$$

is jointly convex in $\boldsymbol{\lambda} \in A$, and hence the objective function for the social optimization problem,

$$\mathcal{U}(\boldsymbol{\lambda}) = \sum_{i \in M} U_i(\lambda_i) - \sum_{i \in M} \lambda_i h_i W_i^\phi(\boldsymbol{\lambda}) ,$$

is jointly concave in $\boldsymbol{\lambda} \in A$. In this case, the KKT conditions, (4.101), (4.102), have a unique solution, which is the vector of socially optimal arrival rates, $\boldsymbol{\lambda}^s = (\lambda_1^s, \dots, \lambda_m^s)$.

4.7.3.3 Application: M/GI/1 Model with Nonpreemptive Priority Discipline

Consider a multiclass $M/GI/1$ queue with a strict nonpreemptive priority discipline, in which the priority ordering is according to the h-μ rule. Let Φ_m

denote the set of all nonanticipative, nonidling, regenerative scheduling rules which are nonpreemptive and service-time independent within each class. It is shown in Section A.3.2 of Appendix A that strong conservation laws hold for the vector, $(W_1, \ldots, W_m)$, for all $\phi \in \Phi_m$, where

$$W_i = \mathrm{E}[\boldsymbol{W}_i] \,,$$

the expected steady-state waiting time in the system for class i, $i \in M$. Hence it follows from Corollary 4.10 that the h-μ rule minimizes the total waiting cost per unit time,

$$\sum_{i \in M} \lambda_i h_i W_i^\phi(\boldsymbol{\lambda}) \,,$$

over all $\phi \in \Phi_m$, for all arrival-rate vectors, $\boldsymbol{\lambda}$.

In this case we have

$$\alpha_S = \left(\frac{\sum_{i \in S} \rho_i}{1 - \sum_{i \in S} \rho_i} \right) \sum_{i \in M} \frac{\rho_i \beta_i}{\mu_i} + \sum_{i \in S} \frac{\rho_i}{\mu_i} \,, \quad S \subseteq M \,. \tag{4.106}$$

In order to apply Theorem 4.12, we must verify that α_S is jointly convex in $\boldsymbol{\lambda}$ for all $S \subseteq M$. It is easily verified that, although α_M is jointly convex in $\boldsymbol{\lambda}$, α_S is *not* jointly convex for any $S \subset M$.

4.7.3.4 Application: M/GI/1 Model with Preemptive-Resume Discipline and Exponential Work Requirements

Consider a multiclass $M/GI/1$ queue with a strict preemptive-resume priority discipline, in which the priority ordering is according to the h-μ rule. Suppose the work requirements in each class are exponentially distributed. Let Φ_m denote the set of all nonanticipative, nonidling, regenerative scheduling rules which are preemptive-resume and service-time independent within each class. It is shown in Section A.4.0.2 of Appendix A that strong conservation laws hold for the vector, $(W_1, \ldots, W_m)$, for all $\phi \in \Phi_m$, where

$$W_i = \mathrm{E}[\boldsymbol{W}_i] \,,$$

the expected steady-state waiting time in the system for class i, $i \in M$. Hence it follows from Corollary 4.10 that the h-μ rule minimizes the total waiting cost per unit time,

$$\sum_{i \in M} \lambda_i h_i W_i^\phi(\boldsymbol{\lambda}) \,,$$

over all $\phi \in \Phi_m$, for all arrival-rate vectors, $\boldsymbol{\lambda}$.

In this case we have

$$\alpha_S = \left(\frac{1}{1 - \sum_{i \in S} \rho_i} \right) \sum_{i \in S} \frac{\rho_i}{\mu_i} \,, \quad S \subseteq M \,. \tag{4.107}$$

In order to apply Theorem 4.12, we must verify that α_S is jointly convex in $\boldsymbol{\lambda}$ for all $S \subseteq M$. Since $(\rho_1, \ldots, \rho_m)$ is linear in $\boldsymbol{\lambda}$, it suffices to show that the expression (4.107) for α_S is jointly convex in $(\rho_1, \ldots, \rho_m)$. This is easily done for the special case in which μ_i does not depend on i. We therefore have the following theorem.

Theorem 4.13 *Consider a multiclass* M/GI/1 *queue operating under the nonpreemptive h-μ rule. Suppose the work requirements in each class are exponentially distributed and that the mean μ_i^{-1} does not depend on $i \in M$: $\mu_i = \mu$, $i \in M$. Then the total-waiting-cost function,*

$$\sum_{i \in M} \lambda_i h_i W_i^\phi(\boldsymbol{\lambda}) \,,$$

is jointly convex in $\boldsymbol{\lambda} \in A$, and hence the objective function for the social optimization problem,

$$\mathcal{U}(\boldsymbol{\lambda}) = \sum_{i \in M} U_i(\lambda_i) - \sum_{i \in M} \lambda_i h_i W_i^\phi(\boldsymbol{\lambda}) \,,$$

is jointly concave in $\boldsymbol{\lambda} \in A$.

4.7.3.5 Numerical Examples

We now give several numerical examples for a multiclass queue operating under a priority discipline. We shall focus exclusively on a two-class M/GI/1 queue with linear waiting costs, a preemptive-resume discipline, and exponential work requirements. We assume equal service rates ($\mu_i = \mu$, $i \in M$), so that we have a multiclass M/M/1 queue. As usual we shall restrict attention to the case $m = 2$. Without loss of generality, we assume that $\mu = 1$.

Our examples confirm the results in the previous section: that the objective function for social optimization is always jointly concave in $\boldsymbol{\lambda}$ when the h-μ rule is used. We also give examples in which the priority discipline in effect is *not* the h-μ rule and show that in this case the objective function for social optimization may not be jointly concave in $\boldsymbol{\lambda}$.

We shall restrict attention to the case of linear utility functions. Suppose $U_i(\lambda_i) = r_i \lambda_i$, where $r_i \geq 0$, $i = 1, 2$.

The social optimization problem takes the form:

$$\max_{\{\lambda_1, \lambda_2\}} \quad r_1 \lambda_1 - \frac{h_1 \lambda_1}{1 - \lambda_1} + r_2 \lambda_2 - \frac{h_2 \lambda_2}{(1 - \lambda_1)(1 - \lambda_1 - \lambda_2)}$$

$$\text{s.t.} \quad \lambda_1 + \lambda_2 < 1$$

$$\lambda_1 \geq 0 \,, \ \lambda_2 \geq 0 \,.$$

For our first example, we shall take $r_1 = 16$, $r_2 = 9$, $h_1 = 4$, and $h_2 = 1$. Figure 4.5 shows the graph of the response surface, $\{\mathcal{U}(\lambda_1, \lambda_2), \lambda_1 \geq 0, \lambda_2 \geq 0\}$. In this case, since $h_1 > h_2$, the h-μ rule is in effect and Figure 4.5 confirms what we know already (from Theorem 4.13): the objective function, $\mathcal{U}(\lambda_1, \lambda_2)$, is jointly concave in $\boldsymbol{\lambda} = (\lambda_1, \lambda_2)$.

Compare this example to the *FIFO* example with the same parameters, presented in Section 4.6.1. As we saw there (cf. Figure 4.3), the objective function, $\mathcal{U}(\lambda_1, \lambda_2)$, is *not* jointly concave in $\boldsymbol{\lambda} = (\lambda_1, \lambda_2)$, under the *FIFO* queue discipline.

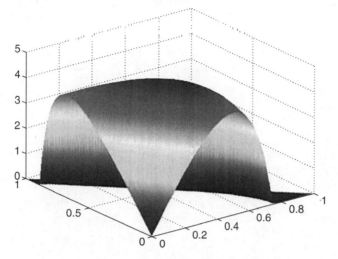

Figure 4.5 *Linear Utility Functions:* $\quad \mathcal{U}(\lambda_1, \lambda_2) = 16\lambda_1 - 4\lambda_1/(1 - \lambda_1) + 9\lambda_2 - \lambda_2/((1 - \lambda_1)(1 - \lambda_1 - \lambda_2))$

For our second example, we shall take $r_1 = 4$, $r_2 = 6$, $h_1 = .4$, and $h_2 = 1$. Figure 4.6 shows the graph of the response surface, $\{\mathcal{U}(\lambda_1, \lambda_2), \lambda_1 \geq 0, \lambda_2 \geq 0\}$. In this case, since $h_1 < h_2$, the h-μ rule is *not* in effect and Figure 4.5 demonstrates that the objective function, $\{\mathcal{U}(\lambda_1, \lambda_2)\}$, is *not* jointly concave in $\boldsymbol{\lambda} = (\lambda_1, \lambda_2)$.

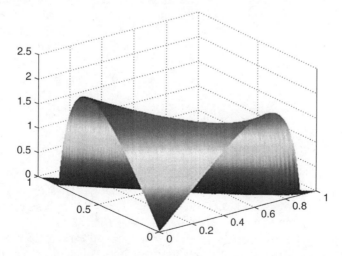

Figure 4.6 *Linear Utility Functions:* $\quad \mathcal{U}(\lambda_1, \lambda_2) = 4\lambda_1 - .4\lambda_1/(1 - \lambda_1) + 6\lambda_2 - \lambda_2/((1 - \lambda_1)(1 - \lambda_1 - \lambda_2))$

Thus we see that using a (preemptive-resume) priority discipline leads to a jointly concave objective function (in contrast to the *FIFO* case), if the priorities are generated by the h-μ rule. On the other hand, if the priorities are not generated by the h-μ rule, then the objective function may not be jointly concave.

4.8 Endnotes

Section 4.1.1

See Gross and Harris [79], Kleinrock [113], or Heyman and Sobel [92] for derivations of the formula (4.1) for the delay in an $M/GI/1$ queue with *FIFO* discipline (the Pollaczek-Khintchine formula), the formula (4.6) for the delay in an $M/GI/1$ queue with a nonpreemptive priority queue, and the formula (4.10) for the delay in an $M/GI/1$ queue with a preemptive-resume priority queue.

Section 4.2.2.1

The analysis of this section is based on Rump and Stidham [169, 171] and uses the approach of Rosen [161] to establish the existence and uniqueness of the fixed point equilibrium. This approach depends on the concavity of the objective function and the convexity of the feasible region for each class. Alternative methods use monotonicity and/or submodularity of the objective functions: see, e.g., Topkis [193], Yao [199].

The uniqueness of individually optimal and class optimal solutions both depend rather delicately on the model assumptions. Assumption 4 is particularly crucial, in that it ensures that an equilibrium cannot occur at a point where the system capacity is fully utilized. This assumption is mild in the context of classical queueing models. By contrast, models in the theory of road traffic flow often allow full utilization of capacity. In such a setting there may be more than one individually optimal equilibrium and/or more than one class optimal equilibrium. See, e.g., Marcotte and Wynter [140]. Sheffi [178] is a good general reference on the theory of road traffic flow. See also Nagourney [150] for a more mathematical treatment, emphasizing equilibria and variational inequalities.

Section 4.4.3

Fridgeirsdottir and Akella [68] give an example with a nonconcave objective function in the context of product portfolio selection. See also Fridgeirsdottir and Chiu [69] for a discussion of when the waiting cost is a convex function of the arrival rate in queueing models.

The phenomenon of class dominance was apparently first discussed in the queueing theory literature by Balachandran and colleagues (see, e.g., Balachandran and Schaefer [14]).

The net utility (utility minus waiting cost) in our model is a special case of the utility function proposed by Mackie-Mason and Varian [136] for a model of a communication network. The analysis in [136] relies entirely on an implicit assumption that the optimal flow allocation is an interior solution to the first-order optimality conditions. The authors do not consider the fact that this assumption may not be valid. Indeed, our analysis shows that in an important subclass of models it is *never* valid.

Section 4.5.2

The material in this section is based on Rump and Stidham [169, 171] and Zhang and Douligeris [200].

Section 4.7.3.4

The optimality of the h-μ rule is a folk theorem with a long history in queueing theory. An early proof, based on the explicit solution for the expected waiting times in a steady-state $M/M/1$ queue with fixed priorities, is in Cox and Smith [45]. Harrison [83], [82] generalized this result to the case of discounted costs. A particularly powerful approach for establishing the optimality of index rules, such as the h-μ rule, is the achievable-region method based on strong conservation laws (cf. the Appendix in this volume and the survey by Bertsimas [21]). For a recent comprehensive text on scheduling, see Pinedo and Chao [157].

Mendelson and Whang [144] consider optimal design of an $M/M/1$ priority queue. They give conditions under which the optimal prices are *incentive compatible*, in the sense that customers in a particular class find it in their own interest to declare correctly their class membership. Kim and Mannino [110] extend these results to an $M/GI/1$ priority queue.

Argon and Ziya [7] also consider optimal design of a priority queue. In their model the service provider has imperfect knowledge of the characteristics of each customer, represented by *signals*. The service provider uses these signals to determine priority levels for the customers with the objective of minimizing waiting costs.

4.9 Figures for *FIFO* Examples

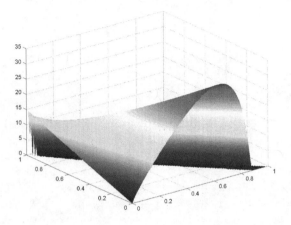

Figure 4.7 *Square-Root Utility Functions:* $\mathcal{U}(\lambda_1, \lambda_2) = 64\lambda_1 + 8\sqrt{\lambda_1} - 9\lambda_1/(1 - \lambda_1 - \lambda_2) + 15\lambda_2$

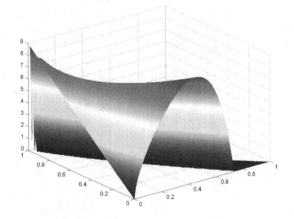

Figure 4.8 *Square-Root Utility Functions:* $\mathcal{U}(\lambda_1, \lambda_2) = 24\lambda_1 + 8\sqrt{\lambda_1} - 9\lambda_1/(1 - \lambda_1 - \lambda_2) + 9\lambda_2$

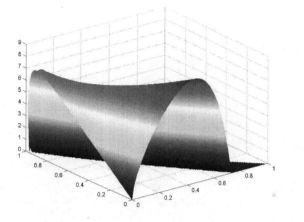

Figure 4.9 *Square-Root Utility Functions:* $\mathcal{U}(\lambda_1, \lambda_2) = 24\lambda_1 + 8\sqrt{\lambda_1} - 9\lambda_1/(1 - \lambda_1 - \lambda_2) + 9\lambda_2 - 0.1\lambda_2/(1 - \lambda_1 - \lambda_2)$

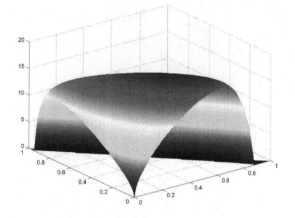

Figure 4.10 *Square-Root Utility Functions:* $\mathcal{U}(\lambda_1, \lambda_2) = 16\lambda_1 + 16\sqrt{\lambda_1} - 4\lambda_1/(1 - \lambda_1 - \lambda_2) + 9\lambda_2 + 9\sqrt{\lambda_2} - \lambda_2/(1 - \lambda_1 - \lambda_2)$

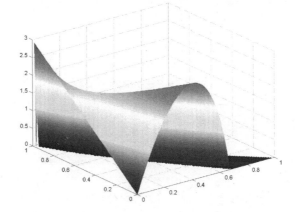

Figure 4.11 *Logarithmic Utility Functions:* $\quad \mathcal{U}(\lambda_1, \lambda_2) = 16\log(1 + \lambda_1) - 4\lambda_1/(1 - \lambda_1 - \lambda_2) + 3\lambda_2$

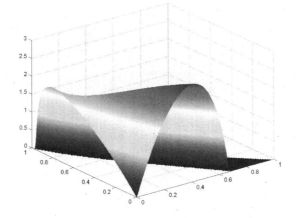

Figure 4.12 *Logarithmic Utility Functions:* $\quad \mathcal{U}(\lambda_1, \lambda_2) = 16\log(1 + \lambda_1) - 4\lambda_1/(1 - \lambda_1 - \lambda_2) + 4\log(1 + \lambda_2) - 0.1\lambda_2/(1 - \lambda_1 - \lambda_2)$

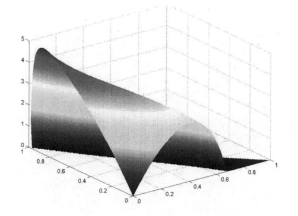

Figure 4.13 *Logarithmic Utility Functions:* $\mathcal{U}(\lambda_1, \lambda_2) = 16\log(1 + \lambda_1) - 4\lambda_1/(1 - \lambda_1 - \lambda_2) + 9\log(1 + \lambda_2) - 0.1\lambda_2/(1 - \lambda_1 - \lambda_2)$

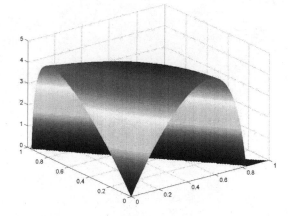

Figure 4.14 *Logarithmic Utility Functions:* $\mathcal{U}(\lambda_1, \lambda_2) = 16\log(1 + \lambda_1) - 2\lambda_1/(1 - \lambda_1 - \lambda_2) + 9\log(1 + \lambda_2) - 0.25\lambda_2/(1 - \lambda_1 - \lambda_2)$

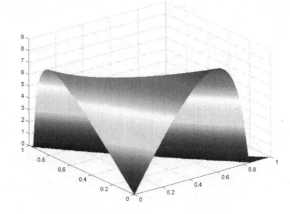

Figure 4.15 *Quadratic Utility Functions:* $\mathcal{U}(\lambda_1, \lambda_2) = 75\lambda_1 - \lambda_1^2 - 4\lambda_1/(1 - \lambda_1 - \lambda_2) + 14\lambda_2 - 0.05\lambda_2^2 - 0.5\lambda_2/(1 - \lambda_1 - \lambda_2)$

CHAPTER 5

Optimal Service Rates in a Single-Class Queue

In this chapter we turn our attention to models in which the service rate is a decision variable. Typically it is the operator of the service facility who is able to choose the service rate, so we shall usually speak of the decision maker and the facility operator as one and the same. The models differ according to whether or not the arrival rate is a decision variable (directly or indirectly) .

In Section 5.1 we introduce our general model, in which both the arrival rate and the service rate may be decision variables. This model may be construed as a generalization (to allow a variable service rate) of the model of Chapter 2 for selecting the optimal arrival rate. From the development of the model of Chapter 2, we know that the choice of λ for any fixed value of μ can be implemented in a setting of individually optimizing customers by an appropriate choice of an admission fee or toll, δ. Thus, an alternate version of our model is one in which the system controller chooses both the service rate and the admission fee.

One may call the problem of choosing the service rate the *long-run* problem, and the optimal-arrival-rate problem of Chapter 2 (in which the service rate is fixed) the *short-run* problem. The motivation for this distinction is that in applications the time scale on which adjustments in service capacity are possible is usually (much) longer than that required for adjustments in the arrival rate, for example by changing the toll charged for admission.

In subsequent sections we consider different versions of the general model, depending on whether the arrival rate and/or the toll are fixed or variable. We also consider different optimality criteria, specifically, social optimality, in which the decision maker wishes to maximize the total net utility, and facility optimality, in which a facility operator is interested only in maximizing his own net profit – the difference between the admission fees received and the service cost incurred per unit time.

First (in Sections 5.2 and 5.3) we consider models in which the arrival rate is fixed and the decision maker is either a social optimizer or a profit maximizer. Then we consider models with a variable arrival rate. In the first of these models (Section 5.4) the admission fee is fixed and the service rate is a decision variable. The arrival rate is the equilibrium arrival rate associated with individually optimizing customers, given the toll and the service rate. Again the solution differs depending on whether the decision maker is a social optimizer or a profit maximizer. We compare these two solutions and show that the resulting service rate is smaller in the case of profit maximization, as expected. Moreover, because the admission fee is fixed, the socially optimal solution will in general under-perform relative to the socially optimal solution in which the admission fee is a decision variable.

Finally (in Section 5.5) we consider the problem in which both the service rate and the admission fee are decision variables. We consider both social and facility optimality in this case. We introduced a simple version of this model in Section 1.3 of Chapter 1, in which we observed that the objective function for social optimality may not be jointly concave in the arrival rate and service rate and therefore may have multiple local maxima. We investigate this phenomenon in more detail in this section.

5.1 The Basic Model

As in Chapter 2, we consider a service facility operating over a finite or infinite time interval. Once again, rather than specify a particular queueing model, we shall describe the system in general terms, keeping structural and stochastic assumptions at a minimum. The basic ingredients are:

- the arrival rate λ – the average number of customers entering the system per unit time (a possible decision variable);

- the average gross utility per unit time, $U(\lambda)$;

- a single server, who serves customers one at a time, and is never idle when customers are present;

- the service rate μ – the average number of services completed per unit time while the server is busy (a decision variable);

- the average waiting cost per customer, $G(\lambda, \mu)$;

- the admission fee or toll, δ, paid by each entering customer (a possible decision variable);

- the average service cost per unit time, $c(\mu)$.

As usual, the meaning of the word "average" depends on the specific model context. For example, it may mean a sample-path time average or (in the case of an infinite time period) the expectation of a steady-state random variable.

The arrival rate λ measures the average number of customers arriving and joining the system per unit time during the period of interest. Initially we shall assume that λ is fixed, but later we shall consider models in which λ is a decision variable, as well as μ.

As usual, to measure the benefit to the system of having a higher through-put, we assume the existence of a utility function, $U(\lambda)$, that measures the average gross value received per unit time by the system as a function of the arrival rate λ, $\lambda \in A$. Our default assumption is that $A = [0, \infty)$ and $U(\lambda)$ is nondecreasing, differentiable, and concave in $\lambda \in A$.

The variable μ measures the average number of services completed per unit time while the server is busy. For example, when the service facility is a

$GI/GI/1$ queue, μ is the service rate of the single server (the reciprocal of the expected service time). Throughout this chapter, μ will be a decision variable, chosen from a feasible region $B \subseteq [0, \infty)$. Our default assumption is that $B = [0, \infty)$.

Balanced against the benefit of throughput is the cost to customers caused by the time they spend in the system. For given λ and μ, $G(\lambda, \mu)$ denotes the average waiting cost of a customer, averaged over all customers who arrive during the period in question. For example, it might be that $G(\lambda, \mu) = E[h(\boldsymbol{W}(\lambda, \mu))]$, where $h(t)$ is the waiting cost incurred by a job that spends a length of time t in the system and $\boldsymbol{W}(\lambda, \mu)$ has the steady-state distribution of waiting time in the system for the queueing system induced by λ and μ. We shall assume that $G(\lambda, \mu)$ is well defined and finite and differentiable in (λ, μ), $0 \le \lambda < \mu$, with $G(\lambda, \mu) = \infty$ for $\lambda \ge \mu \ge 0$. We also shall assume that

(i) for each fixed $\mu > 0$, $G(\lambda, \mu)$ is strictly increasing in λ, $0 \le \lambda < \mu$, and $G(\lambda, \mu) \to \infty$ as $\lambda \uparrow \mu$;

(ii) for each fixed $\lambda \ge 0$, $G(\lambda, \mu)$ is strictly decreasing in μ, $\mu > \lambda$, and $G(\lambda, \mu) \to \infty$ as $\mu \downarrow \lambda$.

For example, in an $M/M/1$ queue operating in steady state with a linear waiting cost $h(t) = h \cdot t$, we have

$$G(\lambda, \mu) = \frac{h}{\mu - \lambda}, \ 0 \le \lambda < \mu,$$

$$G(\lambda, \mu) = \infty, \ \lambda \ge \mu \ge 0.$$

In this case it is easy to verify that $G(\lambda, \mu)$ satisfies the above assumptions.

The toll, or admission fee, δ, is charged to each entering customer. As usual, the decision maker may be able to control the arrival rate indirectly by charging an appropriate toll, assuming individually optimizing customers.

Finally, the service-cost rate $c(\mu)$ measures the average cost per unit time associated with operating the server at rate μ. We assume that $c(\cdot)$ is a nondecreasing and differentiable function of $\mu \in A \subseteq [0, \infty)$, with $c(0) = 0$ and $c(\mu) \to \infty$ as $\mu \to \infty$. Our default assumption is that $c(\mu)$ is convex in $\mu \in A$, but we also consider the case of a concave service-cost function (cf. Section 5.5.2.3).

Remark 1   Let $H(\lambda, \mu) := \lambda G(\lambda, \mu)$, the average waiting cost per unit time. It follows from the assumptions about $G(\lambda, \mu)$ that $G(0, 0) = \infty$, and hence $H(0, 0)$ is indeterminate (taking the form $0 \cdot \infty$). Indeed, since $G(\lambda, \mu) < \infty$ for all $0 \le \lambda < \mu$ and $G(\lambda, \mu)$ decreases as $\lambda \downarrow 0$, we have $\lim_{\lambda \to 0} H(\lambda, \mu) = 0$ for all $\mu > 0$, so that

$$\lim_{\mu \to 0} \lim_{\lambda \to 0} H(\lambda, \mu) = 0.$$

On the other hand, since $\lim_{\mu\downarrow\lambda} G(\lambda,\mu) = \infty$ for each fixed $\lambda \geq 0$, we have $\lim_{\mu\downarrow\lambda} H(\lambda,\mu) = \infty$ for each fixed $\lambda \geq 0$, so that

$$\lim_{\lambda\to 0}\lim_{\mu\downarrow\lambda} H(\lambda,\mu) = \infty .$$

In terms of its economic interpretation, it makes sense to let $H(0,0) := 0$, and this is what we shall do. But note that, inevitably, $H(\lambda,\mu)$ has a discontinuity at $(0,0)$. Indeed, $H(\lambda,\mu)$ assumes arbitrarily large values in every neighborhood of $(0,0)$ in $\{(\lambda,\mu) : 0 \leq \lambda < \mu\}$.

These properties are exhibited graphically in our simplest example – an *M/M/1* queue with linear waiting cost function – in which

$$H(\lambda,\mu) = \frac{h\lambda}{\mu - \lambda} , \; 0 \leq \lambda < \mu .$$

In this case, we can write $H(\lambda,\mu) = f(\lambda/\mu)$, where $f(\rho) := h\rho/(1 - \rho)$, $0 \leq \rho < 1$. For any fixed ρ, $0 \leq \rho < 1$, the function $H(\lambda,\mu)$ is constant and equals $f(\rho)$ along the line $\mu = \lambda/\rho$ in the λ-μ plane. The limit of $H(\lambda,\mu)$ as $\lambda \to 0$ along this line is therefore also equal to $f(\rho)$. But as ρ increases from 0 to 1, $f(\rho)$ assumes every value between 0 and ∞. Therefore, in every neighborhood of $(0,0)$, $H(\lambda,\mu)$ assumes every value between 0 and ∞. (See Figure 5.1.)

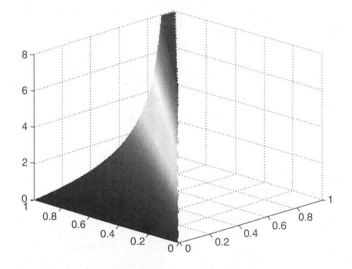

Figure 5.1 M/M/1 *Queue: Graph of* $H(\lambda,\mu)$ *(h = 1)*

Of course, this property is not limited to the *M/M/1* example, but rather holds for any system in which $H(\lambda,\mu) = f(\lambda/\mu)$, where $f(\rho)$ is a continuous, increasing function of $\rho \in [0, 1)$, with $f(0) = 0$ and $\lim_{\rho\to 1} f(\rho) = \infty$ and $f(\rho) = \infty$ for $\rho \geq 1$. In particular, it holds for many queueing systems with a linear waiting-cost function in which λ and μ are scale factors. In such systems, $H(\lambda,\mu) = h \cdot L(\lambda,\mu)$, where $L(\lambda,\mu)$ is the average number of customers in the system, which is independent of the time unit. An example is a $GI/GI/1$ queue with interarrival times distributed as X/λ ($E[X] = 1$) and service times distributed as Y/μ ($E[Y] = 1$).

Another important – and perhaps surprising – property of such systems is that $H(\lambda, \mu)$ is *never* jointly convex in (λ, μ), even when $f(\rho)$ is convex in ρ and even though λ/μ is convex in λ and convex in μ. This fact (proved in Lemma 5.1 below) complicates the analysis of the socially optimal solution when both the toll and the arrival rate (as well as the service rate) are decision variables (see Section 5.5), as there may be multiple solutions to the necessary *KKT* conditions.

Lemma 5.1 *Suppose $H(\lambda, \mu) = f(\lambda/\mu)$, where $f(\rho)$ is differentiable, convex, and strictly increasing in ρ, $\rho \in [0, 1)$. Then $H(\lambda, \mu)$ is differentiable, convex, and strictly increasing in λ and differentiable, convex, and strictly decreasing in μ. However, $H(\lambda, \mu)$ is not jointly convex in (λ, μ).*

Proof   Let $g(\lambda, \mu) := \lambda/\mu$, $0 \le \lambda < \mu$. Then $H(\lambda, \mu) = f(g(\lambda/\mu))$ and we have:

$$\frac{\partial H}{\partial \lambda} = f' \cdot \frac{\partial g}{\partial \lambda} ; \quad \frac{\partial^2 H}{\partial \lambda^2} = f'' \cdot \left(\frac{\partial g}{\partial \lambda}\right)^2 + f' \cdot \frac{\partial^2 g}{\partial \lambda^2} ;$$

$$\frac{\partial H}{\partial \mu} = f' \cdot \frac{\partial g}{\partial \mu} ; \quad \frac{\partial^2 H}{\partial \mu^2} = f'' \cdot \left(\frac{\partial g}{\partial \mu}\right)^2 + f' \cdot \frac{\partial^2 g}{\partial \mu^2} ;$$

$$\frac{\partial^2 H}{\partial \lambda \partial \mu} = f'' \cdot \left(\frac{\partial g}{\partial \lambda}\right)\left(\frac{\partial g}{\partial \mu}\right) + f' \cdot \frac{\partial^2 g}{\partial \lambda \partial \mu}$$

where

$$\frac{\partial g}{\partial \lambda} = \frac{1}{\mu} \ge 0 , \quad \frac{\partial^2 g}{\partial \lambda^2} = 0$$

$$\frac{\partial g}{\partial \mu} = -\frac{\lambda}{\mu^2} \le 0 , \quad \frac{\partial^2 g}{\partial \mu^2} = \frac{2\lambda}{\mu^3} \ge 0 ,$$

$$\frac{\partial^2 g}{\partial \lambda \partial \mu} = -\frac{1}{\mu^4} .$$

Thus, $g(\lambda, \mu) = \lambda/\mu$ is convex in λ and convex in μ, but not jointly convex in (λ, μ), because

$$\left(\frac{\partial^2 g}{\partial \lambda^2}\right)\left(\frac{\partial^2 g}{\partial \mu^2}\right) - \left(\frac{\partial^2 g}{\partial \lambda \partial \mu}\right)^2 = -\frac{1}{\mu^4} < 0 . \tag{5.1}$$

Moreover,

$$\frac{\partial H}{\partial \lambda} > 0 , \quad \frac{\partial^2 H}{\partial \lambda^2} \ge 0 ,$$

$$\frac{\partial H}{\partial \mu} < 0 \quad \frac{\partial^2 H}{\partial \mu^2} \ge 0 .$$

Hence, $H(\lambda, \mu)$ is strictly increasing and convex in λ and strictly decreasing and convex in μ. Moreover,

$$
\begin{aligned}
\Delta \; := \; & \left(\frac{\partial^2 H}{\partial \lambda^2}\right)\left(\frac{\partial^2 H}{\partial \lambda^2}\right) - \left(\frac{\partial^2 H}{\partial \lambda \partial \mu}\right)^2 \\
= \; & \left(f'' \cdot \left(\frac{\partial g}{\partial \lambda}\right)^2 + f' \cdot \frac{\partial^2 g}{\partial \lambda^2}\right)\left(f'' \cdot \left(\frac{\partial g}{\partial \mu}\right)^2 + f' \cdot \frac{\partial^2 g}{\partial \mu^2}\right) \\
& - \left(f'' \cdot \left(\frac{\partial g}{\partial \lambda}\right)\left(\frac{\partial g}{\partial \mu}\right) + f' \cdot \frac{\partial^2 g}{\partial \lambda \partial \mu}\right)^2 \\
= \; & (f'' \cdot f')\left[\left(\frac{\partial g}{\partial \lambda}\right)^2 \frac{\partial^2 g}{\partial \mu^2} - 2\left(\frac{\partial g}{\partial \mu}\right)\left(\frac{\partial g}{\partial \lambda}\right)\frac{\partial^2 g}{\partial \lambda \partial \mu} + \frac{\partial^2 g}{\partial \lambda^2}\left(\frac{\partial g}{\partial \mu}\right)^2\right] + \\
& (f')^2 \left[\left(\frac{\partial^2 g}{\partial \lambda^2}\right)\left(\frac{\partial^2 g}{\partial \mu^2}\right) - \left(\frac{\partial^2 g}{\partial \lambda \partial \mu}\right)^2\right].
\end{aligned}
$$

The first term in brackets equals zero, whereas the second term in brackets equals $-1/\mu^4 < 0$ (cf. (5.1). Thus, $H(\lambda, \mu)$ fails the condition for joint convexity at all points, (λ, μ), $0 \le \lambda < \mu$. ∎

Note that it is the failure of the function $g(\lambda, \mu) = \lambda/\mu$ to be jointly convex that leads to $H(\lambda, \mu)$ failing to be jointly convex, regardless of the properties of the function $f(\rho)$.

5.2 Models with Fixed Toll and Fixed Arrival Rate

The model we study in this section is a generalization of the model introduced in Section 1.1 of the introductory chapter. The arrival rate, λ, and the toll, δ, are both fixed ($\lambda > 0$, $\delta \ge 0$), and the service rate μ is a decision variable. As usual, the solution to the decision problem depends on who is making the decision and what costs and benefits are taken into account. Since the arrival rate λ is fixed, so is the gross utility, $U(\lambda)$, received per unit time by all customers. The relevant economic factors are thus the service-cost rate $c(\mu)$ and the waiting cost $G(\lambda, \mu)$.

5.2.1 Individual Optimality

Since the arrival rate and toll are fixed, the criterion of individual optimality is not relevant in the present context. (Later, however, we shall discuss how the case of a fixed arrival rate and fixed toll can arise as a special case of a problem with fixed toll and variable arrival rate and individually optimizing customers, through appropriate choice of the utility function, $U(\lambda)$. See Section 5.4.1.)

5.2.2 Social Optimality

When the decision criterion is social optimality, the objective function, to be minimized, is given by

$$C(\mu) := c(\mu) + \lambda G(\lambda, \mu) \ , \ \mu \geq 0 \ .$$

That is, $C(\mu)$ measures the expected steady-state total cost per unit time, when service rate μ is chosen. Under our basic convexity and differentiability assumptions, it follows that $C(\mu)$ is convex and differentiable in $\mu > \lambda$ and that $C(\mu) \to \infty$ as $\mu \downarrow \lambda$ as as $\mu \uparrow \infty$. Therefore, the first-order optimality condition, $C'(\mu) = 0$, is both necessary and sufficient for μ to be socially optimal. In other words, the socially optimal service rate, μ^s, is the unique solution in (λ, ∞), to the equation

$$C'(\mu) = c'(\mu) + \lambda \frac{\partial}{\partial \mu} G(\lambda, \mu) = 0 \ . \tag{5.2}$$

5.2.2.1 Example: Simple Linear-Cost Model

Consider again the simple linear-cost model introduced in Section 1.1 of Chapter 1, in which we have an $M/M/1$ queue with fixed arrival rate λ and variable service rate μ. In this model the service-cost rate is linear in the service rate,

$$c(\mu) = c \cdot \mu \ , \ \mu \geq 0 \ ,$$

and the waiting cost per customer is linear in the time spent in the system, so that

$$G(\lambda, \mu) = \frac{h}{\mu - \lambda} \ , \ 0 \leq \lambda < \mu \ .$$

Then

$$C(\mu) = c \cdot \mu + h \cdot \left(\frac{\lambda}{\mu - \lambda} \right) \ ,$$

and the optimization problem takes the form:

$$\min_{\mu : \mu > \lambda} C(\mu) = \min_{\mu : \mu > \lambda} \left\{ c \cdot \mu + h \cdot \left(\frac{\lambda}{\mu - \lambda} \right) \right\} \ . \tag{5.3}$$

In this case equation (5.2) takes the form,

$$C'(\mu) = c - h \left(\frac{\lambda}{(\mu - \lambda)^2} \right) = 0 \ . \tag{5.4}$$

This yields the following expression for the unique socially optimal value of the service rate, μ^s:

$$\mu^s = \lambda + \sqrt{\frac{\lambda h}{c}} \ . \tag{5.5}$$

5.2.3 Facility Optimality

When the decision criterion is facility optimality, the decision maker (facility operator) wants to maximize the following objective function:

$$\lambda\delta - c(\mu) \ .$$

Since both λ and δ are fixed, so is the gross revenue, $\lambda\delta$, earned by the facility. Thus there is no incentive for the facility operator to increase the service rate μ. In fact, the facility operator can increase net revenue by decreasing μ, subject to the feasibility constraint, $\mu > \lambda$. Thus, there is no attainable optimal value of μ. Rather, the objective function is bounded above by $\lambda\delta$ and the decision maker can come arbitrarily close to attaining this upper bound by choosing μ arbitrarily close to λ.

Obviously this is not a particularly interesting or realistic model, inasmuch as it leads to counterintuitive behavior – behavior that can result in arbitrarily large waiting costs, $\lambda G(\lambda, \mu)$, since $G(\lambda, \mu) \to \infty$ as $\mu \downarrow \lambda$.

5.3 Models with Variable Toll and Fixed Arrival Rate

The model we study in this section has a fixed arrival rate, $\lambda > 0$, and a variable toll, $\delta \geq 0$. As usual, the service rate μ is a decision variable. Once again, the solution to the decision problem depends on who is making the decision and what costs and benefits are taken into account. As with the model of the previous section, since the arrival rate λ is fixed, so is the gross utility, $U(\lambda)$, received per unit time by all customers. The relevant economic factors are thus the service-cost rate $c(\mu)$ and the waiting cost $G(\lambda, \mu)$.

5.3.1 Individual Optimality

Since the arrival rate is fixed, the criterion of individual optimality is again not relevant.

5.3.2 Social Optimality

When the decision criterion is social optimality, the objective function, to be minimized, is given by

$$C(\mu) := c(\mu) + \lambda G(\lambda, \mu) \ .$$

The toll does not appear in this objective function and thus is irrelevant to social optimality. (One could argue that in this case the decision maker should set the toll equal to zero, since there is no economic benefit to the aggregate of all customers from charging a positive toll.) It follows that the analysis of the previous section still applies and the socially optimal service rate, μ^s, is the unique solution in (λ, ∞) to the equation

$$C'(\mu) = c'(\mu) + \lambda\frac{\partial}{\partial\mu}G(\lambda, \mu) = 0 \ .$$

5.3.3 Facility Optimality

When the decision criterion is facility optimality, the decision maker (facility operator) wants to maximize the following objective function:

$$\lambda\delta - c(\mu) \ .$$

Since λ is fixed, the gross revenue, $\lambda\delta$, can be made arbitrarily large by choosing an arbitrarily large value of δ. As in the model of the previous section, there is also no incentive for the facility operator to increase the service rate μ above its lower bound, λ.

Again this is not a particularly interesting or realistic model, since $G(\lambda, \mu) \uparrow \infty$ as $\mu \downarrow \lambda$.

5.4 Models with Fixed Toll and Variable Arrival Rate

The model we study in this section has a variable arrival rate, λ, and a fixed toll, δ. As usual, the service rate μ is a decision variable.

5.4.1 Individual Optimality

We now review the characterization of the equilibrium arrival rate, which is the arrival rate induced by individually optimizing customers, for the given fixed toll. (This material has been discussed in more detail in Section 2.1.1 of Chapter 2.)

For a given value of the service rate μ, the full price of admission, $\pi(\lambda)$, is given by

$$\pi(\lambda) = \delta + G(\lambda, \mu) \ , \ 0 \le \lambda < \mu \ .$$

Therefore, the equilibrium conditions, (2.5), (2.5), satisfied by the individually optimal arrival rate, λ^e, take the form

$$U'(\lambda) \ \le \ \delta + G(\lambda, \mu) \ , \text{ and } \lambda \ge 0 \ , \tag{5.6}$$
$$U'(\lambda) \ = \ \delta + G(\lambda, \mu) \ , \text{ if } \lambda > 0 \ . \tag{5.7}$$

It follows from our assumptions about $G(\lambda, \mu)$ that $\pi(\lambda)$ is strictly increasing and continuous. Therefore the equilibrium conditions have a unique solution, for any given μ.

We are interested in looking at the problem from the point of view of a decision maker who can choose the service rate μ. This choice, together with the fixed toll, δ, determines the value of the equilibrium arrival rate – the solution of equations (5.6) and (5.7). For example, since $G(\lambda, \mu)$ is a decreasing function of μ, increasing μ will result in an increase in the equilibrium arrival rate.

5.4.2 Social Optimality

In the case of social optimization with a fixed admission fee δ, variable arrival rate λ, and variable service rate μ, the problem takes the following form:

$$\max_{\{\lambda,\mu\}} \; \mathcal{U}(\lambda,\mu) \quad := \quad U(\lambda) - \lambda G(\lambda,\mu) - c(\mu)$$

$$\text{s.t.} \qquad \delta + G(\lambda,\mu) - U'(\lambda) \geq 0 \;,$$

$$\delta + G(\lambda,\mu) - U'(\lambda) = 0 \;,\; \text{if } \lambda > 0 \;,$$

$$\lambda \geq 0 \;.$$

One approach to solving this problem (formally) is to use the constraints to solve for λ in terms of μ. Let

$$\mu_0 := \inf\{\mu > 0 : U'(0) - G(0,\mu) > \delta\} \;.$$

We consider two cases.

Case 1 $\mu > \mu_0$: We have $U'(0) - G(0,\mu) > \delta$ and therefore, since $U'(\lambda) - G(\lambda,\mu)$ is continuous and strictly decreasing in λ and approaches $-\infty$ as $\lambda \uparrow \mu$, there exists a unique solution, $\lambda \in (0,\mu)$, to the equation,

$$U'(\lambda) - G(\lambda,\mu) = \delta \;.$$

Let $\lambda(\mu)$ denote this solution.

Case 2 $0 \leq \mu \leq \mu_0$: We have $U'(0) - G(0,\mu) \leq \delta$ and therefore, since $U'(\lambda) - G(\lambda,\mu)$ is strictly decreasing in λ, $U'(\lambda) - G(\lambda,\mu) < \delta$ for all $\lambda > 0$. In this case, let $\lambda(\mu) = 0$.

It is easy to verify that $\lambda(\mu)$, so defined, uniquely satisfies the constraints of the social optimization problem, for all $\mu \geq 0$.

Substituting $\lambda(\mu)$ for λ in the objective function, $\mathcal{U}(\lambda,\mu)$, for social optimization leads to the following optimization problem, with the single decision variable, μ:

$$\max_{\{\mu \geq 0\}} \psi(\mu) := \mathcal{U}(\lambda(\mu),\mu) = U(\lambda(\mu)) - \lambda(\mu)G(\lambda(\mu),\mu) - c(\mu) \;.$$

For $0 < \mu \leq \mu_0$, we have

$$\psi(\mu) = -U(0) - 0 \cdot G(0,\mu) - c(\mu) = -c(\mu) \leq -c(0) = 0 = \psi(0) \;.$$

(Recall that, by convention, $H(\lambda,\mu) = \lambda G(\lambda,\mu)$ equals zero at $\lambda = \mu = 0$.) Therefore, the maximum of $\psi(\mu)$ over $\mu \in [0,\mu_0]$ occurs at $\mu = 0$, at which $\psi(\mu) = 0$. It remains to consider the maximum of $\psi(\mu)$ over $\mu \in (\mu_0, \infty)$.

Let us therefore consider the problem:

$$\max_{\mu > \mu_0} \psi(\mu) = \mathcal{U}(\lambda(\mu),\mu) = U(\lambda(\mu)) - \lambda(\mu)G(\lambda(\mu),\mu) - c(\mu) \;.$$

Recall that, over the interval (μ_0, ∞), $\lambda(\mu)$ is the unique solution, $0 < \lambda < \mu$, to the equation, $U'(\lambda(\mu)) - G(\lambda(\mu),\mu) = \delta$. Differentiating $\psi(\mu)$ with respect to μ and setting the derivative equal to zero leads to the following necessary condition for the socially optimal value of μ over (μ_0, ∞)

$$\psi'(\mu) = U'(\lambda(\mu))\lambda'(\mu) - \lambda'(\mu)G(\lambda(\mu), \mu)$$
$$-\lambda(\mu)\left[\frac{\partial}{\partial\lambda}G(\lambda(\mu), \mu)\lambda'(\mu) + \frac{\partial}{\partial\mu}G(\lambda(\mu), \mu)\right] - c'(\mu) = 0 .$$

Collecting terms yields

$$\left[U'(\lambda(\mu)) - G(\lambda(\mu), \mu) - \lambda(\mu)\frac{\partial}{\partial\lambda}G(\lambda(\mu), \mu)\right]\lambda'(\mu)$$
$$-\lambda(\mu)\frac{\partial}{\partial\mu}G(\lambda(\mu), \mu) - c'(\mu) = 0 .$$

Finally, since $U'(\lambda(\mu)) - G(\lambda(\mu), \mu) = \delta$, an equivalent form for the optimality condition is

$$\left[\delta - \lambda(\mu)\frac{\partial}{\partial\lambda}G(\lambda(\mu), \mu)\right]\lambda'(\mu) = \lambda(\mu)\frac{\partial}{\partial\mu}G(\lambda(\mu), \mu) + c'(\mu) . \qquad (5.8)$$

We can derive an expression for $\lambda'(\mu)$ in terms of μ and $\lambda(\mu)$ by differentiating both sides of the equation which defines $\lambda(\mu)$, namely,

$$U'(\lambda(\mu)) - G(\lambda(\mu), \mu) = \delta ,$$

with respect to μ. We thus obtain the equation

$$\left[U''(\lambda(\mu)) - \frac{\partial}{\partial\lambda}G(\lambda(\mu), \mu)\right]\lambda'(\mu) - \frac{\partial}{\partial\mu}G(\lambda(\mu), \mu) = 0 ,$$

which, when solved for $\lambda'(\mu)$, yields

$$\lambda'(\mu) = f(\lambda(\mu), \mu) ,$$

where

$$f(\lambda, \mu) := \frac{-\frac{\partial}{\partial\mu}G(\lambda, \mu)}{\frac{\partial}{\partial\lambda}G(\lambda, \mu) - U''(\lambda)} . \qquad (5.9)$$

Note that $f(\lambda, \mu) > 0$, since $G(\lambda, \mu)$ is increasing in λ and decreasing in μ and $U(\lambda)$ is concave.

From this result and equation (5.8) we see that a socially optimal allocation (λ^s, μ^s), with $\mu^s > \mu_0$, satisfies the following necessary conditions:

$$\left[\delta - \lambda\frac{\partial}{\partial\lambda}G(\lambda, \mu)\right]f(\lambda, \mu) = \lambda\frac{\partial}{\partial\mu}G(\lambda, \mu) + c'(\mu) , \qquad (5.10)$$
$$U'(\lambda) - G(\lambda, \mu) = \delta . \qquad (5.11)$$

From our analysis (in Chapter 2) of the problem of finding the optimal arrival rate for a given μ, we recognize that the term,

$$\lambda\frac{\partial}{\partial\lambda}G(\lambda, \mu) ,$$

represents the external effect incurred when the arrival rate is λ. As such, it is the toll that *ought to be* charged each entering customer in order that individually optimizing customers will enter in a manner that results in an arrival rate that is socially optimal (for that particular value of μ). Thus, the term in brackets in equation (5.10),

$$\left[\delta - \lambda \frac{\partial}{\partial \lambda} G(\lambda, \mu) \right] ,$$

measures the discrepancy between the fixed toll, δ, and the external effect at (λ, μ). On the other hand, the necessary and sufficient condition for μ to be socially optimal, for a fixed value of λ, is that μ satisfy the equation

$$\lambda \frac{\partial}{\partial \mu} G(\lambda, \mu) + c'(\mu) = 0 .$$

But, since $G(\lambda, \mu)$ and $c(\mu)$ are both convex in μ, $\lambda \frac{\partial}{\partial \mu} G(\lambda, \mu) + c'(\mu) < 0 \ (= 0, > 0)$ if and only if μ is strictly less than (equal to, greater than) the socially optimal value of μ associated with λ.

Now, from equation (5.10) (and the fact that $f(\lambda, \mu) > 0$) we conclude that, at a socially optimal solution (λ^s, μ^s) to the problem with a fixed toll δ, with $\mu^s > \mu_0$, the following implications hold:

$$\lambda^s \frac{\partial}{\partial \mu} G(\lambda^s, \mu^s) + c'(\mu^s) \begin{Bmatrix} < \\ = \\ > \end{Bmatrix} 0 \Leftrightarrow \delta \begin{Bmatrix} < \\ = \\ > \end{Bmatrix} \lambda^s \frac{\partial}{\partial \lambda} G(\lambda^s, \mu^s) .$$

From these implications and the above observations, we have the following theorem.

Theorem 5.2 *Suppose $\mu^s > \mu_0$. Then μ^s (the socially optimal value of μ for a fixed toll δ) is strictly less than (equal to, greater than) the socially optimal value of μ with λ fixed at λ^s if and only if the fixed toll δ is strictly less than (equal to, greater than) the external effect at (λ^s, μ^s).*

For example, if the fixed toll is less than the external effect (in particular, if no toll is charged at all), then the facility operator will select a smaller service rate than would be socially optimal if customers were charged a toll equal to the external effect. As a result, the system will be underutilized from the point of view of social optimization.

On the other hand, when the fixed toll is equal to the external effect, then the optimal solution (λ^s, μ^s) satisfies the necessary conditions for a socially optimal solution to the problem with a variable toll (cf. Section 5.5 below). As we shall see, however, in that problem there may be multiple solutions to the necessary conditions for optimality, only one of which is globally optimal.

An equivalent representation for condition (5.10) may be obtained by substituting for $f(\lambda, \mu)$ from (5.9) and simplifying:

$$(\lambda U''(\lambda) - \delta) \frac{\partial}{\partial \mu} G(\lambda, \mu) = \left(\frac{\partial}{\partial \lambda} G(\lambda, \mu) - U''(\lambda) \right) c'(\mu) . \tag{5.12}$$

We shall find this version useful in the special case of a linear utility function.

5.4.2.1 Linear Utility Function

Consider the case of a linear utility function: $U(\lambda) = r \cdot \lambda$. To avoid trivialities, assume $r > \delta$. In this case, μ_0 is the unique solution to the equation

$$G(0, \mu) = r - \delta .$$

Let us first consider the optimal solution over the range, $\mu \in (\mu_0, \infty)$. In this case $U''(\lambda) = 0$ and

$$\lambda'(\mu) = f(\lambda(\mu), \mu) = \left(\frac{-\frac{\partial}{\partial \mu} G(\lambda(\mu), \mu)}{\frac{\partial}{\partial \lambda} G(\lambda(\mu), \mu)} \right) .$$

Therefore condition (5.12) implies

$$\delta \lambda'(\mu) = c'(\mu) . \tag{5.13}$$

Moreover, the other necessary condition, (5.11), which defines $\lambda(\mu)$, implies

$$G(\lambda(\mu), \mu) = r - \delta . \tag{5.14}$$

(Note the interesting implication of this equilibrium condition in this case: the waiting cost per customer is constant in μ.)

Assume for the moment that the necessary conditions, (5.13) and (5.14), are also sufficient. (Presently we shall exhibit examples in which they are *not* sufficient.) In principle, then, we can solve (5.14) to obtain an expression for $\lambda(\mu)$, then differentiate and substitute the resulting expression for $\lambda'(\mu)$ into (5.13) and solve for μ^s, the socially optimal value of μ. This value can then be substituted into the expression for $\lambda(\mu)$ to find λ^s, the socially optimal value of λ.

Example We now illustrate this approach in the special case of an *M/M/1* queue with a linear waiting cost, in which

$$G(\lambda, \mu) = \frac{h}{\mu - \lambda} .$$

In this case (still assuming $\mu > \mu_0$) it follows from (5.14) that

$$\lambda(\mu) = \mu - \frac{h}{r - \delta} ,$$

and hence $\lambda'(\mu) = 1$. Then it follows from (5.13) that an interior solution, $\mu_0 < \mu^s < \infty$, satisfies the equation

$$c'(\mu) = \delta .$$

Since $c'(\mu)$ is nondecreasing, we have three cases.

Case 1 $c'(\mu_0) \geq \delta$. In this case,

$$\psi'(\mu) = \frac{\partial}{\partial \lambda} G(\lambda(\mu), \mu)(\delta - c'(\mu)) \leq 0 , \text{ for all } \mu > \mu_0 ,$$

and hence the objective function, $\psi(\mu)$, for social optimality is nonincreasing for all $\mu > \mu_0$, so that $\mu^s = 0$ (and hence $\lambda^s = 0$).

Case 2 $c'(\mu_0) < \delta < \lim_{\mu \to \infty} c'(\mu)$. In this case, there exists a value of μ, $\mu_0 < \mu < \infty$, such that $c'(\mu) = \delta$. (We may resolve ties by taking the smallest value of μ such that $c'(\mu) = \delta$. Because the utility function, $U(\lambda(\mu))$, and the waiting cost function, $G(\lambda(\mu), \mu)$, are both constant in μ, all values of μ satisfying this equation yield the same value of the objective function for social optimality.) If the value of the objective function at this value of μ is positive, then this value of μ is optimal. Otherwise, $\mu^s = 0$ and hence $\lambda^s = 0$ as well.

Case 3. $\lim_{\mu \to \infty} c'(\mu) \leq \delta$. In this case,

$$\psi'(\mu) = \frac{\partial}{\partial \lambda} G(\lambda(\mu), \mu)(\delta - c'(\mu)) \geq 0 \; , \; \text{for all } \mu \geq 0 \; ,$$

and hence the objective function, $\psi(\mu)$, for social optimality is nondecreasing for all $\mu \geq 0$. If $c'(\mu) = \delta$ for all sufficiently large μ, then let μ_1 be the smallest value of μ such that $c'(\mu) = \delta$. On the other hand, if $c'(\mu) < \delta$, for all $\mu_0 < \mu < \infty$, then let $\mu_1 := \infty$. If $\psi(\mu_1) > 0$, then $\mu^s = \mu_1$. Otherwise, $\mu^s = 0$ and hence $\lambda^s = 0$ as well.

Figure 5.2 illustrates Case 2, with $c(\mu) = 0.01\mu^2$, $\delta = 1$, $h/(r - \delta) = 20$.

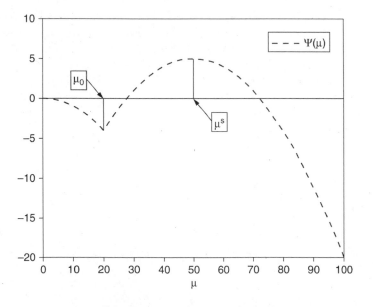

Figure 5.2 M/M/1 *Queue: Graph of* $\psi(\mu)$

Finally, we note that in the case of a linear service cost, $c(\mu) = c \cdot \mu$, we have (for $\mu > \mu_0$)

$$\psi(\mu) = (\delta - c)\mu - \frac{\delta h}{r - \delta} \; .$$

Thus, there are just two cases. For $c \geq \delta$, $\psi(\mu)$ is nonincreasing for $\mu > \mu_0$, so that $\mu^s = 0$ (and hence $\lambda^s = 0$). For $c < \delta$, $\psi(\mu)$ is strictly linearly increasing for $\mu > \mu_0$, so that $\mu^s = \infty$ (and hence $\lambda^s = \infty$). Note that there is never an interior optimal solution.

We now return to the general model with a linear utility function. As promised, we consider the question of when the necessary conditions, (5.13) and (5.14), are sufficient for a pair $(\lambda(\mu), \mu)$ to be socially optimal. Recall that in this case, the objective function for social optimality is

$$\psi(\mu) = \lambda(\mu)\delta - c(\mu) .$$

Therefore, we see that the necessary condition, (5.13), will be sufficient if $\lambda(\mu)$ is concave in $\mu > \mu_0$, where $\lambda(\mu)$ is uniquely defined by the other necessary condition, (5.14). (In this case, $\psi(\mu)$ will be concave and therefore will attain its maximum over $\mu > \mu_0$ at a point where $\psi'(\mu) = 0$, that is, where (5.13) holds.)

To find conditions under which $\lambda(\mu)$ is concave, we shall focus on the case where the waiting-cost function, $G(\lambda, \mu)$, satisfies the following assumption.

Assumption 1   The waiting-cost function, $G(\lambda, \mu)$, takes the form,

$$G(\lambda, \mu) = g(\lambda/\mu)/\mu , \ 0 \leq \lambda < \mu ,$$

where $g(\rho)$ is an increasing, continuous, and convex function of ρ, $0 \leq \rho \leq 1$, with $g(0) \geq 0$ and $g(\rho) = \infty$ for $\rho > 1$.

When this assumption is satisfied, the waiting cost per unit time, $H(\lambda, \mu) = \lambda G(\lambda, \mu)$, depends on λ and μ only through the ratio, λ/μ. That is,

$$H(\lambda, \mu) = f(\lambda/\mu) ,$$

where $f(\rho)$ is an increasing, convex function of $0 \leq \rho < 1$, with $f(0) = 0$. (We have $f(\rho) = \rho g(\rho)$.) We observed early in this chapter (cf. Section 5.1) that this property holds for many queueing systems with a linear waiting-cost function in which λ and μ are scale factors, including the example of an *M/M/1* queue.

Theorem 5.3   *Suppose $G(\lambda, \mu)$ satisfies Assumption 1 and $1/g(\rho)$ is concave (convex) in $\rho \in [0, 1)$. Then $\lambda(\mu)$ is concave (convex) in $\mu > \mu_0 = (r - \delta)^{-1}g(0)$.*

Proof   First note that Assumption 1 implies that

$$\mu > \mu_0 \Leftrightarrow \frac{g(0)}{\mu} < r - \delta \Leftrightarrow \mu > \frac{g(0)}{r - \delta} ,$$

so that $\mu_0 = (r - \delta)^{-1}g(0)$. Let $g^{-1}(x)$ be the inverse of $g(\rho)$, $x \geq g(0)$. Since $g(\rho)$ is strictly increasing and continuous, $g^{-1}(x)$ equals the unique value of ρ such that $g(\rho) = x$, $x \geq g(0)$. For $\mu > \mu_0$, $\lambda(\mu)$ equals the unique value of λ such that $G(\lambda, \mu) = r - \delta$, or, equivalently, the unique value of λ such that

$$g(\lambda/\mu) = (r - \delta)\mu .$$

Therefore,
$$\lambda(\mu) = \mu g^{-1}((r - \delta)\mu) \, ,$$

for $\mu > \mu_0$. It follows that

$$
\begin{aligned}
\lambda'(\mu) &= g^{-1}((r - \delta)\mu) + (r - \delta)\mu \cdot (g^{-1})'((r - \delta)\mu) \\
\lambda''(\mu) &= (r - \delta) \left[2(g^{-1})'((r - \delta)\mu) + (r - \delta)\mu \cdot (g^{-1})''((r - \delta)\mu) \right] \\
&= (r - \delta) \left[2(g^{-1})'(x) + x(g^{-1})''(x) \right] \, ,
\end{aligned}
$$

for $x = (r - \delta)\mu > (r - \delta)\mu_0 = g(0)$. Now

$$(g^{-1})'(x) = \frac{1}{g'(g^{-1}(x))} \; ; \; (g^{-1})''(x) = \frac{-g''(g^{-1}(x))}{(g'(g^{-1}(x)))^3} \, .$$

Therefore,

$$
\begin{aligned}
(r - \delta)^{-1}\lambda''(\mu) &= \frac{2}{g'(g^{-1}(x))} - \frac{xg''(g^{-1}(x))}{(g'(g^{-1}(x)))^3} \\
&= \frac{2(g'(g^{-1}(x)))^2 - xg''(g^{-1}(x))}{(g'(g^{-1}(x)))^3} \, .
\end{aligned}
$$

It follows that $\lambda(\mu)$ is concave in $\mu > \mu_0$ if and only if

$$2(g'(g^{-1}(x)))^2 \le xg''(g^{-1}(x)) \, ,$$

for all $x = (r - \delta)\mu > (r - \delta)\mu_0 = g(0)$. But, as x ranges from $g(0)$ to ∞, $g^{-1}(x) = \rho$, where $g(\rho) = x$ and ρ ranges from 0 to 1. Therefore, $\lambda(\mu)$ is concave in $\mu > \mu_0$ if and only if

$$2(g'(\rho)^2 \le g(\rho)g''(\rho) \, , \text{ for all } \rho \, , \; 0 \le \rho < 1 \qquad (5.15)$$

for all ρ, $0 \le \rho < 1$. But (5.15) holds if and only if $1/g(\rho)$ is concave in ρ, $0 \le \rho < 1$. The proof that $\lambda(\mu)$ is convex if and only if $1/g(\rho)$ is convex is symmetric.   ∎

Since an affine function is both concave and convex, we note that $\lambda(\mu)$ is an affine function of μ if and only if $1/g(\rho)$ is an affine function of ρ. Recall that this was the case in the example of an *M/M/1* queue with a linear waiting-cost function, in which $g(\rho) = h/(1 - \rho)$ and $1/g(\rho) = (1 - \rho)/h$ and

$$\lambda(\mu) = \mu - \frac{h}{r - \delta} \, .$$

When $\lambda(\mu)$ is not concave, the necessary condition for a socially optimal μ, $\mu > \mu_0$, namely,

$$\lambda'(\mu)\delta = c'(\mu) \, ,$$

may not be sufficient. It may have several solutions, some of which are local (not global) maxima, or even local minima.

Example Suppose Assumption 1 is satisfied, with $g(\rho) = a + b\rho$, $0 \le \rho \le 1$, and $g(\rho) = \infty$, $\rho > 1$, where $a > 0$, $b > 0$. Suppose $c(\mu) = c \cdot \mu^2$, $\mu \ge 0$. Then $\mu_0 = (r - \delta)^{-1}a$. Let $\mu_1 := (r - \delta)^{-1}(a + b)$. For $\mu \le \mu_0$, $\lambda(\mu) = 0$

and $\psi(\mu) \leq 0 = \psi(0)$, whereas for $\mu > \mu_1$, $\lambda(\mu) - \mu$ and $\psi(\mu) < \psi(\mu_1)$. For $\mu_0 < \mu < \mu_1$, $\lambda(\mu)$ is the unique solution to

$$a + b\left(\frac{\lambda}{\mu}\right) = (r - \delta)\mu \ .$$

Therefore,

$$\lambda(\mu) = \frac{\mu((r - \delta)\mu - a)}{b} \ ,$$

which is convex in $\mu_0 < \mu \leq \mu_1$. Thus, for $\mu_0 < \mu \leq \mu_1$, the objective function for social optimization takes the form,

$$\begin{aligned}
\psi(\mu) &= \lambda(\mu)\delta - c(\mu) \\
&= \left(\frac{\delta(r - \delta)}{b} - c\right)\mu^2 - \left(\frac{a\delta}{b}\right)\mu \ .
\end{aligned}$$

We see that $\psi(\mu)$ is concave if $\delta(r-\delta) \leq bc$ and (strictly) convex if $\delta(r-\delta) > bc$. In the latter case, the necessary condition for an interior optimum,

$$2\left(\frac{\delta(r - \delta)}{b} - c\right)\mu - \left(\frac{a\delta}{b}\right) = 0 \ ,$$

has a unique solution,

$$\hat{\mu} := \frac{a\delta}{2\left(\delta(r - \delta) - bc\right)} \ .$$

We have $\hat{\mu} > \mu_0$ if and only if $\delta(r - \delta) < 2bc$, and $\hat{\mu} \leq \mu_1$ if and only if $(a\delta + 2b)(r - \delta) \geq 2(a + b)bc$.

It follows that if

$$\begin{aligned}
bc < \delta(r - \delta) &\leq 2bc \ , \text{ and} \\
(a\delta + 2b)(r - \delta) &\geq 2(a + b)bc \ ,
\end{aligned}$$

then $\hat{\mu}$ is in fact the global *minimum* of $\psi(\mu)$ over $\mu \in (\mu_0, \mu_1]$, not the sought-for global maximum. Moreover, we have

$$\psi(\hat{\mu}) = -\frac{(a\delta)^2}{4b(\delta(r - \delta) - bc)} < 0 \ .$$

That is, at the unique solution to the necessary condition for an interior optimum, the objective function is actually strictly negative, even though there may be an interval of values of μ (namely, $(2\hat{\mu}, \mu_1]$) over which $\psi(\mu) > 0$! This interval is nonempty if and only if $(a + b)c < \delta(r - \delta)$.

Combining these results, we see that if

$$(a + b)c < \delta(r - \delta) \leq 2bc \ ,$$

then:

(i) the unique solution to the necessary condition for an interior optimum, $\hat{\mu}$, is actually the global *minimum* of the objective function, $\psi(\mu)$, for social optimization;

(ii) the objective function takes on a negative value at $\hat{\mu}$; whereas

(iii) $\psi(\mu)$ is positive and increasing over the interval, $(a(\delta(r-\delta)-bc)^{-1}, (a+b)(r-\delta)^{-1}]$; and hence

(iv) the global maximum actually occurs at $\mu^s = \mu_1 = (a+b)(r-\delta)^{-1}$, with $\lambda^s = \lambda(\mu_1) = \mu_1$. (See Figure 5.3.)

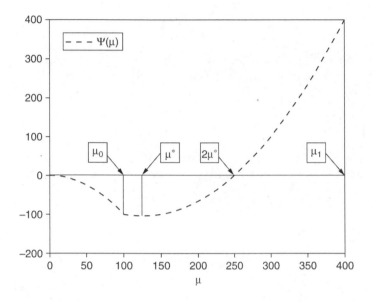

Figure 5.3 *Example with Convex Objective Function,* $\mu > \mu_0$

5.4.3 Facility Optimality

Now consider the problem from the point of view of a facility operator who wishes to maximize profit. Once again the admission fee δ is fixed and the service rate μ is the decision variable. The facility receives gross revenue in the form of the fees received per unit time, $\lambda\delta$, and incurs service cost at rate $c(\mu)$ per unit time. The profit of the service facility is therefore

$$\lambda\delta - c(\mu) .$$

The arrival rate λ is the equilibrium arrival rate uniquely determined by equations (5.6) and (5.7).

Note that the waiting cost is now experienced by the customers, but not directly by the decision maker (the operator of the facility). In the social optimization models of this chapter, the decision maker (acting on behalf of the collective of all customers) had an incentive to increase the service rate in order to reduce the waiting cost incurred. Now the incentive to increase the service rate comes from the increase in the fee collected per unit time, which results from the increase in the arrival rate caused by the decrease in waiting cost experienced by the customers. Thus, in both models it is the decrease in waiting cost that serves as an incentive to the decision maker to increase the service rate, but in the present model this waiting-cost decrease is experienced by the facility operator indirectly through the increase on fees collected per unit time.

The profit-maximization problem thus takes the following form:

$$\max_{\{\lambda,\mu\}} \quad \lambda\delta - c(\mu)$$

$$\text{s.t.} \quad \delta + G(\lambda,\mu) - U'(\lambda) \geq 0 \,,$$

$$\delta + G(\lambda,\mu) - U'(\lambda) = 0 \,, \text{ if } \lambda > 0 \,,$$

$$0 \leq \lambda < \mu \,.$$

As in the case of social optimality, we consider separately the two cases $0 \leq \mu \leq \mu_0$, and $\mu > \mu_0$, where $\mu_0 = \inf\{\mu > 0 : U'(0) - G(0,\mu) > \delta\}$. For $0 \leq \mu \leq \mu_0$, the unique solution to the constraints is given by $\lambda = 0$. Thus the maximum profit over this range of values of μ occurs at $\mu = 0$, at which the profit equals zero.

Over the range $\mu > \mu_0$, λ satisfies the constraints if and only if $U'(\lambda) = \delta + G(\lambda,\mu)$. Thus we have the following equivalent form for the problem with μ restricted to (μ_0, ∞):

$$\max_{\{\lambda,\mu:\mu>\mu_0\}} \quad \lambda\delta - c(\mu)$$

(F) $\qquad \text{s.t.} \quad U'(\lambda) - G(\lambda,\mu) = \delta \,,$

$$0 \leq \lambda < \mu \,.$$

Following the approach suggested in the analysis of social optimality, let $\lambda(\mu)$ denote the (unique) solution of the equilibrium equation, $U'(\lambda) = \delta + G(\lambda,\mu)$, associated with service rate $\mu > \mu_0$. Then the profit maximization problem for this range of values of μ may be rewritten as

$$\max_{\{\mu>\mu_0\}} \quad \lambda(\mu)\delta - c(\mu) \,.$$

The first-order necessary condition for a facility-optimal value of μ (in the range $\mu > \mu_0$) is therefore

$$\lambda'(\mu)\delta = c'(\mu) \,. \tag{5.16}$$

Remark 2   Recall that this condition is also satisfied by the socially optimal value of μ in the special case of a linear utility function: $U(\lambda) = r \cdot \lambda$. This is not surprising, since in that case $U(\lambda) = \lambda U'(\lambda)$ and hence the objective functions for social and facility optimality coincide. Since they share the same constraint,

$$U'(\lambda) - G(\lambda,\mu) = \delta \,,$$

it follows that the two problems are equivalent in this case. Hence we can apply the results of Section 5.4.2.1. In particular, the conditions given there for $\lambda(\mu)$ to be concave are also relevant here, and under these conditions equation (5.16) is sufficient as well as necessary for a facility-optimal value of μ.

When the utility function is not linear, a different approach is called for. Let us return to Problem **(F)**, in which the search for a facility-optimal pair, (λ, μ), in the range $\mu > \mu_0$ is formulated as a nonlinear program (NLP). First note that, since $U'(\lambda)$ is nonincreasing and $G(\lambda, \mu)$ is nondecreasing in λ and nonincreasing in μ, we can replace the "=" in the constraint, $U'(\lambda) - G(\lambda, \mu) = \delta$, with a "≥" without loss of optimality. Thus we can replace Problem **(F)** with the following equivalent NLP:

$$\max_{\{\lambda, \mu : \mu > \mu_0\}} \quad \lambda\delta - c(\mu)$$

(F′)   s.t. $U'(\lambda) - G(\lambda, \mu) \geq \delta$,

$$0 \leq \lambda < \mu .$$

(For any given values of λ and μ, a feasible solution to this problem in which the constraint is satisfied with a strict inequality can always be replaced with a feasible solution with a strictly larger value of λ and/or a strictly smaller value of μ, resulting in a strictly larger value of the objective function. Thus the ≥ constraint will always be satisfied with equality in an optimal solution.)

Theorem 5.4 *Suppose $U'(\lambda)$ is concave in λ, $G(\lambda, \mu)$ is jointly convex in (λ, μ), and $c(\mu)$ is convex in μ. Then Problem* **(F)** *has a unique global maximum pair, (λ, μ), which is the unique solution to the* KKT *conditions,*

$$\delta + \alpha \left(U''(\lambda) - \frac{\partial}{\partial\lambda} G(\lambda, \mu) \right) = 0 ,$$

$$-c'(\mu) - \alpha \frac{\partial}{\partial\mu} G(\lambda, \mu) = 0 ,$$

$$U'(\lambda) - G(\lambda, \mu) - \delta = 0 .$$

Proof   Since both the objective function and the l.h.s. of the ≥ constraint are jointly concave in (λ, μ), it follows that Problem **(F′)** has a unique global maximum which is the unique solution to the KKT conditions. The equivalence of Problems **(F)** and **(F′)** implies that the inequality KKT conditions are satisfied with equality and hence that the complementary slackness conditions hold.   ∎

Remark 3   Note that the results of this section apply to utility functions $U(\lambda)$ for which $U'(0) = \infty$. In particular, because δ is fixed, they apply without the technical assumption that $\lambda U'(\lambda) \to 0$ as $\lambda \to 0$. This is in contrast to the facility optimization models of Chapter 2 (see, e.g., Sections 2.1.3 and 2.3.3), in which δ was a decision variable. There we saw, for example, that if $\lambda U'(\lambda) \to \infty$, the facility operator could earn an arbitrarily large profit by choosing an arrival rate arbitrarily close to zero (equivalently, choosing an arbitrarily large toll). Recall that the condition, $\lambda U'(\lambda) \to \infty$, can occur only if the distribution of customer rewards, $F(x) = \mathrm{P}\{R \leq x\}$, has a heavy tail.

5.4.3.1 Comparison of Socially Optimal and Facility Optimal Solutions

We now compare the socially optimal and facility optimal solutions to the problem with a fixed admission fee. Recall that when we examined the socially optimal and facility optimal arrival rates in the case of a system with a fixed service rate and a variable admission fee (in Chapter 2) we found that the socially optimal admission fee (arrival rate) is always smaller (larger) than the facility optimal admission fee (arrival rate). This result is consistent with the general property from welfare economics that profit maximization leads to underutilization of an economic system than is optimal from the point of view of overall welfare maximization. We shall see that a similar phenomenon holds in the present context, where the admission fee is constant and control is exercised by varying the service rate.

Since the optimal solution for both social and facility optimization over the range $0 \leq \mu \leq \mu_0$ occurs at $\mu = 0$, with an objective function value equal to zero in both cases, we shall restrict our attention to the range $\mu > \mu_0$.

First observe that, in both social optimization and facility optimization, for any particular value of the service rate in the range $\mu > \mu_0$ the associated arrival rate is determined by the same equation,

$$U'(\lambda) = \delta + G(\lambda, \mu) .$$

Let $\lambda(\mu)$ denote the unique solution, $\lambda \in [0, \mu)$, to this equation, for $\mu > \mu_0$.

Lemma 5.5 $\lambda(\mu)$ *is a (strictly) increasing function of* $\mu > \mu_0$.

Proof   Let $\mu_0 < \mu_1 < \mu_2$ and note that

$$U'(\lambda(\mu_1)) - G(\lambda(\mu_1), \mu_1) = \delta .$$

Since $G(\lambda, \mu)$ is strictly decreasing in μ, it follows that

$$U'(\lambda(\mu_1)) - G(\lambda(\mu_1), \mu_2) > \delta .$$

But $U(\lambda)$ is concave and $G(\lambda, \mu)$ is strictly increasing in λ. Hence $U'(\lambda) - G(\lambda, \mu_2)$ is strictly decreasing in λ. Since

$$U'(\lambda(\mu_2)) - G(\lambda(\mu_2), \mu_2) = \delta ,$$

we conclude that $\lambda(\mu_2) > \lambda(\mu_1)$.   ∎

Let λ^s and μ^s denote the socially optimal arrival rate and service rate, respectively, and let λ^f and μ^f denote the facility optimal arrival rate and service rate, respectively. We shall resolve ties by selecting the largest maximizer. The following theorem expresses the ordering between these rates, alluded to above.

Theorem 5.6 *Both the arrival rate and the service rate are larger in the socially optimal solution than in the facility optimal solution:* $\lambda^s > \lambda^f$ *and* $\mu^s > \mu^f$.

Proof   By Lemma 5.5, it suffices to show that $\mu^s > \mu^f$. Substituting $\lambda(\mu)$ for

λ in the objective functions for social and facility optimization, we can write the two problems in equivalent form as

$$\max_{\mu} B_s(\mu) := U(\lambda(\mu)) - \lambda(\mu)G(\lambda(\mu), \mu) - c(\mu)$$

and

$$\max_{\mu} B_f(\mu) := \lambda(\mu)U'(\lambda(\mu)) - \lambda(\mu)G(\lambda(\mu), \mu) - c(\mu) ,$$

respectively. Now

$$B_s(\mu) - B_f(\mu) = U(\lambda(\mu)) - \lambda(\mu)U'(\lambda(\mu)) .$$

But $\lambda(\mu)$ is nondecreasing in μ by Lemma 5.5 and the concavity of $U(\lambda)$ implies that $U(\lambda) - \lambda U'(\lambda)$ is nondecreasing in λ (see Lemma 2.1 in Chapter 2). Therefore $B_s(\mu) - B_f(\mu)$ is nondecreasing in μ. But this implies that $\mu^s \geq \mu^f$.
∎

Recall that the constraint for both social and facility optimality, which defines $\lambda(\mu)$, is the equilibrium condition for the individually optimal arrival rate when the service rate is μ and the toll is δ:

$$U'(\lambda(\mu)) - G(\lambda(\mu), \mu) = \delta . \qquad (5.17)$$

Since the right-hand side of this equation is constant in μ, we have, upon differentiating both sides with respect to μ,

$$U''(\lambda(\mu))\lambda'(\mu) - \left[\frac{\partial}{\partial \lambda}G(\lambda(\mu), \mu)\lambda'(\mu) + \frac{\partial}{\partial \mu}G(\lambda(\mu), \mu) \right] = 0 ,$$

so that

$$U''(\lambda(\mu))\lambda'(\mu) = \frac{\partial}{\partial \lambda}G(\lambda(\mu), \mu)\lambda'(\mu) + \frac{\partial}{\partial \mu}G(\lambda(\mu), \mu) . \qquad (5.18)$$

Now, in the case of social optimality the first-order condition for the optimality of μ is

$$\frac{d}{d\mu}[U(\lambda(\mu)) - \lambda(\mu)G(\lambda(\mu), \mu) - c(\mu)] = U'(\lambda(\mu))\lambda'(\mu) - \lambda'(\mu)G(\lambda(\mu), \mu)$$

$$-\lambda(\mu)\left[\frac{\partial}{\partial \lambda}G(\lambda(\mu), \mu)\lambda'(\mu) + \frac{\partial}{\partial \mu}G(\lambda(\mu), \mu) \right] = 0$$

Substituting from equations (5.17) and (5.18) and rearranging terms yields the following equation:

$$\lambda'(\mu) \cdot \delta - c'(\mu) = \lambda(\mu)U''(\lambda(\mu))\lambda'(\mu) .$$

Since $U(\lambda)$ is concave and $\lambda(\mu)$ is nonnegative and nondecreasing in μ, the right-hand side of this equation is nonpositive.

In the case of facility optimality, the equation satisfied by the optimal value of μ is

$$\lambda'(\mu) \cdot \delta - c'(\mu) = 0 .$$

The left-hand side of both these equations represents the marginal net revenue of the toll collector, with respect to the service rate μ. We see that the facility optimizer sets this marginal net revenue equal to zero, whereas the social optimizer sets it equal to a nonpositive number, which agrees with our intuition and previous results. In the case of linear utility, $U''(\lambda) = 0$ for all $\lambda \geq 0$, and hence the socially optimal and facility optimal service rates coincide, as expected.

One final observation: it also follows from the above observations and equation (5.18) that

$$\frac{d}{d\mu}G(\lambda(\mu), \mu) = \frac{\partial}{\partial \lambda}G(\lambda(\mu), \mu)\lambda'(\mu) + \frac{\partial}{\partial \mu}G(\lambda(\mu), \mu) \leq 0 \, .$$

That is, the waiting cost per customer does not increase as the service rate increases. Note that this is a direct consequence of the concavity of the utility function. If the utility function is strictly concave, then the waiting cost per customer strictly decreases as μ increases. If the utility function is linear, then the above inequality is replaced by equality and hence (as we have already seen) the waiting cost per customer remains constant as μ increases.

5.5 Models with Variable Toll and Variable Arrival Rate

In this section we consider an optimal design model in which both the toll and the arrival rate – as well as the service rate – are design variables. In the model of the previous section, the toll was fixed and the system designer could influence the arrival rate only by choice of the service rate. By contrast, in this section the system designer is free to choose the toll as well as the service rate. Together the toll and the service rate determine the arrival rate, through the equilibrium condition for individually optimizing customers.

5.5.1 Individual Optimality

For a given value of the toll, δ, and the service rate, μ, the individually optimal arrival rate, λ^e, uniquely satisfies the equilibrium conditions, (5.6) and (5.7), which we rewrite here for convenient reference:

$$
\begin{aligned}
U'(\lambda) &\leq & \delta + G(\lambda, \mu) \, , \\
U'(\lambda) &=& \delta + G(\lambda, \mu) \, , \text{ if } \lambda > 0 \, .
\end{aligned}
$$

5.5.2 Social Optimality

The social optimization problem takes the form

$$\max_{\lambda, \mu} \ \mathcal{U}(\lambda, \mu) \ := \ U(\lambda) - \lambda G(\lambda, \mu) - c(\mu)$$

$$\text{s.t.} \qquad 0 \leq \lambda < \mu \, .$$

The KKT necessary conditions for a maximizing pair, (λ, μ), are:

$$U'(\lambda) \ \leq \ G(\lambda, \mu) + \lambda \frac{\partial}{\partial \lambda}G(\lambda, \mu) \, , \text{ and } \lambda \geq 0 \, , \tag{5.19}$$

$$U'(\lambda) \;=\; G(\lambda,\mu) + \lambda\frac{\partial}{\partial\lambda}G(\lambda,\mu) \;,\; \text{if } \lambda > 0 \;, \tag{5.20}$$

$$c'(\mu) \;\geq\; -\lambda\frac{\partial}{\partial\mu}G(\lambda,\mu) \;,\; \text{and } \mu \geq 0 \;, \tag{5.21}$$

$$c'(\mu) \;=\; -\lambda\frac{\partial}{\partial\mu}G(\lambda,\mu) \;,\; \text{if } \mu > 0 \;;, \tag{5.22}$$

$$0 \;\leq\; \lambda < \mu \;. \tag{5.23}$$

Remark 4   Note that, like $H(\lambda,\mu)$, $\frac{\partial}{\partial\mu}H(\lambda,\mu) = \lambda\frac{\partial}{\partial\mu}G(\lambda,\mu)$ may not be well defined at $\lambda = 0$, $\mu = 0$. We shall adopt the convention that

$$\frac{\partial}{\partial\mu}H(0,0) = \lim_{\mu\to 0}\lim_{\lambda\to 0}\lambda\frac{\partial}{\partial\mu}G(\lambda,\mu) = 0 \;.$$

With this convention and under the default assumption that $c'(\mu) > 0$ for all $\mu \geq 0$, the effect of conditions (5.21) and (5.22) is to require that $\mu = 0$ when $\lambda = 0$.

In the case of an interior maximum, $0 < \lambda < \mu$, the necessary conditions simplify to:

$$U'(\lambda) \;=\; G(\lambda,\mu) + \lambda\frac{\partial}{\partial\lambda}G(\lambda,\mu) \;, \tag{5.24}$$

$$c'(\mu) \;=\; -\lambda\frac{\partial}{\partial\mu}G(\lambda,\mu) \;. \tag{5.25}$$

For any given $\mu > 0$, (5.24) is necessary and sufficient for $\lambda > 0$ to be socially optimal for that particular μ (cf. Chapter 2). Likewise, for any given $\lambda > 0$, (5.25) is necessary and sufficient for $\mu > \lambda$ to be socially optimal for that particular λ (cf. Section 5.2 of this chapter). However, because the function $H(\lambda,\mu) = \lambda G(\lambda,\mu)$ may not be jointly convex in (λ,μ), as we have already noted, the two conditions together may not be sufficient for the pair (λ,μ) to be socially optimal for the problem in which both λ and μ are variables.

As usual, we note that, by charging individually optimizing customers a toll equal to the external effect, namely,

$$\delta^s := \lambda^s\frac{\partial}{\partial\lambda}G(\lambda^s,\mu^s) \;,$$

we can induce the customers to join at a rate that satisfies the necessary condition (5.24) for λ^s.

Remark 5   Consider a solution, (λ,μ), to the necessary conditions, (5.24) and (5.25). For systems in which

$$\frac{\partial}{\partial\lambda}G + \frac{\partial}{\partial\mu}G = 0 \;, \tag{5.26}$$

(which includes all $M/M/1$ systems with linear waiting costs) it follows from (5.24) and (5.25) that

$$\delta^s = c'(\mu^s) \;, \tag{5.27}$$

that is, the optimal toll to charge customers for entering equals the marginal cost of increasing the service rate.

As in the short-run problem, by assuming that there is an interior maximum, $0 < \lambda < \mu$, one is able to ignore the nonnegativity constraints and simply look for solutions to the necessary conditions, (5.24) and (5.25), for an interior maximum. In the long-run problem, however, the assumption that there is an interior maximum is no longer trivial nor easily verified *a priori*, since μ is now also a decision variable. An additional complication comes from the fact that the first-order necessary conditions (5.24) and (5.25) are not in general sufficient, even when an interior point is optimal. They would be sufficient if $H(\lambda, \mu) = \lambda G(\lambda, \mu)$ were jointly convex in (λ, μ), but this is not true in general, as we have noted. We shall illustrate these complications in Section 5.5.2.1 below in the context of an *M/GI/1* model with linear waiting cost. To analyze this and other examples, we shall sometimes find it convenient to work with the following alternative formulation of the long-run problem.

First observe that for any fixed choice of $\lambda > 0$, the cost component of $\mathcal{U}(\lambda, \mu)$, namely $\lambda G(\lambda, \mu) + c(\mu)$, is convex in μ. Thus we can find the associated optimal value $\mu(\lambda)$ of μ by differentiating with respect to μ and setting the derivative equal to zero. In other words, $\mu(\lambda)$ satisfies the first-order condition (5.25). For $\lambda = 0$, we define $\mu(\lambda) := 0$. The socially optimal value, λ^s, of λ can then be found by substituting $\mu(\lambda)$ for μ in $\mathcal{U}(\lambda, \mu)$ and maximizing the resulting function of λ, namely,

$$\phi(\lambda) := \mathcal{U}(\lambda, \mu(\lambda)) = U(\lambda) - \lambda G(\lambda, \mu(\lambda)) - c(\mu(\lambda)) , \qquad (5.28)$$

over $\lambda \geq 0$. Provided that $\mu(\lambda)$ is a differentiable function of λ, an interior maximum λ^s will satisfy the first-order necessary condition obtained by differentiating $\phi(\lambda)$ and setting the derivative equal to zero:

$$
\begin{aligned}
\phi'(\lambda) &= U'(\lambda) - G(\lambda, \mu(\lambda)) - \lambda \frac{\partial}{\partial \lambda} G(\lambda, \mu(\lambda)) \\
&\quad - \lambda \frac{\partial}{\partial \mu} G(\lambda, \mu(\lambda)) \mu'(\lambda) - c'(\mu(\lambda) \mu'(\lambda) \\
&= U'(\lambda) - G(\lambda, \mu(\lambda)) - \lambda \frac{\partial}{\partial \lambda} G(\lambda, \mu(\lambda)) = 0 . \qquad (5.29)
\end{aligned}
$$

(The second equality follows from (5.25).)

Each solution λ to (5.29), together with the associated optimal service rate $\mu = \mu(\lambda)$, constitutes a solution to the first-order conditions (5.24) and (5.25) for an interior maximum. Therefore, the question of the existence of multiple solutions to (5.24) and (5.25)) can be answered by examining the solutions to (5.29).

5.5.2.1 An Example with a Nonconcave Objective Function

In this subsection we show that the objective function in (5.28), $\phi(\lambda) = U(\lambda) - \lambda G(\lambda, \mu(\lambda)) - c(\mu(\lambda))$, is not in general concave and that its nonconcavity can give rise to multiple solutions to the first-order optimality conditions. We shall give an example in which the cost component of $\phi(\lambda)$, namely, $C(\lambda) := \lambda G(\lambda, \mu(\lambda)) + c(\mu(\lambda))$, is itself a concave function, so that $\phi(\lambda)$ is the difference of two concave functions and hence not in general concave nor even unimodal.

Our example is a system in which $H(\lambda, \mu) = f(\lambda/\mu)$, where $f(\rho)$ is a continuous, increasing function of $\rho \in [0, 1)$, with $f(0) = 0$ and $f(\rho) = \infty$ for $\rho \geq 1$. (See Remark 1 in Section 5.1.) Recall that this property holds for many queueing systems with a linear waiting-cost function in which λ and μ are scale factors. In this case we have $H(\lambda, \mu) = \lambda G(\lambda, \mu) = h\mathrm{E}[\boldsymbol{L}(\lambda, \mu)]$, where $\boldsymbol{L}(\lambda, \mu)$ has the steady-state distribution of the number of jobs in the system. An example is an $M/GI/1$ queue with linear waiting cost, in which the server performs work at rate μ and customers bring i.i.d. amounts of work to be served. (See Remark 6 below.)

We also assume that the service-cost function, $c(\mu)$, is linear: $c(\mu) = c \cdot \mu$, $\mu \geq 0$. Thus

$$C(\lambda) = H(\lambda, \mu(\lambda)) + c\mu(\lambda) \,, \tag{5.30}$$

where $\mu(\lambda)$ is a solution to

$$-\frac{\partial}{\partial \mu} H(\lambda, \mu) = c \,. \tag{5.31}$$

To resolve ties, we shall assume that $\mu(\lambda)$ is the smallest such solution. From the assumption that $H(\lambda, \mu) = f(\lambda/\mu)$, it follows that

$$\lambda \frac{\partial}{\partial \lambda} H(\lambda, \mu) + \mu \frac{\partial}{\partial \mu} H(\lambda, \mu) = 0 \,. \tag{5.32}$$

We wish to show that $C'(\lambda)$ is nonincreasing, that is, $C(\lambda)$ is concave, where $C(\lambda)$ is given by (5.30). Now

$$\begin{aligned}
C'(\lambda) &= \frac{\partial}{\partial \lambda} H(\lambda, \mu(\lambda)) \\
&\quad + \frac{\partial}{\partial \mu} H(\lambda, \mu(\lambda)) \mu'(\lambda) + c\mu'(\lambda) \\
&= \frac{\partial}{\partial \lambda} H(\lambda, \mu(\lambda)) \\
&= -\left(\frac{\mu(\lambda)}{\lambda}\right) \frac{\partial}{\partial \mu} H(\lambda, \mu(\lambda)) \\
&= c\left(\frac{\mu(\lambda)}{\lambda}\right) \,,
\end{aligned}$$

where the second and fourth equalities follow from (5.31) and the third equality from (5.32).

Thus it suffices to show that $\mu(\lambda)/\lambda$ is nonincreasing in λ, or, equivalently, that $\rho(\lambda) := \lambda/\mu(\lambda)$ is nondecreasing in λ. Now for each λ, $\rho(\lambda)$ is the smallest minimizer of the function $f(\rho) + c\lambda/\rho$, which is easily seen to be submodular in (λ, ρ), so that $\rho(\lambda)$ is nondecreasing in λ, the desired result (Topkis [192], Heyman and Sobel [92]).

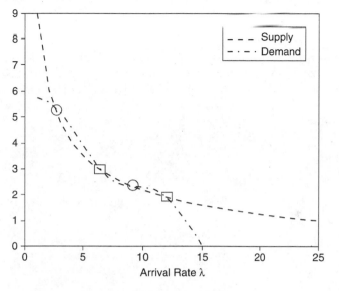

Figure 5.4 *Long-Run Demand and Supply Curves*

Remark 6   In the case of a $M/GI/1$ queue with linear waiting costs, we have $H(\lambda, \mu) = \lambda G(\lambda, \mu) = h \cdot \mathrm{E}[\boldsymbol{L}(\lambda, \mu)] = f(\rho)$, where

$$f(\rho) = h \left[\rho + \frac{\rho^2 \beta}{1 - \rho} \right] .$$

Here $\beta := \mathrm{E}[\boldsymbol{S}^2]/2$, where $\boldsymbol{S}$ is the generic random amount of work brought to the system by a customer. (We assume w.l.o.g. that $\mathrm{E}[\boldsymbol{S}] = 1$.)

Marginal economic analysis of the long-run problem suggests that we interpret $U'(\lambda)$ as the demand curve and $C'(\lambda)$ as the supply curve for the service facility (cf. the analysis of the short-run problem in Section 2.4 of Chapter 2). Thus we see that our example has a downward-sloping supply curve. That is, if the service rate μ is a design variable and the service (capacity) cost is a linear function of μ, then the service facility exhibits economies of scale with respect to increasing load λ, when both waiting cost and capacity cost are considered.

When both the demand and the supply curves are downward sloping, they may intersect at more than one point. That is, there may be multiple solutions to the equality $U'(\lambda) = C'(\lambda)$. Figure 5.4 illustrates what might happen.

The solutions to $\phi'(\lambda) = 0$ are the points where the demand curve (the graph of $U'(\lambda)$) and the supply curve (the graph of $C'(\lambda)$) intersect. Of these, some are relative maxima (indicated by a square) and some are relative minima (indicated by a circle) of the objective function $\phi(\lambda)$. Only one point, of course, can be the sought-for global maximum of $\phi(\lambda)$.

It should be noted that the value distribution associated with the demand curve in Figure 5.4 has a bimodal density, which roughly corresponds to the case of two classes of customers with distinctly different value distributions.

Example We now provide a simple concrete example in which either there are multiple solutions to the first-order conditions or the optimal solution lies on the boundary of the feasible region. Consider an $M/M/1$ queueing system in which potential customers arrive at a fixed rate Λ, with a linear waiting cost and a customer value distribution that is uniform on the interval $S = [d, a]$, where $0 \leq d < a < \infty$. That is, $F(x) = 0$, $x < d$; $F(x) = (x - d)/(a - d)$, $d \leq x < a$; $F(x) = 1$, $x \geq a$. In this case it is readily verified that

$$U'(\lambda) = \bar{F}^{-1}(\lambda/\Lambda) = [d\lambda + a(\Lambda - \lambda)]/\Lambda ,$$

$0 \leq \lambda \leq \Lambda$. The cost component, $C(\lambda)$, of the objective function, $\phi(\lambda)$, is given by

$$C(\lambda) = h\lambda/(\mu(\lambda) - \lambda) + c\mu(\lambda) , \tag{5.33}$$

where $\mu(\lambda) = \lambda + (\lambda h/c)^{1/2}$. Substituting for $\mu(\lambda)$ in (5.33) and differentiating with respect to λ, we obtain

$$C'(\lambda) = c + (hc/\lambda)^{1/2} .$$

Recall that the solutions to the first-order necessary conditions for an interior maximum are given by $(\lambda, \mu(\lambda))$, where λ is a solution to $U'(\lambda) = C'(\lambda)$, that is, a point at which the demand and supply curves intersect. We consider two cases.

Case 1: $d \leq c + (hc/\Lambda)^{1/2}$

In this case it can happen that the demand and supply curves do not intersect at all in $[0, \Lambda]$. This occurs if $U'(\lambda) < C'(\lambda)$ for all $0 \leq \lambda \leq \Lambda$, in which case $\lambda^s = \mu^s = 0$. Otherwise, there are exactly two solutions in $[0, \Lambda]$ to the equation $U'(\lambda) = C'(\lambda)$, which we shall denote λ^1 and λ^2, where $0 < \lambda^1 < \lambda^2 \leq \Lambda$. These are the two points of intersection of the demand and supply curves in $[0, \Lambda]$, as illustrated in Figure 5.5.

The smaller of these, λ^1, is a relative minimum of $\phi(\lambda)$ (at which $\phi(\lambda) < 0$). The larger, λ^2, is a relative maximum, which is the global maximum, λ^s, over $0 \leq \lambda \leq \Lambda$ if and only if $\phi(\lambda^2) \geq 0$. (If not, $\lambda^s = 0$.)

Case 2: $d > c + (hc/\Lambda)^{1/2}$

In this case there is exactly one solution in $[0, \Lambda]$ to the equation $U'(\lambda) = C'(\lambda)$, which we shall denote λ^1, where $0 < \lambda^1 < \Lambda$. This is the only point of intersection of the demand and supply curves in $[0, \Lambda]$, as illustrated in Figure 5.6.

It is a relative minimum of the objective function $\phi(\lambda)$ and again yields a negative value, $\phi(\lambda^1) < 0$. The global maximum λ^s of $\phi(\lambda)$ occurs at Λ if $\phi(\Lambda) \geq 0$. Otherwise, $\lambda^s = 0$.

Thus we have an example in which the unique solution $(\lambda^1, \mu(\lambda^1))$ to the

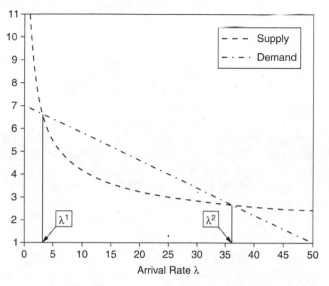

Figure 5.5 *Uniform [d, a] Value Distribution Long-Run Demand and Supply Curves, Case 1*

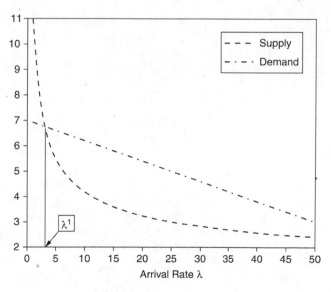

Figure 5.6 *Uniform [d, a] Value Distribution Long-Run Demand and Supply Curves, Case 2*

first-order necessary condition for an interior maximum of $f(\lambda, \mu)$ is in fact a local minimum and yields a negative value of the objective function. The sought-for global maximum (λ^s, μ^s) always occurs on the boundary of the feasible region.

Remark 7 Note that the case of a deterministic reward (which corresponds to a linear utility function) is the limiting case of the uniform value distribution as $d \to a$. We introduced this example in Chapter 1, in which we focused on Case 2. In this case the supply and demand curves intersect at exactly one point in the interior of the feasible region, $\{\lambda : 0 < \lambda < \Lambda\}$. This point is the global *minimum* of the objective function, $\phi(\lambda)$. The global maximum occurs at one of the extreme points, $\lambda = 0$ or $\lambda = \Lambda$. We return to the case of a linear utility function in Section 5.5.2.3 below, where we use an alternative approach which allows us to extend this result to concave service-cost functions, $c(\mu)$.

In the next subsection we shall further examine the $M/M/1$ model with uniform value distribution for the case $d = 0$. Our interest there will be in an iterative algorithms for computing the optimal solution (λ^s, μ^s) for the long-run problem.

5.5.2.2 Dynamic Adaptive Algorithm

Recall that in the long-run problem we are trying to find the value of λ, denoted λ^s, that maximizes the objective function $\phi(\lambda) = U(\lambda) - C(\lambda)$ over $0 \le \lambda \le \Lambda$. The first-order necessary condition for an interior maximum, $0 < \lambda^s < \Lambda$, is given by $\phi'(\lambda) = U'(\lambda) - C'(\lambda) = 0$, which in this case reduces to:

$$a(1 - \lambda/\Lambda) = c + (ch/\lambda)^{1/2} . \tag{5.34}$$

When $d = 0$, Case 2 cannot occur. As we saw in the previous section, in Case 1 this equation has either no solutions or two solutions in the feasible region, $0 \le \lambda \le \Lambda$. Figure 5.7 illustrates the case where there are two solutions, λ^1 and λ^2 $(0 < \lambda^1 < \lambda^2 < \Lambda)$ to (5.34).

For numerical computation of λ^s several iterative schemes are possible. We shall focus attention on a class of schemes that exploit the equilibrium-seeking behavior of the customers who use the facility. To motivate these schemes we shall once again consider a discrete-time, dynamic version of our problem, as we did in the case of the short-run problem (cf. Chapters 2 and 3).

Suppose the system operates over a succession of time periods, labeled $n = 0, 1, \ldots$. During time period n the arrival rate is fixed at a particular value λ_n and each time period is assumed to last long enough for the system to attain steady state (approximately). At the beginning of period n the system operator fixes the service rate at

$$\mu_n = \mu(\lambda_n) = \lambda_n + (\lambda_n h/c)^{1/2} ; \tag{5.35}$$

that is, μ_n is the optimal value of μ associated with arrival rate λ_n. Next the system operator chooses a toll δ to charge for entry during period n. Together, the toll δ, the arrival rate λ_n, and the service rate μ_n induce a

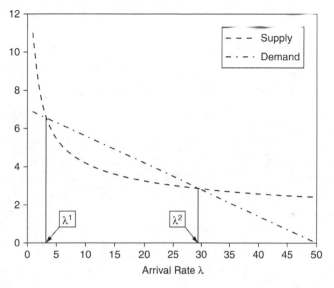

Figure 5.7 *Long-Run Demand and Supply Curves; Uniform $[0, a]$ Value Distribution*

cost per customer equal to $h/(\mu_n - \lambda_n) + \delta$ during period n. The customers react collectively to this perceived cost by adjusting the arrival rate during the next period so as to equate this cost with the marginal value received per unit time. That is, λ_{n+1} is the unique solution in $[0, \Lambda]$ of the equation $a(1 - \lambda_{n+1}/\Lambda) = h/(\mu_n - \lambda_n) + \delta$. (Note that this algorithm corresponds to the relaxed discrete-time algorithm presented in Section 3.3 of Chapter 3 in the extreme case, $\omega = 1$. This is sometimes called *best reply*.)

Now if (λ_n, μ_n) were the optimal pair, (λ^s, μ^s), then the optimal arrival rate $\lambda_{n+1} = \lambda^s$ could be induced by charging entering customers the optimal toll $\delta = c$ (see Remark 5). Hence it makes sense for the system operator to set $\delta = c$ at each period n. Thus λ_{n+1} is the unique solution in $[0, \Lambda]$ of the equation

$$a(1 - \lambda_{n+1}/\Lambda) = h/(\mu_n - \lambda_n) + c . \tag{5.36}$$

If $\lambda_{n+1} = \lambda_n$, then it follows from (5.35) and (5.36) that (λ_n, μ_n) satisfy both first-order conditions (5.24) and (5.25) for the optimal pair (λ^s, μ^s). If $\lambda_{n+1} \neq \lambda_n$, then we replace n by $n + 1$ and repeat the process.

Does this algorithm converge? If so, does it converge to the correct solution to (5.34)? To answer these questions, first consider the case $\lambda^2 < \lambda_0 \leq \Lambda$. Suppose we are at iteration $n \geq 0$, with $\lambda^2 < \lambda_n \leq \Lambda$. Then, since $a(1 - \lambda/\Lambda)$ is nonincreasing in λ and $c + (bh/\lambda_n)^{1/2} > a(1 - \lambda_n/\Lambda)$, it follows from (5.34) and (5.35) that $\lambda^2 < \lambda_{n+1} < \lambda_n$. Thus the sequence, λ_n, $n = 0, 1, \ldots$, converges to a limit $\tilde{\lambda} \geq \lambda^2$. Suppose $\tilde{\lambda} > \lambda^2$. Then, with $\lambda_0 = \tilde{\lambda}$, the above argument with $n = 0$ implies that $\lambda^2 < \lambda_1 < \lambda_0$. Now consider the original sequence, which converges to $\tilde{\lambda}$ from above. By choosing n sufficiently large, we can ensure that λ_n is arbitrarily close to Λ. Since the functions appearing in (5.34)

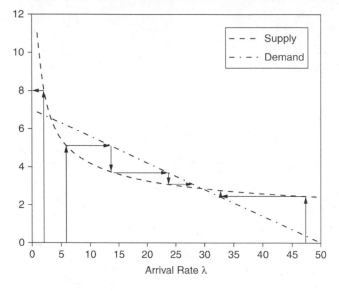

Figure 5.8 *Convergence of Iterative Algorithm for Case of Uniform $[0, a]$ Demand*

are continuous, we conclude that $\lambda_{n+1} < \tilde{\lambda}$ for sufficiently large n. But this contradicts our hypothesis that the sequence, λ_n, $n = 0, 1, \ldots$, converges to $\tilde{\lambda} > \lambda^2$. Therefore, it must converge to λ^2.

A similar argument shows that the sequence, λ_n, $n = 0, 1, \ldots$, is monotonically nondecreasing and converges to λ^2 for the case $\lambda^1 < \lambda_0 < \lambda^2$. Finally, if $0 < \lambda_0 < \lambda^1$, then the sequence, λ_n, $n = 0, 1, \ldots$, converges to 0. Figure 5.8 illustrates each of these cases.

The algorithm we have described is open to two different interpretations. On the one hand, we may imagine a situation in which the designer of a facility not yet in operation wishes to find the optimal service capacity μ^s. The designer knows the utility function $U(\lambda)$ of the customers as well as the waiting-cost coefficient h and the service-cost coefficient c. Then the algorithm may be used offline as an iterative technique for finding μ^s. On the other hand, suppose the facility designer (or operator) does not know the utility function of the customers. Then the algorithm can be used online and adaptively to find μ^s. In this case the dynamic scenario described above provides a rough description of the actual evolution of the system from period to period, where a "period" is a time interval long enough for steady state to be attained (approximately) and during which the arrival rate and service rate remain constant.

An implicit assumption underlying this interpretation of the algorithm is that the facility operator is able to adjust the service rate at the beginning of each period. That is, the service rate can be changed every time the arrival rate changes. In practice this may not be a feasible alternative. For example, in the applications to computer systems changing the service rate may correspond to replacing one computer with another.

What if it is not feasible to change the service rate so often? To take an extreme case, suppose that after the facility operator selects the service rate μ_n for period n, enough time elapses that a new equilibrium arrival rate is attained before the next opportunity to change the service rate. Then λ_{n+1} satisfies

$$a(1 - \lambda_{n+1}/\Lambda) = h/(\mu_n - \lambda_{n+1}) + c . \tag{5.37}$$

That is, λ_{n+1} is the equilibrium arrival rate associated with service rate $\mu = \mu_n$ and toll $\delta = c$. The analysis and results of Section 2.4.1 of Chapter 2 thus apply. In particular, λ_{n+1} is a stable equilibrium if and only if $\mu_n > [(a - c)/a]\Lambda$. If λ_{n+1} is not stable for some n, then, after the service rate is changed to $\mu_n = \mu(\lambda_n)$, the arrival rate will not converge to the new equilibrium value λ_{n+1}, and hence our proposed adaptive, dynamic scheme for finding the optimal pair (λ^s, μ^s) will not work.

Note that, even if $\mu_0 > [(a-c)/a]\Lambda$, after several iterations of the algorithm, it may be that $\mu_n \leq [(a - c)/a]\Lambda$, so that the algorithm fails to converge beyond this iteration. This will be the case, for example, if $\lambda^s = \lambda^2$ and $\mu(\lambda^2) \leq [(a - c)/a]\Lambda$. Conversely, if $\lambda^s = \lambda^2$ and $\mu(\lambda^2) > [(a - c)/a]\Lambda$, then the algorithm will always converge to $\lambda^s = \lambda^2$, provided that $\lambda_0 > \lambda^2$. The next theorem (see Stidham [187] for a proof) gives a sufficient condition for this to be true.

Theorem 5.7 *Suppose*

$$a > 2c + (h/2\Lambda)[1 + (1 + 8c\Lambda/h)^{1/2}] . \tag{5.38}$$

Then the proposed modified algorithm converges to λ^2, provided that $\lambda_0 > \lambda^2$.

Proof. From the preceding remark, we see that it suffices to show that $\mu(\lambda^2) > \Lambda' = [(a - c)/a]\Lambda$. Let $\underline{\lambda}$ be the unique solution in $[0, \Lambda']$ to

$$\mu(\lambda) = \lambda + (\lambda h/c)^{1/2} = \Lambda' .$$

That is, $\underline{\lambda} = \Lambda' - (h/2c)[(1 + 4\Lambda'c/h)^{1/2} - 1]$. Then to show that $\mu(\lambda^2) > \Lambda'$ it suffices to show that $\lambda^2 > \underline{\lambda}$. But this is true if $c + (ch/\underline{\lambda})^{1/2} < a(1 - \underline{\lambda}/\Lambda)$ or, equivalently, if

$$\begin{aligned} (bh/\underline{\lambda})^{1/2} \quad &< \quad (a'/\Lambda')(\Lambda' - \underline{\lambda}) \\ &= \quad (a'/\Lambda')(\underline{\lambda}h/c)^{1/2} , \end{aligned}$$

that is, if $\underline{\lambda}/\Lambda' > c/a'$. Substituting for $\underline{\lambda}$ we obtain (after some algebraic manipulations) the equivalent inequality

$$a' > c + (h/2\Lambda')[1 + (1 + 4c\Lambda'/h)^{1/2}] ,$$

which in turn is equivalent to (5.38).

Remark 8 For the long-run problem, in which μ as well as λ and δ are design variables, we have observed that a socially optimal solution may not be uniquely characterized by the first-order necessary (KKT) conditions. The KKT conditions typically have multiple solutions, some of which may not be relative maxima and may produce a negative value of the objective function.

Indeed, we gave an example in which the *KKT* conditions have a unique interior solution which is in fact a relative *minimum* with a negative value of the objective function. In this example, the global maximum occurs on the boundary of the feasible region.

Using the case of an *M/M/1* queue with linear waiting cost and a uniform distribution for the customer's value of service for illustration, we examined the convergence of a dynamic, adaptive algorithm for finding the optimal arrival rate and service rate. The algorithm always converges to a relative maximum of the objective function if the service rate can be adjusted every time the arrival rate changes. Otherwise, the algorithm will diverge if and when the service rate ever falls below the threshold value associated with stability of the equilibrium arrival rate.

Although our results are for simple queueing models, they illustrate the type of economic behavior that can arise in the design and operation of a service facility when both the arrival rate and service rate are design variables. They indicate the need for caution in applying economic insights gained only from examination of the first-order necessary conditions for an interior maximum. For example, economic analysis of a service facility may indicate that the arrival rate is in equilibrium and that the service rate is optimal for that arrival rate, and yet this pair of arrival and service rates may not be even locally, let alone globally, optimal.

5.5.2.3 Linear Utility Function: Optimality of an Extreme-Point Solution

In this section we restrict attention to the special case in which the utility function is linear: $U(\lambda) = r \cdot \lambda$. In this case we are able to extend some of the above results for a linear service-cost function to a concave service-cost function, $c(\mu)$. In particular we show that the objective function for social optimality is not in general concave and that the socially optimal pair, (λ^s, μ^s), may occur at an extreme point of the feasible region.

The social optimization problem now takes the form

$$\max_{\lambda, \mu} \mathcal{U}(\lambda, \mu) \quad := \quad r \cdot \lambda - \lambda G(\lambda, \mu) - c(\mu)$$

$$\text{s.t.} \qquad 0 \leq \lambda < \mu \leq \bar{\mu} \,. \tag{5.39}$$

Note that we are assuming that the feasible region for μ is $[0, \bar{\mu}]$, where $\bar{\mu} \leq \infty$.
We also make the following assumptions.

Assumption 2 The service cost, $c(\mu)$, is a concave, strictly increasing, and differentiable function of $\mu \in [0, \bar{\mu}]$.

Assumption 3 The waiting cost per unit time, $H(\lambda, \mu) = \lambda G(\lambda, \mu)$, takes the form

$$H(\lambda, \mu) = f(\lambda/\mu) \,,$$

where $f(\rho)$ is a convex, strictly increasing, and differentiable function of $\rho \in [0, 1)$, with $f(0) = 0$ and $\lim_{\rho \to 1} f(\rho) = \infty$ and $f(\rho) = \infty$ for $\rho \geq 1$.

Note that in Assumption 2 we depart from our default assumption that $c(\mu)$ is convex. Recall that Assumption 3 holds for many queueing systems with a linear waiting-cost function in which λ and μ are scale factors, for example, a steady-state $M/GI/1$ queue with a linear waiting-cost function. (See Remark 1 in Section 5.1.

For a given value of μ, let $\lambda(\mu)$ denote the value of λ that maximizes the objective function, $\mathcal{U}(\lambda, \mu)$, for social optimization. Let μ_0 be the unique solution to $r = G(0, \mu)\}$. Then for $0 \leq \mu \leq \mu_0$, we have $\lambda(\mu) = 0$. For $\mu > \mu_0$, $\lambda(\mu)$ maximizes the function, $r \cdot \lambda - f(\lambda/\mu)$, and is therefore the unique solution to the optimality condition,

$$r = \left(\frac{1}{\mu}\right) f'(\lambda/\mu) . \tag{5.40}$$

Let $\psi(\mu) := \mathcal{U}(\lambda(\mu), \mu)$, $\mu \in [0, \bar{\mu}]$. Then

$$\psi(\mu) = r \cdot \lambda(\mu) - f(\lambda(\mu)/\mu) - c(\mu) .$$

The following theorem shows that $\psi(\mu)$ is not concave. Indeed, it is a *convex* function of μ and therefore attains its maximum at a boundary point of the feasible region for μ.

Theorem 5.8 *The objective function, $\psi(\mu) = \mathcal{U}(\lambda(\mu), \mu)$, is convex in $\mu \geq 0$. Hence, the socially optimal pair (λ^s, μ^s) is either $(0, 0)$ or $(\lambda(\bar{\mu}), \bar{\mu})$.*

Proof Differentiating $\psi(\mu)$ with respect to μ yields

$$\begin{aligned}
\psi'(\mu) &= r \cdot \lambda'(\mu) - f'(\lambda/\mu)\left(\frac{\lambda'(\mu)}{\mu}\right) - f'(\lambda/\mu)\left(\frac{-\lambda(\mu)}{\mu^2}\right) - c'(\mu) \\
&= f'(\lambda/\mu)\left(\frac{\lambda(\mu)}{\mu^2}\right) - c'(\mu) \\
&= \left(\frac{\lambda(\mu)}{\mu}\right) r - c'(\mu) .
\end{aligned}$$

The second and third equalities both follow from (5.40). Let $\rho(\mu) := \lambda(\mu)/\mu$. Thus we have

$$\psi'(\mu) = \rho(\mu) r - c'(\mu) .$$

It follows from (5.40) that $\rho(\mu)$ is the unique solution to the equation

$$r\mu = f'(\rho(\mu)) .$$

Since $f'(\cdot)$ is nondecreasing, $\rho(\mu)$ is nondecreasing in μ. This fact, together with the assumption that $c(\mu)$ is concave, implies that $\psi'(\mu)$ is nondecreasing in μ, that is, $\psi(\mu)$ is convex in μ. ∎

Remark 9 It is important to note that, because $\psi(\mu)$ is a strictly convex function of μ, there will be (at most) one interior point, $\mu = \tilde{\mu} \in (0, \bar{\mu})$, at which $\psi'(\mu) = 0$. The pair, $(\lambda(\tilde{\mu}), \tilde{\mu})$, will then solve the necessary *KKT* conditions for a social optimum. But in fact $\tilde{\mu}$ will be a global *minimum*, not a maximum, of $\psi(\mu)$. Not only that, but the objective function for social optimization will assume a *negative* value at $(\lambda(\tilde\mu), \tilde\mu)$. We saw an example of this phenomenon in Section 1.3 of Chapter 1.

The above results hold for all values of the reward parameter, r, in the linear utility function, but they require Assumptions 2 and 3. We now show that the extreme-point solution $(0,0)$ is always optimal for sufficiently small r, under a weak assumption.

Assumption 4   The service cost $c(\mu)$ satisfies the following asymptotic condition:

$$\liminf_{\mu \to \infty} \frac{c(\mu)}{\mu} > 0 . \tag{5.41}$$

Assumption 4 says that $c(\mu)$ grows asymptotically at least linearly. This assumption holds trivially when $c(\mu)$ is convex. It may, but does not necessarily, hold when $c(\mu)$ is concave. Note that the case of a finite upper bound, $\bar\mu$, on μ may be accommodated by setting $c(\mu) = \infty$ for $\mu > \bar\mu$, which implies that Assumption 4 is automatically satisfied when $\bar\mu < \infty$.

Theorem 5.9 *Suppose Assumption 4 holds. Then there exists $r_0 > 0$ such that for $0 \leq r \leq r_0$, the socially optimal arrival and service rates for the problem with linear utility function, $U(\lambda) = r \cdot \lambda$, given by (5.39), are $\lambda^s = \mu^s = 0$.*

Proof   As a function of the parameter r, let $\mu_0(r)$ denote the unique solution to

$$r = G(0, \mu) .$$

Let $\lambda(\mu)$ denote the socially optimal value of λ corresponding to the service rate μ. Then $\lambda(\mu)$ is the unique solution to the optimality conditions

$$r \leq G(\lambda, \mu) + \lambda \frac{\partial}{\partial \lambda} G(\lambda, \mu) ,$$

$$r = G(\lambda, \mu) + \lambda \frac{\partial}{\partial \lambda} G(\lambda, \mu) , \text{ if } \lambda > 0 .$$

Since $G(0, \mu)$ is strictly decreasing in μ, it follows that $\lambda(\mu) = 0$ for $0 \leq \mu < \mu_0(r)$. Moreover, since $\lim_{\mu \to \infty} G(0, \mu) = 0$, we have

$$\mu_0(r) \to \infty , \text{ as } r \to 0 .$$

Let

$$\psi(\mu) := \max_{0 \leq \lambda < \mu} \{ r \cdot \lambda - \lambda G(\lambda, \mu) - c(\mu)$$

$$= r \cdot \lambda(\mu) - \lambda(\mu) G(\lambda(\mu), \mu) - c(\mu) .$$

For $0 \leq \mu < \mu_0(r)$ we have $\psi(\mu) = -c(\mu) \leq 0$, whereas for $\mu \geq \mu_0(r)$,

$$\psi(\mu) \leq r \cdot \mu - c(\mu) .$$

But Assumption 4 implies that there exists a $\underline\mu$ such that, for all sufficiently small r, $c(\mu) \geq r \cdot \mu$ for all $\mu \geq \underline\mu$. Now let r_0 be sufficiently small that this condition holds and also that $\mu_0(r_0) \geq \mu$. Then, for all $0 \leq r < r_0$, we have $\psi'(\mu) \leq 0$ for all $\mu > 0$ and hence $\mu^s = 0$, and therefore $\lambda^s = \lambda(\mu^s) = 0$ as well. ∎

5.5.3 Facility Optimality

Now consider the problem from the point of view of a facility operator who wishes to maximize its profit. Both the admission fee δ and the service rate μ are decision variables, and the arrival rate is determined by the equilibrium condition for individually optimizing customers. The facility receives gross revenue in the form of the fees received per unit time, $\lambda \delta$, and incurs service cost at rate $c(\mu)$ per unit time. The profit of the service facility is therefore

$$\lambda \delta - c(\mu) .$$

Given δ and $\mu \geq 0$, the arrival rate λ is the individually optimal equilibrium arrival rate uniquely determined by equations (5.6) and (5.7), which we rewrite here for convenience:

$$\delta + G(\lambda, \mu) - U'(\lambda) \geq 0 ,$$
$$\delta + G(\lambda, \mu) - U'(\lambda) = 0 , \text{ if } \lambda > 0 .$$

The profit-maximization problem thus takes the following form:

$$\max_{\{\lambda, \mu, \delta\}} \quad \mathcal{U}(\lambda, \mu, \delta) := \lambda \delta - c(\mu)$$
$$\text{s.t.} \quad \delta + G(\lambda, \mu) - U'(\lambda) \geq 0 ,$$
$$\delta + G(\lambda, \mu) - U'(\lambda) = 0 , \text{ if } \lambda > 0 ,$$
$$\mu \geq 0 , \ 0 \leq \lambda < \mu .$$

Let $\mu_0 := \inf\{\mu \geq 0 : U'(0) - G(0, \mu) > 0\}$. For $0 < \mu \leq \mu_0$ and any $\lambda > 0$, it follows from the constraints that $\delta = U'(\lambda) - G(\lambda, \mu) < U'(0) - G(0, \mu) \leq 0$. Hence $\lambda \delta < 0$ and for such a feasible solution we have $\mathcal{U}(\lambda, \mu, \delta) \leq \mathcal{U}(0, \mu, \delta') = -c(\mu) < 0 = \mathcal{U}(0, \delta', 0)$, where $\delta' \geq U'(0) - G(0, \mu)$. Therefore, over the range $0 < \mu \leq \mu_0$, the profit-maximizing service rate is $\mu = 0$ (with $\lambda = 0$ and toll δ').

It remains to characterize the facility-optimal solution over the range $\mu > \mu_0$, with $\lambda > 0$. In this case the individually optimal equilibrium constraints hold if and only if $\delta = U'(\lambda) - G(\lambda, \mu)$. Therefore, over this range the profit-maximization problem takes the form

$$\max_{\{\lambda, \mu, \delta\}} \quad \mathcal{U}(\lambda, \mu, \delta) = \lambda \delta - c(\mu)$$
$$\text{s.t.} \quad U'(\lambda) - G(\lambda, \mu) = \delta ,$$
$$\mu \geq \mu_0 , \ 0 < \lambda < \mu .$$

Substituting for δ in the objective function gives the following equivalent form for the problem with μ restricted to (μ_0, ∞):

$$\max_{\{\lambda,\mu,\delta\}} \quad \lambda U'(\lambda) - \lambda G(\lambda, \mu) - c(\mu)$$

$$\text{s.t.} \quad U'(\lambda) - G(\lambda, \mu) = \delta \,,$$

$$\mu > \mu_0 \,, \ 0 < \lambda < \mu \,.$$

Thus we see that the profit maximization problem takes exactly the same form as the social optimization problem, with the utility function $U(\lambda)$ replaced by $\lambda U'(\lambda)$. (The equality constraint now simply serves as a definition of the toll, δ.)

We leave it to the reader to complete the analysis of the facility optimization problem, following the approach used above for the social optimization problem. We note here that the facility optimization problem is subject not only to all the difficulties associated with the social optimization problem, brought on by the possible nonconvexity of the cost component of the objective function, $\lambda G(\lambda, \mu) + c(\mu)$, but also to additional complexities caused by the possible nonconcavity of $\lambda U'(\lambda)$. The latter complexities were discussed at length in Chapter 2.

One point worth mentioning, however, is the effect of the behavior of $\lambda U'(\lambda)$ as $\lambda \to 0$ on the behavior of the objective function for facility optimization,

$$\tilde{\mathcal{U}}(\lambda, \mu) := \lambda U'(\lambda) - \lambda G(\lambda, \mu) - c(\mu) \,.$$

For any fixed value of μ, a value of λ maximizes $\tilde{\mathcal{U}}(\lambda, \mu)$ if and only if it maximizes $\lambda U'(\lambda) - \lambda G(\lambda, \mu)$. This problem was covered in Section 2.1.3 of Chapter 2. If $\lambda U'(\lambda) \to 0$ as $\lambda \to 0$, then the search for the maximizing value λ may be confined to the interior of the feasible region for λ, namely, $(0, \mu)$. In this case, λ must satisfy the first-order necessary condition,

$$U'(\lambda) + \lambda U''(\lambda) - G(\lambda, \mu) - \lambda G'(\lambda, \mu) = 0 \,. \tag{5.42}$$

As noted above, if $\lambda U'(\lambda)$ is not concave, then this equation may have multiple solutions, some of which are local maxima or minima, and only one of which can be the global maximum.

Suppose $\lambda U'(\lambda) \to \kappa$ as $\lambda \to 0$, where $0 < \kappa < \infty$. In this case, $\tilde{\mathcal{U}}(\lambda, \mu)$ has a discontinuity at $\lambda = 0$, since (by convention) $\tilde{\mathcal{U}}(0, \mu) = -c(\mu)$, whereas $\tilde{\mathcal{U}}(0+, \mu) = \kappa - c(\mu) > \tilde{\mathcal{U}}(0, \mu)$. For a particular value of $\mu > 0$, we must compare $\tilde{\mathcal{U}}(0+, \mu)$ to the value of $\tilde{\mathcal{U}}(\lambda, \mu)$ at each of the positive values of λ that satisfy the optimality condition (5.42). (Again, see Section 2.1.3 of Chapter 2.)

Finally, suppose $\lambda U'(\lambda) \to \infty$ as $\lambda \to 0$. Then, for any $\mu > 0$, the profit-maximizing facility operator can achieve an arbitrarily large profit by choosing an arbitrarily small arrival rate, λ (equivalently, charging an arbitrarily large toll, δ). Since this can be done regardless of the value of μ, as long as $\mu > \lambda$, it follows that there is a sequence of values of λ and μ, with $0 < \lambda < \mu$, approaching zero along which the facility operator can earn larger and larger profits, with the profit equalling ∞ in the limit.

5.6 Endnotes

Dewan and Mendelson [54] considered the combined selection of the arrival rate and the service rate in the context of a model for a computer facility within an organization. They introduced the distinction between the *long-run* problem (selecting the service rate) and the *short-run* problem (selecting the arrival rate).

Section 5.2.2

As mentioned in Chapter 1, the optimal-service-rate problem was introduced by Hillier [93] in the context of a steady-state $M/M/1$ queue with deterministic reward and linear waiting cost function. A recent reference is George and Harrison [73], which compares the solution to the optimal design problem to the solution to the corresponding optimal control problem, in which the service rate is allowed to depend on the number of customers in the system.

Section 5.4.3

For a reference on the theory of nonlinear programming, see, e.g., Bazaraa et al. [16].

Section 5.5.2

The material in this section is based largely on Stidham [187].

Dewan and Mendelson [54]) were apparently the first to observe (in the context of a linear service-cost function) that equation (5.27) holds at a socially optimal solution, provided that equation (5.26) is satisfied. That is, the optimal toll to charge customers for entering equals the marginal cost of increasing the service rate, which Dewan and Mendelson called the "allocated fixed cost."

Section 5.5.2.1

For references on submodularity and its properties, see, e.g., Topkis [192] and Heyman and Sobel [92].

Section 5.5.2.2

Masuda and Whang [142] analyzed the problem of selecting optimal arrival rates and service rates in a Jackson network of queues. Their paper may be regarded as an extension of the $M/M/1$ model of this section to a network of queues.

CHAPTER 6

Multi-Facility Queueing Systems: Parallel Queues

In this chapter we show how some of the models and techniques presented in the previous chapters can be extended to multi-facility queueing systems: in particular, systems consisting of parallel service facilities fed by one or more arrival processes. Such systems are the simplest examples of networks of queues, which will also be the subject of the next two chapters.

We begin (Section 6.1) by looking at a system of n parallel facilities in which the arrival rates are decision variables. The starting point for all our models is the competitive equilibrium set of arrival rates, for a given set of entrance fees or tolls, which results when customers are individual optimizers. We then examine how the tolls may be chosen by the facility operator(s) to achieve various objectives, including social optimization and facility optimization (profit maximization). For the latter criterion we examine both the cooperative case, in which the facility operators act together to maximize total profit, and the competitive case, in which the facility operators act independently, each with the goal of maximizing the profit at the facility in question.

The next topic (Section 6.2) deals with the case where the service rates are the decision variables and the arrival rates are fixed. Finally (Section 6.3) we consider a model in which both the arrival rates (and/or the tolls) and service rates at the facilities are decision variables.

6.1 Optimal Arrival Rates

In this section we present a general model for selection of arrival rates to a system of parallel facilities. This model is a multi-facility version of the general single-facility model introduced in Chapter 2. It is also a generalization of the simple parallel-facility model introduced in Section 1.5 of Chapter 1. Later in this chapter we shall revisit this model and other special cases in which explicit solutions for the arrival rates are available.

As with the model introduced in Section 1.5 of Chapter 1, we consider a system with n independent parallel queues with fixed service rates and variable arrival rates, λ_j, $j \in J := \{1, \ldots, n\}$. Now, however, the total arrival rate, $\lambda = \sum_{j \in J} \lambda_j$, is a decision variable, as in the single-facility model of Chapter 2. The set of feasible values for λ_j is denoted by A_j. Our default assumption will be that $A_j = [0, \infty)$, $j \in J$. The system earns a utility, $U(\lambda)$, per unit time when the total arrival rate is λ. As usual, we assume that $U(\lambda)$ is nondecreasing, differentiable, and concave in $\lambda \geq 0$.

Also in contrast to the parallel-facility model introduced in Chapter 1, the waiting cost per customer may differ from facility to facility and is no longer required to be linear in the waiting time at the facility. As in the general single-facility model of Chapter 2, we assume that at each facility the average waiting cost per customer is a function of the arrival rate at that facility. For a given value of the arrival rate, λ_j, let $G_j(\lambda_j)$ denote the average waiting cost of a job at facility j, averaged over all customers who arrive during the period in question. We make the same assumptions about each $G_j(\lambda_j)$ as we made for the single-facility model; we repeat them here for convenient reference.

We assume that $G_j(\lambda_j)$ takes values in $[0, \infty]$ and is strictly increasing and differentiable in $\lambda_j \in A_j$. We allow infinite values for $G_j(\lambda_j)$ to accommodate, for example, a single-server system with service rate μ_j, in which we typically have $G_j(\lambda_j) = \infty$ for $\lambda_j \geq \mu_j$.

For example, in the case of an infinite time period, it might be that

$$G_j(\lambda_j) = \mathrm{E}[h_j(\boldsymbol{W}_j(\lambda_j))] \,, \tag{6.1}$$

where $h_j(t)$ is the waiting cost incurred by a job that spends a length of time t at facility j and, for each $\lambda_j \geq 0$, $\boldsymbol{W}_j(\lambda_j)$ is the steady-state random waiting time in the system for the queueing system induced by λ_j.

Let $H_j(\lambda_j) = \lambda_j G_j(\lambda_j)$: the average waiting cost per unit time at facility j, $j \in J$. We assume that $H_j(\lambda_j)$ is convex in $\lambda_j \geq 0$, $j \in J$.

There may also be an admission fee (or toll) δ_j which is charged to each customer who joins facility j, $j \in J$.

As usual, the solution to the decision problem depends on who is making the decision and what criteria are being used. The decision may be made by the individual customers, each concerned only with its own net utility (*individual optimality*), by an agent for the customers as a whole, who might be interested in maximizing the aggregate net utility to all customers (*social optimality*), or by a system operator interested in maximizing profit (*facility optimality*).

6.1.1 Individually Optimal Arrival Rates

As in the single-facility model, a natural starting point is the notion of individually optimal arrival rates. In this case (roughly speaking) the arrival rates are chosen so as to balance the marginal utility per unit time with the full price (sum of admission fee and waiting cost) at each facility that is used.

It is useful to separate the decision about the total arrival rate λ from the decision about the arrival rates λ_j at each facility j. As in the single-facility model, the total arrival rate is found by equating the marginal utility to the full price of admission, $\pi(\lambda)$. Determining the appropriate value for $\pi(\lambda)$ is more complicated than in the single-facility model; we shall deal with this issue presently. For the time being, we just assume (as usual) that $\pi(\lambda)$ is strictly increasing in λ and that $\lim_{\lambda \to \infty}(U'(\lambda) - \pi(\lambda)) < 0$. (We shall see that these properties hold for $\pi(\lambda)$ in the present model.) Thus the equilibrium arrival rate, λ^e, for the system as a whole satisfies the equilibrium equation,

$$U'(\lambda) = \pi(\lambda) \,,$$

if this equation has a solution in $A = [0, \infty)$. This is the case if and only if $U'(0) \geq \pi(0)$. If $U'(0) < \pi(0)$, then no user has any incentive to join and we set $\lambda^e := 0$. If $U'(0) \geq \pi(0)$, then there is a unique solution to this equation in A. Thus we have

$$
\begin{aligned}
U'(\lambda) &\leq \pi(\lambda) \,, \text{ and } \lambda \geq 0 \,, \\
U'(\lambda) &= \pi(\lambda) \,, \text{ if } \lambda > 0 \,.
\end{aligned}
$$

These are the equilibrium conditions which uniquely define the individually optimal arrival rate, λ^e.

Equivalent equilibrium conditions are the following:

$$
\begin{aligned}
\pi(\lambda) - U'(\lambda) &\geq 0 \,, \\
\lambda &\geq 0 \,, \\
\lambda(\pi(\lambda) - U'(\lambda)) &= 0 \,.
\end{aligned}
$$

Note that, if we know the full-price function, $\pi(\lambda)$, the process of finding the overall equilibrium arrival rate λ^e is identical to that for the single-facility model.

It remains to show how to calculate $\pi(\lambda)$ in the case of parallel facilities and to characterize the equilibrium arrival rate, λ_j^e, at each facility $j \in J$. Of course, if $\lambda^e = 0$, then $\lambda_j^e = 0$ for all $j \in J$. Suppose $\lambda^e > 0$, in which case λ^e is the unique solution to $U'(\lambda) = \pi(\lambda)$.

Now, for a given value of the total arrival rate, λ, let us consider the behavior of a marginal user who joins the system. At equilibrium, such a user will choose to go to a facility which offers the minimum full price, which implies that

$$\pi(\lambda) = \min_{\{j \in J\}} \{\delta_j + G_j(\lambda_j)\} \,.$$

If, to the contrary, a facility with a larger price has positive flow, then such a solution cannot be an equilibrium, since there is an incentive to divert some of this flow to a facility that achieves the minimum price. Thus $\lambda_j^e > 0$ only if $\delta_j + G_j(\lambda_j) = \pi(\lambda)$.

We can summarize these observations as follows. An allocation of flows, $(\lambda_j, j \in J)$, is individually optimal (and denoted $(\lambda_j^e, j \in J)$) if and only if

$$
\begin{aligned}
U'(\lambda) &\leq \pi \,, & (6.2) \\
\lambda(\pi - U'(\lambda)) &= 0 \,, & (6.3) \\
\sum_{j \in J} \lambda_j &= \lambda \,, & (6.4) \\
G_j(\lambda_j) + \delta_j &\geq \pi \,, \ j \in J \,, & (6.5) \\
\lambda_j(G_j(\lambda_j) + \delta_j - \pi) &= 0 \,, \ j \in J \,, & (6.6) \\
\lambda_j &\geq 0 \,, \ j \in J \,. & (6.7)
\end{aligned}
$$

Together with conditions (6.2) and (6.7), the complementary-slackness condition (6.3) ensures that either (i) the total flow λ is positive and the marginal utility, $U'(\lambda)$, equals the price, π, or (ii) $\lambda = 0$ and $U'(\lambda) \leq \pi$. Together with conditions (6.5) and (6.7), the complementary-slackness conditions (6.6) ensure that $\pi = \min_{j \in J}\{\delta_j + G_j(\lambda_j)\}$ and that only the facilities with the minimal price have positive flow. Note that it will be typical for an equilibrium solution to have more than one facility sharing the minimal price and, therefore, having positive flow.

Before proceeding, we need to answer a basic question: do the equilibrium conditions have a unique solution? The answer is "yes" and we shall verify this by showing that the equilibrium conditions are the (necessary and sufficient) optimality conditions for a related maximization problem with a concave objective function and linear constraints. The maximization problem is the following:

$$\max_{\{\lambda, \lambda_j, j \in J\}} \quad U(\lambda) - \sum_{j \in J} \int_0^{\lambda_j} (\delta_j + G_j(\nu))d\nu$$

$$\text{s.t.} \quad \sum_{j \in J} \lambda_j = \lambda \,,$$

$$\lambda_j \geq 0 \,, \ j \in J \,.$$

Since the objective function is jointly concave in $(\lambda, \lambda_j, \ j \in J)$, the *KKT* conditions are necessary and sufficient for a global maximum to this problem. These conditions have a unique solution and it is easily verified that they are identical to the equilibrium conditions, (6.2)-(6.7), for an individually optimal solution.

6.1.1.1 Individual Optimization with a Fixed Total Arrival Rate

Now consider the case in which the total arrival rate, λ, is fixed. An allocation of arrival rates, λ_j, $j \in J$, subject to the constraint, $\sum_{j \in J} \lambda_j = \lambda$, will be individually optimal if the full price is the same (say equal to π) at every facility with a positive arrival rate and at least π at every facility with zero arrival rate. More formally, as a function of the parameter π, let $\lambda_j^e(\pi)$, $j \in J$, denote the unique solution to the equilibrium conditions,

$$\pi \leq G_j(\lambda_j) + \delta_j \,, \ \lambda_j \geq 0 \,, \tag{6.8}$$

$$\pi = G_j(\lambda_j) + \delta_j \,, \ \text{if } \lambda_j > 0 \,. \tag{6.9}$$

The arrival rates, $\lambda_j^e(\pi)$, $j \in J$, will be individually optimal for the toll-free individual optimization problem if π is chosen so that the constraint, $\sum_{j \in J} \lambda_j^e(\pi) = \lambda$, is satisfied.

Note that $\lambda_j^e(\pi) = 0$ for $\pi < G_j(0) + \delta_j$ and $\lambda_j^e(\pi)$ is continuous and strictly increasing in $\pi \geq G_j(0) + \delta_j$, for all $j \in J$. It follows that $\lambda^e(\pi) := \sum_{j \in J} \lambda_j^e(\pi)$ is continuous and strictly increasing in $\pi \geq \min_{j \in J}\{G_j(0) + \delta_j\}$. Hence the individual optimization problem with a fixed total arrival rate λ can be solved parametrically by increasing π until the constraint, $\lambda^e(\pi) = \sum_{j \in J} \lambda_j^e(\pi) =$

λ, is (uniquely) satisfied. (Recall that we used this parametric approach in Section 1.5 of Chapter 1 for the special case of a system of parallel steady-state $M/M/1$ facilities with linear waiting costs and zero tolls.)

Remark 1 Note that adding a constant to each of the tolls, δ_j, $j \in J$, has no effect on the individually optimal allocation when the total arrival rate is fixed. The result is simply that the same constant is added to both sides of (6.8) and (6.9). This observation will be relevant when we analyze socially optimal and facility optimal allocations with a fixed total arrival rate and consider implementation of such allocations by charging appropriate tolls to individually optimizing customers. (See Sections 6.1.2.2 and 6.1.3.2.)

6.1.2 Socially Optimal Arrival Rates

Now let us consider the problem from the point of view of social optimization. The objective is to choose the system arrival rate, λ, and arrival rates, λ_j, at each facility $j \in J$, to maximize the steady-state expected net benefit (utility minus waiting costs) per unit time. Thus the problem takes the form:

$$\max_{\{\lambda, \lambda_j, j \in J\}} \quad U(\lambda) - \sum_{j \in J} \lambda_j G_j(\lambda_j)$$

$$\text{s.t.} \quad \sum_{j \in J} \lambda_j = \lambda \,, \tag{6.10}$$

$$\lambda_j \geq 0 \,, \ j \in J \,.$$

Note that the admission fees, δ_j, do not appear in the objective function for social optimality. This is consistent with the single-facility model of Chapter 2 and reflects the fact that they are internal transfer fees, which we assume are redistributed to the collective of all customers and thus do not affect the aggregate net benefit.

We shall use a Lagrange multiplier to eliminate the constraint on the total arrival rate. The Lagrangean problem is:

$$\max_{\{\lambda, \lambda_j, j \in J\}} \quad U(\lambda) - \sum_{j \in J} \lambda_j G_j(\lambda_j) + \alpha \left(\sum_{j \in J} \lambda_j - \lambda \right) \tag{6.11}$$

$$\text{s.t.} \quad \lambda_j \geq 0 \,, \ j \in J \,.$$

The solution is parameterized by α, which can be interpreted as an imputed cost per unit time per unit of arrival rate. Rewriting the objective function in equivalent form,

$$U(\lambda) - \alpha\lambda + \sum_{j \in J} \lambda_j(\alpha - G_j(\lambda_j)) \,,$$

reveals that it is separable in λ and λ_j, $j \in J$, for any fixed value of α. Thus we can maximize the objective function separately with respect to the arrival rate λ for the system as a whole and with respect to the arrival rate λ_j for each facility j.

For the system as a whole the problem takes the form of a single-facility arrival-rate-optimization problem (cf. Chapter 2), with utility function $U(\lambda)$ and constant waiting-cost function, $G(\lambda) \equiv \alpha$. It is clear that λ maximizes $U(\lambda) - \alpha\lambda$ if (and only if) either (i) $\lambda > 0$ and $U'(\lambda) = \alpha$, or (ii) $\lambda = 0$ and $U'(0) \leq \alpha$.

The problem for each facility j also takes the form of a single-facility arrival-rate-optimization problem, in this case with a linear utility function $\alpha\lambda_j$, and waiting-cost function $G_j(\lambda_j)$. The optimal value of λ_j for the Lagrangean problem therefore has the following characterization:

$$\alpha \leq G_j(\lambda_j) + \lambda_j \frac{\partial}{\partial\lambda_j} G_j(\lambda_j) , \; \lambda_j \geq 0 ,$$

$$\alpha = G_j(\lambda_j) + \lambda_j \frac{\partial}{\partial\lambda_j} G_j(\lambda_j) , \text{ if } \lambda_j > 0 .$$

A solution λ, λ_j, $j \in J$, to these conditions will be optimal for the original problem if α is chosen so that the constraint, $\sum_{j\in J} \lambda_j = \lambda$, is satisfied.

Thus we see that a socially optimal allocation (denoted λ^s, λ_j^s, $j \in J$) is characterized by the following (necessary and sufficient) optimality conditions:

$$U'(\lambda) \leq \alpha , \tag{6.12}$$

$$\lambda(\alpha - U'(\lambda)) = 0 , \tag{6.13}$$

$$\sum_{j\in J} \lambda_j = \lambda , \tag{6.14}$$

$$G_j(\lambda_j) + \lambda_j \frac{\partial}{\partial\lambda_j} G_j(\lambda_j) - \alpha \geq 0 , \; j \in J , \tag{6.15}$$

$$\lambda_j \left(G_j(\lambda_j) + \lambda_j \frac{\partial}{\partial\lambda_j} G_j(\lambda_j) - \alpha \right) = 0 , \; j \in J , \tag{6.16}$$

$$\lambda_j \geq 0 , \; j \in J . \tag{6.17}$$

These are the *KKT* conditions for the social optimization problem.

As in the single-facility model, we can interpret the term $G_j(\lambda_j)$ in (6.15) and (6.16) as the *internal effect* of a marginal increase in the arrival rate λ_j at facility j. It is the portion of the marginal increase in aggregate waiting cost that is borne by the "marginal" arriving customer when the arrival rate is λ_j. Similarly, we can interpret the term $\lambda_j \frac{\partial}{\partial\lambda_j} G_j(\lambda_j)$ as the *external effect*: the rate of increase in waiting cost borne by all users as a result of a marginal increase in the arrival rate λ_j.

Remark 2   An intuitive interpretation of the Lagrangean approach might be instructive. From the point of view of a decision maker setting the overall arrival rate, λ, the Lagrange multiplier, α, represents a price paid per unit of flow. By contrast, from the point of view of a decision maker setting the arrival rate, λ_j, at facility j, the Lagrange multiplier, α, represents a reward received per unit of flow. Thus, as α increases from zero, the value of λ which satisfies the optimality conditions, (6.12) and (6.13), will at first be large, but decreasing. On the other hand, the values of λ_j which satisfy the optimality conditions, (6.15), (6.16), and (6.17), will at first be small, but increasing, as will the value of the sum, $\sum_{j \in J} \lambda_j$. Thus, at first we will have $\lambda > \sum_{j \in J} \lambda_j$. Eventually, however, as α increases, we will reach a value at which $\lambda = \sum_{j \in J} \lambda_j$. At this point, all the optimality conditions for the original problem will be satisfied.

6.1.2.1 Comparison of Socially Optimal and Toll-Free Individually Optimal Solutions

Note that that the conditions for social optimality are of the same form as those for the competitive equilibrium. Choosing the admission fee, δ_j, at facility j to equal the external effect, $\lambda_j \frac{\partial}{\partial \lambda_j} G_j(\lambda_j)$, makes the individually optimal and socially optimal allocations coincide: $\lambda_j^e = \lambda_j^s$, $j \in J$. The Lagrange multiplier α then equals the full price π paid by each individually optimizing customer who enters the system.

As in the single-facility model, it is possible to make some strong statements about the relationship between the socially optimal and the individually optimal flow allocations, when no tolls are charged in the individually optimal case. The strongest of these statements concerns the relation between the total arrival rates under the two optimality criteria. Using the single-facility model as a paradigm, we expect that the total arrival rate will be smaller in the socially optimal allocation than in the individually optimal allocation without tolls. This indeed turns out to be true. The relationship between the arrival rates to a particular facility under the two allocations is more complicated, however.

To compare the total arrival rates under the two criteria, it is convenient once again to consider separately the problem of setting the total arrival rate and the problem of choosing arrival rates to the various facilities. From the optimality conditions for the two problems, we see that in both cases the total arrival rate is found by equating the marginal utility to a price parameter. In the case of social optimization, the equation is

$$U'(\lambda) = \alpha \,, \tag{6.18}$$

whereas in the case of individual optimization, the equation is

$$U'(\lambda) = \pi \,. \tag{6.19}$$

The following lemma will help us make the connection between the problem of allocating flows among the facilities and the problem of selecting the total arrival rate.

Lemma 6.1 $\lambda^s \leq \lambda^e$ if and only if $\alpha \geq \pi$. $\lambda^s < \lambda^e$ if and only if $\alpha > \pi$.

Proof An immediate consequence of the concavity of $U(\lambda)$ and the defining equations, (6.18) and (6.19). ∎

The difference between the two problems, of course, lies in how the two parameters, α and π, are calculated. Examining the two sets of optimality conditions, we see that

$$\alpha = \min_{\{j \in J\}} \{G_j(\lambda_j) + \lambda_j G'_j(\lambda_j)\}$$

$$\text{s.t.} \quad \lambda_j(G_j(\lambda_j) + \lambda_j G'_j(\lambda_j) - \alpha) = 0 \,, \; j \in J \,,$$

$$\sum_{j \in J} \lambda_j = \lambda \,,$$

$$\lambda_j \geq 0 \,, \; j \in J \,,$$

whereas (when, as we are now assuming, $\delta_j = 0$ for all j)

$$\pi = \min_{\{j \in J\}} \{G_j(\lambda_j)\}$$

$$\text{s.t.} \quad \lambda_j(G_j(\lambda_j) - \pi) = 0 \,, \; j \in J \,,$$

$$\sum_{j \in J} \lambda_j = \lambda \,,$$

$$\lambda_j \geq 0 \,, \; j \in J \,.$$

For any given value of λ, these formulations tell us how to allocate a total arrival rate of λ to the individual facilities under the socially optimal and the (toll-free) individually optimal solutions, respectively.

Remark 3   Allocating a fixed total arrival among parallel facilities is an interesting problem in its own right, which we shall study presently (see Section 6.1.2.2 below). We introduced a special case of this problem in Chapter 1, in which each facility was an *M/M/1* queue and the waiting-cost function $G_j(\lambda_j)$ was the same linear function at each facility. So the two formulations above are natural generalizations of this problem.

Our goal now is to show that $\alpha \geq \pi$. Note that the two formulations differ only in the presence of the extra term, $\lambda_j G'_j(\lambda_j)$ (the external effect), in the minimand and the complementary-slackness condition in the socially optimal formulation. This term is always nonnegative, since the waiting-cost function, $G_j(\lambda_j)$, is nondecreasing. This observation makes it intuitively plausible that $\alpha \geq \pi$.

Following the above intuition, let us consider two individually optimal allocations, for a fixed total arrival rate λ. The waiting-cost functions in the first allocation are $G_j(\lambda_j)$, $j \in J$. In the second allocation, the waiting-cost functions are $\tilde{G}_j(\lambda_j)$, $j \in J$. We assume that $\tilde{G}_j(\lambda_j) \geq G_j(\lambda_j)$, for all $\lambda_j \geq 0$, $j \in J$. Let $\pi(\lambda)$ ($\tilde{\pi}(\lambda)$) denote the full price of admission for the problem with waiting-cost functions $G_j(\lambda_j)$ ($\tilde{G}_j(\lambda_j)$), $j \in J$. The full price, $\pi(\lambda)$, for the former problem may be found by solving the following mathematical program (denoted Problem (**P**)):

$$\pi = \min_{\{j \in J\}} \{G_j(\lambda_j)\}$$

$$(\mathbf{P}) \quad \text{s.t.} \quad \lambda_j(G_j(\lambda_j) - \pi) = 0 \,, \; j \in J \,,$$

$$\sum_{j \in J} \lambda_j = \lambda \,,$$

$$\lambda_j \geq 0 \,, \; j \in J \,.$$

For the latter problem (denoted Problem $(\tilde{\mathbf{P}})$), the same mathematical program with $G_j(\lambda_j)$ replaced by $\tilde{G}_j(\lambda_j)$, $j \in J$, may be used to solve for $\tilde{\pi}(\lambda)$.

Theorem 6.2 *For all $\lambda > 0$, $\tilde{\pi}(\lambda) \geq \pi(\lambda)$.*

Proof The proof is by contradiction. Let $\lambda > 0$ be given and let $\pi := \pi(\lambda)$ and $\tilde{\pi} := \tilde{\pi}(\lambda)$. Suppose $\tilde{\pi} < \pi$. Let λ_j denote the arrival rate at facility j in the solution to Problem $(\mathbf{P})$ and let $\tilde{\lambda}_j$ denote the arrival rate at facility j in the solution to Problem $(\tilde{\mathbf{P}})$. Now $\tilde{G}_j(\tilde{\lambda}_j) \geq \tilde{\pi}$ for all $j \in J$, with equality for j such that $\tilde{\lambda}_j > 0$. Thus for all j such that $\tilde{\lambda}_j > 0$ we have

$$\tilde{G}_j(\tilde{\lambda}_j) = \tilde{\pi} < \pi \leq G_j(\lambda_j) \leq \tilde{G}_j(\lambda_j) \,.$$

Since $\tilde{G}_j$ is a nondecreasing function, it follows that $\tilde{\lambda}_j < \lambda_j$ for all j such that $\tilde{\lambda}_j > 0$. In particular this implies that $\lambda_j > 0$ if $\tilde{\lambda}_j > 0$, or equivalently, $1\{\lambda_j > 0\} \geq 1\{\tilde{\lambda}_j > 0\}$. That is, a facility that has positive flow in Problem $(\tilde{\mathbf{P}})$ also has positive flow in Problem $(\mathbf{P})$, under the assumption that $\tilde{\pi} < \pi$. Therefore, we have

$$
\begin{aligned}
\sum_{j \in J} \lambda_j 1\{\tilde{\lambda}_j > 0\} \quad &> \quad \sum_{j \in J} \tilde{\lambda}_j 1\{\tilde{\lambda}_j > 0\} \\
&= \quad \lambda \\
&= \quad \sum_{j \in J} \lambda_j 1\{\lambda_j > 0\} \\
&\geq \quad \sum_{j \in J} \lambda_j 1\{\tilde{\lambda}_j > 0\} \,,
\end{aligned}
$$

which is a contradiction. Thus $\tilde{\pi} \geq \pi$. $\blacksquare$

Remark 4 Theorem 6.2 is a kind of monotonicity result. It says that the full price of admission to the system under a Nash equilibrium allocation of flows is monotonically increasing in the waiting-cost function at each facility. This is an intuitively plausible property. After all, the users are all trying to find a path through the system that has least cost. It makes sense that increasing any particular cost should lead to an increase in the cost of the path(s) they follow in an equilibrium solution. As we shall see in Section 7.4.1 (Braess's Paradox) in Chapter 7, however, this monotonicity property may not hold in more general networks, in which users choose between paths, each of which may consist of a sequence of facilities. In such cases the topology of the network may lead to complicated interactions between different customers, whose chosen paths may or may not share particular facilities. If we were concerned with socially optimal allocations, then the monotonicity property would hold trivially. The fact that customers are self-optimizing, however, can lead to surprising and nonintuitive behavior.

As a consequence of Lemma 6.1 and Theorem 6.2, we have the following result:

Corollary 6.3 $\lambda^s \leq \lambda^e$

6.1.2.2 Social Optimization with a Fixed Total Arrival Rate

In this section we consider the case in which the total arrival rate, λ, is fixed. For social optimization, the objective is to choose the arrival rates, λ_j, at each facility $j \in J$, to minimize the steady-state expected total waiting cost per unit time, subject to the constraint that the arrival rates at the facilities sum to λ. Thus the problem takes the form:

$$\min_{\{\lambda_j, j \in J\}} \quad \sum_{j \in J} \lambda_j G_j(\lambda_j)$$

$$\text{s.t.} \quad \sum_{j \in J} \lambda_j = \lambda \,, \tag{6.20}$$

$$\lambda_j \geq 0 \,, \ j \in J \,.$$

In Chapter 1 we considered the special case of this model in which each facility is an *M/M/1* queue with the same linear waiting-cost function. In that case an equivalent objective is to minimize the steady-state total expected number of customers in the system.

As usual, we shall use a Lagrange multiplier to eliminate the constraint on the total arrival rate. The Lagrangian problem is:

$$\min_{\{\lambda_j, j \in J\}} \quad \sum_{j \in J} \lambda_j G_j(\lambda_j) - \alpha \left(\sum_{j \in J} \lambda_j - \lambda \right) \tag{6.21}$$

$$\text{s.t.} \quad \lambda_j \geq 0 \,, \ j \in J \,.$$

Rewriting the objective function in equivalent form,

$$\sum_{j \in J} \lambda_j (G_j(\lambda_j) - \alpha) + \alpha \lambda \,,$$

reveals that it is separable in λ_j, $j \in J$, for any fixed value of α. Thus we can minimize the objective function separately with respect to the arrival rate λ_j for each facility j.

The optimal value of λ_j for the Lagrangian problem therefore has the following characterization:

$$\alpha \ \leq \ G_j(\lambda_j) + \lambda_j \frac{\partial}{\partial \lambda_j} G_j(\lambda_j) \,, \ \lambda_j \geq 0 \,, \tag{6.22}$$

$$\alpha \ = \ G_j(\lambda_j) + \lambda_j \frac{\partial}{\partial \lambda_j} G_j(\lambda_j) \,, \ \text{if } \lambda_j > 0 \,. \tag{6.23}$$

Let $\lambda_j^s(\alpha)$, $j \in J$, denote the unique solution to these conditions. The arrival rates, $\lambda_j^s(\alpha)$, $j \in J$, will be optimal for the original social optimization problem if α is chosen so that the constraint, $\sum_{j \in J} \lambda_j^s(\alpha) = \lambda$, is satisfied.

6.1.2.3 Comparison of Socially Optimal and Toll-Free Individually Optimal Solutions with a Fixed Total Arrival Rate

Now consider the individual optimization problem with zero tolls. Here we apply the parametric analysis of Section 6.1.1.1 to the special case with $\delta_j = 0$, $j \in J$. (To facilitate comparison with social optimization, we shall denote the parameter as α rather than π.) Let $\lambda_j^e(\alpha)$, $j \in J$, denote the unique solution to the equilibrium conditions (6.8) and (6.9), which now take the form:

$$\alpha \leq G_j(\lambda_j), \; \lambda_j \geq 0, \tag{6.24}$$

$$\alpha = G_j(\lambda_j), \; \text{if } \lambda_j > 0. \tag{6.25}$$

The arrival rates, $\lambda_j^e(\alpha)$, $j \in J$, will be individually optimal for the toll-free individual optimization problem if α is chosen so that the constraint, $\sum_{j \in J} \lambda_j^e(\alpha) = \lambda$, is satisfied. (In equilibrium, all facilities with positive arrival rates have equal waiting costs per customer, and all facilities with zero arrival rates have waiting costs at least this large.)

The following lemma is an immediate consequence of the assumption that $G_j(\cdot)$ is strictly increasing.

Lemma 6.4 For $0 < \alpha \leq G_j(0)$, $\lambda_j^s(\alpha) = \lambda_j^e(\alpha) = 0$. For $\alpha > G_j(0)$, $0 < \lambda_j^s(\alpha) < \lambda_j^e(\alpha)$.

Remark 5 The conditions satisfied by $\lambda_j^s(\alpha)$ and $\lambda^e(\alpha)$ are equivalent to those satisfied by the socially optimal and individually optimal arrival rates, respectively, in a single-facility model with linear utility function, $U(\lambda_j) = \alpha\lambda_j$, and waiting-cost function, $G_j(\lambda_j)$. Therefore, Lemma 6.4 is in fact a direct consequence of the general result for single-facility systems, proved in Chapter 2, that the socially optimal arrival rate is always smaller than the individually optimal arrival rate, when both are positive.

Let α^s denote the value of α for which $\sum_{j \in J} \lambda^s(\alpha) = \lambda$, and let α^e denote the value of α for which $\sum_{j \in J} \lambda_j^e(\alpha) = \lambda$. Then $\lambda_j^s = \lambda_j^s(\alpha^s)$, $j \in J$, and $\lambda_j^e = \lambda_j^e(\alpha^e)$, $j \in J$, solve the original social and individual optimization problems, respectively, with fixed arrival rate λ.

Lemma 6.5 $\alpha^s > \alpha^e$.

Proof Suppose $\alpha^s \leq \alpha^e$. Then it follows from (6.22)-(6.23) and (6.24)-(6.25) that $\lambda_j^e = 0$ implies $\lambda_j^s = 0$. On the other hand, if $\lambda_j^e > 0$, then by Lemma 6.4, $\lambda_j^e = \lambda_j^e(\alpha^e) \geq \lambda_j^e(\alpha^s) > \lambda_j^s(\alpha^s) = \lambda_j^s$. Hence $\sum_{j \in J}(\lambda_j^e - \lambda_j^s)$ is positive, since it contains at least one positive term and no negative terms. But this contradicts the requirement that $\sum_{j \in J} \lambda_j^e = \sum_{j \in J} \lambda_j^s = \lambda$. Thus $\alpha^s > \alpha^e$. ∎

Remark 6 The "weak" version of the inequality proved in Lemma 6.5 – that is, $\alpha^s \geq \alpha^e$ – is a corollary of Theorem 6.2 of the previous section.

M/GI/1 Queues with Linear Waiting Costs. We now consider the special case where each facility consists of an *M/GI/1* queue operating in steady state with a linear waiting-cost function. Specifically, we assume that the arrival process at facility j is Poisson with parameter λ_j, $j \in J$. Successive service times at each facility are i.i.d. with a general distribution. Let S_j be a generic service time at facility j, with $\mathrm{E}[S_j] = 1/\mu_j$ and $\mathrm{E}[S_j^2] = 2\beta/\mu_j^2$, $j \in J$. Let h_j denote the cost per customer per unit of time waiting in the system at facility j, $j \in J$. Further assume that the facilities are numbered so that $h_1/\mu_1 \leq \cdots \leq h_n/\mu_n$.

The expected steady-state waiting time in the system at facility j is

$$W_j(\lambda_j) = \frac{1}{\mu_j} + \frac{\lambda_j \beta}{\mu_j(\mu_j - \lambda_j)} . \tag{6.26}$$

The waiting cost per customer at facility j is given by

$$G_j(\lambda_j) = h_j \cdot W_j(\lambda_j) . \tag{6.27}$$

Remark 7 Note that we are assuming that the coefficient of variation of the service times is the same at all n facilities. (It equals $(2\beta-1)^{1/2}$.) Two possible scenarios which might make this assumption plausible are: (1) the nature of the service process is such that it is reasonable to assume that all service-time distributions come from a parametric family (e.g., the exponential family) with a common coefficient of variation; or (2) the servers process "work" in a deterministic fashion (at rate μ_j at facility $j \in J$); customers require a random duration of the server's attention because they present the server with a random amount of work; customer's work requirements have expected value 1 and variance $(2\beta - 1)$. In Section 1.5 of Chapter 1 we considered a system of parallel $M/M/1$ facilities, in which the assumption of a common coefficient of variation is automatically satisfied, since all exponential distributions have coefficient of variation equal to one (and $\beta = 1$).

For this model we have the following explicit formulas for $\lambda_j^s(\alpha)$ and $\lambda_j^e(\alpha)$:

$$\lambda_j^s(\alpha) = \left(\mu_j - \mu_j \left[\frac{\beta}{\beta + \alpha h_j^{-1}\mu_j - 1} \right]^{1/2} \right)^+ \tag{6.28}$$

$$\lambda_j^s(\alpha) = \left(\mu_j - \mu_j \left[\frac{\beta}{\beta + \alpha h_j^{-1}\mu_j - 1} \right] \right)^+ . \tag{6.29}$$

For each $j \in J$, define $\rho_j^s := \lambda_j^s/\mu_j$, $\rho_j^e := \lambda_j^e/\mu_j$. Dividing both sides of (6.28) and (6.29) by μ_j yields the following result, which is used in the proof of the main theorem.

Lemma 6.6 For $k > j$,

(i) $\lambda_j^s > 0 \Rightarrow \rho_j^s \geq \rho_k^s$ with equality only if $h_k/\mu_k = h_j/\mu_j$,

(ii) $\lambda_j^e > 0 \Rightarrow \rho_j^e \geq \rho_k^e$ with equality only if $h_k/\mu_k = h_j/\mu_j$.

That is, under both solutions, faster servers are busier.

Theorem 6.7 *If $\lambda_j^s \geq \lambda_j^e$, then $\lambda_k^s \geq \lambda_k^e$ for all $k > j$. That is, lower numbered (faster) servers have smaller arrival rates under a socially optimal than an individually optimal allocation; higher numbered (slower) servers have larger arrival rates.*

Proof   Assume $k > j$. First note that $\lambda_j^s = 0$ and $\lambda_j^s \geq \lambda_j^e$ imply $\lambda_j^e = 0$ and hence $\lambda_k^s = \lambda_k^e = 0$ by Lemma 6.6 and thus the Theorem holds. Now suppose $\lambda_j^s > 0$ and $\lambda_j^s \geq \lambda_j^e$, but $\lambda_k^s < \lambda_k^e$. We shall show that this leads to a contradiction. First we observe that Lemmas 6.5 and 6.6 imply that all of λ_j^s, λ_j^e, λ_k^s, and λ_k^e are positive. From (6.23) and (6.25) we have

$$h_j(W_j(\lambda_j^s) + \lambda_j^s W_j'(\lambda_j^s)) \;=\; \alpha^s = h_k(W_k(\lambda_k^s) + \lambda_k^s W_k'(\lambda_k^s)), \quad (6.30)$$

$$h_j W_j(\lambda_j^s) \;=\; \alpha^s = h_k W_k(\lambda_k^s). \quad\quad (6.31)$$

Define

$$f(\rho) := \beta\rho/(1-\rho), \; 0 \leq \rho < 1. \quad\quad (6.32)$$

Note that $f(\rho)$ is nonnegative, differentiable, strictly increasing, and strictly convex in $0 \leq \rho < 1$. We shall also need the following property of $f(\cdot)$, which is easily verified from (6.32) by differentiation:

$$\rho f'(\rho)/f(\rho) \text{ is strictly increasing in } 0 \leq \rho < 1. \quad\quad (6.33)$$

It follows from (6.26) that $W_i(\lambda_i) = (1 + f(\rho_i))/\mu_i$ and $W_i(\lambda_i) + \lambda_i W_i'(\lambda_i) = (1 + f(\rho_i) + \rho_i f'(\rho_i))/\mu_i$, and , where $\rho_i := \lambda_i/\mu_i$, $i \in J$. Hence we can rewrite (6.30) and (6.31) in equivalent form:

$$(h_j\mu_j^{-1})(1 + f(\rho_j^s) + \rho_j^s f'(\rho_j^s))$$
$$= \alpha^s \;=\; (h_k\mu_k^{-1})(1 + f(\rho_k^s) + \rho_k^s f'(\rho_k^s)) (6.34)$$
$$(h_j\mu_j^{-1})(1 + f(\rho_j^e)) = \alpha^e \;=\; (h_k\mu_k^{-1})(1 + f(\rho_k^e)). \quad (6.35)$$

By the hypotheses, $\rho_j^s \geq \rho_j^e$ and $\rho_k^s < \rho_k^e$. It follows by subtracting (6.35) from (6.34) and using the fact that $f(\cdot)$ is strictly increasing and strictly convex that

$$(h_j\mu_j^{-1})\rho_j^e f'(\rho_j^e) \leq \alpha^s - \alpha^e < (h_k\mu_k^{-1})\rho_k^e f'(\rho_k^e). \quad (6.36)$$

But (6.35) and the fact that $h_j\mu_j^{-1} \leq h_k\mu_k^{-1}$ imply that

$$(h_j\mu_j^{-1})f(\rho_j^e) \geq (h_k\mu_k^{-1})f(\rho_k^e). \quad\quad (6.37)$$

Dividing (6.36) by (6.37) yields

$$\rho_j^e f'(\rho_j^e)/f(\rho_j^e) < \rho_k^e f'(\rho_k^e)/f(\rho_k^e),$$

from which it follows, using property (6.33), that $\rho_j^e < \rho_k^e$, which is a contradiction of Lemma 6.6. Therefore it cannot be true that both $\rho_j^s \geq \rho_j^e$ and $\rho_k^s < \rho_k^e$. We conclude that $\rho_j^s \geq \rho_j^e$ implies $\rho_k^s \geq \rho_k^e$, which is the desired result.   ∎

Remark 8   Note that $f(\rho)$ as defined by (6.32) equals the average delay (waiting time in the queue), expressed in units of mean service time. Theorem 6.7 uses only the fact that $\rho f'(\rho)/f(\rho)$ is nondecreasing in ρ; thus it extends to any queueing system with this property. A sufficient condition for $\rho f'(\rho)/f(\rho)$ to be nondecreasing is that $f(\rho)$ be log convex.

We studied the special case of a system of parallel $M/M/1$ facilities in Section 1.5 of Chapter 1 and graphically displayed the individually and socially optimal arrival rates for an example with three facilities. Figures 1.7, 1.9, and 1.10 illustrate the behavior predicted by Theorem 6.7.

Heavy Traffic Limits. By definition a socially optimal allocation has a lower total cost, $C^s := \sum_{j \in J} \lambda_j^s h_j W_j(\lambda_j^s)$, than that of an equilibrium allocation, $C^e := \sum_{j \in J} \lambda_j^e h_j W_j(\lambda_j^e)$. How much lower? That is, what can we say about the ratio, C_e/C_s? facilities.

Theorem 6.8 *Suppose each facility operates as an* M/M/1 *queue in steady state. Assume* $h_j = 1$, $j \in J$. *Let* $\mu := \sum_{j \in J} \mu_j$. *Then*

$$\lim_{\lambda \uparrow \mu} C_e/C_s = n\mu/(\sum_{j \in J} \sqrt{\mu_j})^2 \leq n \ .$$

Proof   (See Exercise 1 below.)   ∎

Note that C_e/C_s attains its lower bound, 1, in the symmetric case: $\mu_j = \mu/n$, $j = 1, \ldots, n$. (Of course, in this case the individually optimal and socially optimal allocations coincide at all levels of traffic: $\lambda_j^s = \lambda_j^e = \lambda/n$.) The upper bound, n, is the least upper bound, since it is the limit as $\epsilon \to 0$ of $n\mu/(\sum_{j \in J} \sqrt{\mu_j})^2$ for the (very unsymmetric) case $\mu_1 = \mu - (n-1)\epsilon$, $\mu_j = \epsilon$, $j = 2, \ldots, n$.

Exercise 1   Prove Theorem 6.8. (**Hint:** As $\lambda \uparrow \mu$, all facilities will eventually be open in both the individually optimal and the socially optimal allocations.)

6.1.3 Facility Optimal Arrival Rates and Tolls

As in the case of a single-facility queueing system, one may also consider the system from the point of view of a facility operator whose goal is to set tolls that will maximize its revenue. Now, however, we have multiple facilities, each of which may have its own operator. Or there may be a single operator who manages (and sets prices for) all the facilities. Naturally the resulting arrival rates differ according to which of these scenarios prevails. We shall refer to the former scenario as *competing* facilities and the latter as *cooperating* facilities. We shall consider only the case of cooperating facilities.

6.1.3.1 Cooperating Facilities

Suppose a single operator (or a team of cooperating operators) sets the tolls, δ_j, for all facilities $j \in J$. The goal is to maximize the total profit. We assume that the costs of operating the facilities are fixed. (Later (cf. Section 6.3.3) we shall examine models in which the service rates, μ_j, and hence the operating costs, at the facilities are variable.) Thus maximizing the total profit is equivalent to maximizing the total revenue,

$$\sum_{j \in J} \lambda_j \delta_j \, ,$$

received from the tolls paid by the entering customers.

Assuming (as usual) that the arriving customers are individual optimizers, any particular set of values δ_j for the tolls will result in arrival rates λ_j that satisfy the equilibrium conditions, (6.2)-(6.7). To keep the exposition simple, let us focus on the case where the total arrival rate, λ, is positive. Then the equilibrium conditions may be written in the following form:

$$U'(\lambda) = \pi \, ,$$

$$\sum_{j \in J} \lambda_j = \lambda \, ,$$

$$G_j(\lambda_j) + \delta_j \geq \pi \, , \, j \in J$$

$$;,$$

$$\lambda_j(G_j(\lambda_j) + \delta_j - \pi) = 0 \, , \, j \in J \, ,$$

$$\lambda_j \geq 0 \, , \, j \in J \, .$$

Thus, we have the following formulation of the facility optimization problem:

$$\max_{\{\pi, \lambda, \lambda_j, \delta_j, j \in J\}} \quad \sum_{j \in J} \lambda_j \delta_j$$

$$\text{s.t.} \quad \lambda_j(G_j(\lambda_j) + \delta_j - \pi) = 0 \, , \, j \in J \, ,$$

$$\delta_j \geq \pi + G_j(\lambda_j) \, , \, j \in J \, ,$$

$$\sum_{j \in J} \lambda_j = \lambda \, ,$$

$$U'(\lambda) = \pi \, ,$$

$$\lambda_j \geq 0 \, , \, j \in J \, .$$

Summing the first set of constraints (the complementary-slackness conditions) over $j \in J$, subtracting this sum (which equals zero) from the objective function, and simplifying leads to the following equivalent problem:

$$\max_{\{\pi, \lambda, \lambda_j, \delta_j, j \in J\}} \quad \sum_{j \in J} \lambda_j(\pi - G_j(\lambda_j))$$

$$\text{s.t.} \quad \delta_j \geq \pi + G_j(\lambda_j) \, , \, j \in J \, ,$$

$$\delta_j = \pi + G_j(\lambda_j) \, , \, \text{if } \lambda_j > 0 \, , \, j \in J \, ,$$

$$\sum_{j \in J} \lambda_j = \lambda \, ,$$

$$U'(\lambda) = \pi \, ,$$

$$\lambda_j \geq 0 \, , \, j \in J \, .$$

Since the tolls, $\delta_j, j \in J$, no longer appear in the objective function, we can solve this problem by first solving the problem,

$$\max_{\{\pi,\lambda,\lambda_j,j\in J\}} \quad \sum_{j\in J}\lambda_j(\pi - G_j(\lambda_j))$$

$$\text{s.t.} \quad \sum_{j\in J}\lambda_j = \lambda \,,$$

$$U'(\lambda) = \pi \,,$$

$$\lambda_j \geq 0 \,, \ j \in J \,,$$

and then choosing δ_j, such that

$$\delta_j \ \geq \ \pi - G_j(\lambda_j) \,, \ j \in J \,, \tag{6.38}$$

$$\delta_j \ = \ \pi - G_j(\lambda_j) \,, \text{ if } \lambda_j > 0 \,, \tag{6.39}$$

for all $j \in J$. (Note that the solution to (6.38) and (6.39) may not be unique, inasmuch as δ_j can be *any* value greater than or equal to $\pi - G_j(\lambda_j)$ when $\lambda_j = 0$.)

Finally, using the constraint, $\sum_{j\in J}\lambda_j = \lambda$, we can rewrite the maximization problem in equivalent form as

$$\max_{\{\pi,\lambda,\lambda_j,j\in J\}} \quad \lambda\pi - \sum_{j\in J}\lambda_j G_j(\lambda_j)$$

$$\textbf{(F)} \qquad \text{s.t.} \quad \sum_{j\in J}\lambda_j = \lambda \,,$$

$$U'(\lambda) = \pi \,,$$

$$\lambda_j \geq 0 \,, \ j \in J \,.$$

Substituting for π in the objective function, we see that Problem **(F)** is equivalent to a social optimization problem in which the utility function $U(\lambda)$ is replaced by $\lambda U'(\lambda)$ (cf. the analysis of the facility optimization problem in the case of a single facility in Section 2.1.3 of Chapter 2). Hence we can apply the analysis of the previous section, bearing in mind that the necessary *KKT* conditions may not have a unique solution, because the function, $\lambda U'(\lambda)$, may not be concave. (It may also fail to be nondecreasing.)

We shall have more to say about this problem below in Section 6.1.3.3, but first we shall consider cooperating facilities when the total arrival rate λ is fixed.

6.1.3.2 Cooperating Facilities: Fixed Total Arrival Rate

In this subsection we consider the cooperative facility optimal solution in the case of a fixed total arrival rate λ. The problem takes the following form:

$$\max_{\{\pi,\lambda_j,\delta_j,j\in J\}} \quad \sum_{j\in J}\lambda_j\delta_j$$

$$\text{s.t.} \quad \pi \leq \delta_j + G_j(\lambda_j) \,, \ j \in J \,,$$

$$\pi = \delta_j + G_j(\lambda_j) \text{ , if } \lambda_j > 0 \text{ , } j \in J \text{ ,}$$

$$\sum_{j \in J} \lambda_j = \lambda \text{ ,}$$

$$\lambda_j \geq 0 \text{ , } j \in J \text{ .}$$

Let us begin by considering the case $n = 2$. To keep the exposition simple, we shall assume that the arrival rates at both facilities are positive in an optimal solution. In this case we can write the problem as

$$\max_{\{\pi, \lambda_j, \delta_j, j=1,2\}} \quad \lambda_1 \delta_1 + \lambda_2 \delta_2$$

$$\text{s.t.} \quad \pi = \delta_1 + G_1(\lambda_1) \text{ ,}$$

$$\pi = \delta_2 + G_2(\lambda_2) \text{ ,}$$

$$\lambda_1 + \lambda_2 = \lambda \text{ ,}$$

$$\lambda_1 \geq 0 \text{ , } \lambda_2 \geq 0 \text{ .}$$

Using the equality constraints to eliminate the variables π and λ_2, we obtain the following equivalent formulation:

$$\max_{\{\lambda_1, \delta_j, j=1,2\}} \quad \lambda_1(\delta_1 - \delta_2) + \lambda \delta_2$$

$$\text{s.t.} \quad \delta_1 - \delta_2 = G_2(\lambda - \lambda_1) - G_1(\lambda_1) \text{ ,}$$

$$\lambda_1 \geq 0 \text{ .}$$

Substituting for $\delta_1 - \delta_2$ in the objective function leads to the following formulation with λ_1 and δ_2 as the only decision variables:

$$\max_{\{\lambda_1, \delta_2\}} \quad \lambda_1(G_2(\lambda - \lambda_1) - G_1(\lambda_1)) + \lambda \delta_2$$

$$\text{s.t.} \quad \lambda_1 \geq 0 \text{ .}$$

Differentiating the objective function with respect to λ_1 yields the first-order optimality condition,

$$G_2(\lambda - \lambda_1) - G_1(\lambda_1) + \lambda_1(-G_2'(\lambda - \lambda_1) - G_1'(\lambda_1)) = 0 \text{ ,}$$

which must be satisfied by an interior maximum, $0 < \lambda_1 < \lambda$. (Note that there may be multiple solutions to this equation, since the objective function is not necessarily concave (or even unimodal) in λ_1.) For any given value of δ_2, the corresponding value of the toll, δ_1, to be charged at facility 1 is given by

$$\delta_1 = \delta_2 + G_2(\lambda - \lambda_1) - G_1(\lambda_1) \text{ .} \tag{6.40}$$

Note that this equation implies that $\delta_1 > \delta_2$, since $G_2(\lambda - \lambda_1) > G_1(\lambda_1)$ in an optimal solution to the given maximization problem.

Recall, however, that the problem is symmetric in facilities 1 and 2, so we could equally well have eliminated λ_1 to obtain an equivalent formulation with λ_2 and δ_1 as the only decision variables:

$$\max_{\{\lambda_2, \delta_1\}} \quad \lambda_2(G_1(\lambda - \lambda_2) - G_2(\lambda_2)) + \lambda \delta_1$$

$$\text{s.t.} \qquad \lambda_2 \geq 0 \,.$$

Differentiating this objective function with respect to λ_2 yields the first-order optimality condition,

$$G_1(\lambda - \lambda_2) - G_2(\lambda_1) + \lambda_2(-G_1'(\lambda - \lambda_2) - G_2'(\lambda_2)) = 0 \,,$$

which must be satisfied by an interior maximum, $0 < \lambda_2 < \lambda$. For any given value of δ_1, the corresponding value of the toll, δ_2, to be charged at facility 2 is given by

$$\delta_2 = \delta_1 + G_1(\lambda - \lambda_2) - G_2(\lambda_2) \,. \tag{6.41}$$

But, by symmetry, $G_1(\lambda - \lambda_2) > G_2(\lambda_2)$ in an optimal solution to the given maximization problem. Hence, $\delta_2 > \delta_1$, a contradiction of our previous conclusion that $\delta_1 > \delta_2$.

We must conclude that there is no set of values $\lambda_j, \delta_j, j = 1, 2$, that simultaneously satisfy the two first-order optimality conditions for λ_1 and λ_2. What has gone wrong here? Looking again at the objective function in the first reformulation of the problem, we see that there is also a term, $\lambda\delta_2$, which we have been neglecting. Since δ_2 is a decision variable, in addition to λ_1, it follows that the objective function can be made arbitrarily large by selecting a suitably large value of δ_2, whatever the value of λ_1. (Symmetrically, in the second reformulation, the objective function can be made arbitrarily large by selecting a suitably large value of δ_1, whatever the value of λ_2.)

This tells us that the objective function in the cooperative facility optimization problem with fixed λ is actually unbounded: the facility operator(s) can obtain an arbitrarily large profit by charging sufficiently large tolls.

Recall that we observed a similar phenomenon in the case of a single facility in which the arrival rate was fixed and the toll (in addition to the service rate) was variable. (See Section 5.3.3 of Chapter 5.) For any fixed value of the service rate, we observed there that the facility operator could earn an arbitrarily large profit by setting the toll at an arbitrarily large value.

Remark 9 Another way to arrive at this conclusion is to imagine a dynamic, sequential toll-setting process along the lines of the dynamical-system models proposed in Chapter 3. Suppose that servers 1 and 2 adjust their tolls in turn, using the first-order optimality conditions given above. That is, for a given value of δ_2, server 1 employs its "best reply" and charges the toll

$$\delta_1 = \delta_2 + G_2(\lambda - \lambda_1) - G_1(\lambda_1) > \delta_2 \,.$$

Server 2 then adjusts its toll, also in accordance with its "best reply," resulting in

$$\delta_2 = \delta_1 + G_1(\lambda - \lambda_2) - G_2(\lambda_2) > \delta_1 \,.$$

This process continues *ad infinitum*, resulting in a sequence of strictly increasing tolls, approaching infinity. It is important to remember that the servers, although acting sequentially, are still cooperating. That is, each is setting its arrival rate and toll in accordance with the objective of maximizing the total value of all tolls collected.

We shall see (in Section 6.1.3.5) that this result – the fact that the facilities, by cooperating, can earn an arbitrarily large profit – does not hold if the facilities compete, that is, if each is concerned only with maximizing its own profit. Moreover, the result does not hold in general when the total arrival rate, λ, is not fixed but instead is determined by maximization of a concave utility function, $U(\lambda)$. A little reflection should convince the reader that, indeed, there is something special, perhaps even peculiar, about the assumption that the total arrival rate is fixed. This assumption requires that the demand be completely *inelastic*: no matter how large the tolls and/or the waiting costs, the total demand on the system remains fixed. With this perspective, it is not so surprising that the facilities, by cooperating, can "get away with" charging exorbitant tolls.

Now, with these insights in mind, let us turn our attention to the problem with more than two facilities ($n > 2$). Using the same analysis that led to the formulation of the problem with a variable total arrival rate as Problem (**F**), we can reformulate the problem with a fixed total arrival rate as one of solving the maximization problem,

$$\max_{\{\pi, \lambda_j, j \in J\}} \quad \lambda\pi - \sum_{j \in J} \lambda_j G_j(\lambda_j)$$

$$\text{s.t.} \quad \sum_{j \in J} \lambda_j = \lambda \,,$$

$$\lambda_j \geq 0 \,, \ j \in J \,,$$

and then choosing δ_j, to satisfy (6.38) and (6.39) for all $j \in J$. Now we see that, whatever the values of λ_j, $j \in J$, the objective function may be made arbitrarily large by choosing a sufficiently large value of π. Thus, the result obtained above for $n = 2$ generalizes to $n > 2$.

Remark 10 In fact, the unboundedness of the facility optimization problem with a fixed total arrival rate was foreshadowed in Remark 1 in Section 6.1.1.1, where we considered the individual optimization problem. There we observed that adding the same constant to the toll at each facility has no effect on the individually optimal arrival rates at the facilities when the total arrival rate is fixed. It follows from this observation that, given *any* set of tolls, δ_j, $j \in J$ (with associated individually optimal arrival rates, λ_j, $j \in J$, such that $\sum_{j \in J} \lambda_j = \lambda$), by adding a positive constant K to each toll we can increase the revenue at facility j by $\lambda_j K$ and the total revenue by λK. Thus, there cannot be a finite optimal solution, δ_j, $j \in J$, to the cooperative facility optimization problem with fixed total arrival rate λ.

Now suppose, for some reason, that we are not free to choose an arbitrarily large value for π. Suppose instead that we are given a fixed value of π and then wish to find the values of λ_j, $j \in J$ (equivalently, the values of the tolls, δ_j, $j \in J$) that will maximize the total revenue received by the cooperating servers. Then it follows from the above formulation (in which π is now fixed, as well as λ) that the problem is solved by finding values of λ_j, $j \in J$, that solve the minimization problem,

$$\min_{\{\lambda_j, j \in J\}} \quad \sum_{j \in J} \lambda_j G_j(\lambda_j)$$

$$\text{s.t.} \quad \sum_{j \in J} \lambda_j = \lambda \, , , ,$$

$$\lambda_j \geq 0 \, , \, j \in J \, .$$

But this minimization problem is just the problem of finding the socially optimal arrival rates at each facility when the total arrival rate is fixed, which we solved in Section 6.1.2. Recall that the socially optimal arrival rates satisfy the optimality conditions,

$$\alpha \leq G_j(\lambda_j) + \lambda_j G'_j(\lambda_j) \, , \, \lambda_j \geq 0 \, ,$$
$$\alpha = G_j(\lambda_j) + \lambda_j G'_j(\lambda_j) \, , \, \text{if } \lambda_j > 0 \, ,$$

for all $j \in J$, where α is chosen so that $\sum_{j \in J} \lambda_j = \lambda$. The facility-optimal tolls then must satisfy (6.38) and (6.39). Note that these tolls coincide with the socially optimal tolls at each facility j with $\lambda_j > 0$, up to addition of a constant. Specifically,

$$\delta_j = \pi - \alpha + \lambda_j G'_j(\lambda_j) \, . \tag{6.42}$$

Recall that the socially optimal toll at facility j equals the external effect, $\lambda_j G'_j(\lambda_J)$. Thus, the facility optimal toll associated with π equals the external effect plus $\pi - \alpha$.

6.1.3.3 Cooperating Facilities: Variable Total Arrival Rate

Now we return to the problem with a variable total arrival rate λ and a (concave) utility function $U(\lambda)$. Recall that this problem can be solved by first solving Problem (**F**) and then choosing tolls, δ_j, to satisfy (6.38) and (6.39) for all $j \in J$. For convenience we rewrite Problem (**F**) here:

$$\max_{\{\pi, \lambda, \lambda_j, j \in J\}} \quad \lambda \pi - \sum_{j \in J} \lambda_j G_j(\lambda_j)$$

(**F**) $\quad$ s.t. $\quad \sum_{j \in J} \lambda_j = \lambda \, ,$

$$U'(\lambda) = \pi \, ,$$

$$\lambda_j \geq 0 \, , \, j \in J \, .$$

We use a Lagrangean approach to solve Problem (**F**). For simplicity we only consider the case in which $\lambda_j > 0$, $j \in J$. Attach Lagrange multipliers, α and β, to the first and second constraints, respectively, and form the Lagrangean:

$$L(\pi, \lambda, \lambda_j, j \in J; \alpha, \beta) \quad := \quad \lambda \pi - \sum_{j \in J} \lambda_j G_j(\lambda_j)$$

$$+ \alpha \left(\sum_{j \in J} \lambda_j - \lambda \right) + \mu(U'(\lambda) - \pi) .$$

Differentiating the Lagrangean with respect to the decision variables, λ_j, $j \in J$, λ, and π, leads to the following first-order optimality conditions:

$$\frac{\partial L}{\partial \lambda_j} = -G_j(\lambda_j) - \lambda_j G'_j(\lambda_j) + \alpha = 0 , \; j \in J ,$$

$$\frac{\partial L}{\partial \lambda} = \pi + \beta U''(\lambda) = 0 ,$$

$$\frac{\partial L}{\partial \pi} = \lambda - \beta .$$

Combining and simplifying, we obtain

$$G_j(\lambda_j) + \lambda_j G'_j(\lambda_j) = \alpha , \; j \in J , \qquad (6.43)$$
$$\pi + \lambda U''(\lambda) = \alpha . \qquad (6.44)$$

Together with the (definitional) constraints

$$\lambda = \sum_{j \in J} \lambda_j ,$$

$$\delta_j = \pi - G_j(\lambda_j) , \; j \in J ,$$

these equations constitute the *KKT* conditions, which are necessary for an optimal solution to the original problem.

Note that these equations differ from the optimality conditions for the problem with fixed λ only in that the parameter α now must be chosen so that $\pi + \lambda U''(\lambda) = \alpha$, where $\lambda = \sum_{j \in J} \lambda_j$ is variable, rather than being chosen so that the latter constraint holds with λ fixed.

Now consider λ as a variable and let $\alpha(\lambda)$ denote the value of α which, together with the equations (6.43) defining λ_j, $j \in J$, yields $\sum_{j \in J} \lambda_j = \lambda$. Likewise, let $\pi(\lambda)$ denote the value of π that satisfies (6.44), that is,

$$\pi(\lambda) = \alpha(\lambda) + \lambda(-U''(\lambda)) .$$

Thus, $\pi(\lambda)$ is the full price of admission, expressed as a function of the total arrival rate λ.

When

$$U'(\lambda) = \pi = \pi(\lambda) , \qquad (6.45)$$

all the necessary conditions for facility optimization are satisfied.

Note that the term $\lambda(-U''(\lambda))$ need not be nondecreasing. Thus it is possible for there to be multiple values of λ that satisfy (6.45), that is, multiple solutions to the necessary conditions for optimality. The situation here is exactly as it was in our analysis of the single-facility problem in Chapter 2, and we refer the reader there for a discussion of classes of utility functions with multiple solutions to the optimality conditions and those for which the solution is unique.

As in the model with fixed arrival rate, we find that the facility-optimal tolls are given by

$$\delta_j = \pi - G_j(\lambda_j) = \pi - \alpha + \lambda_j G'_j(\lambda_j) \, , \; j \in J \, ,$$

where we have used (6.43). Now, however, instead of being arbitrary, π must satisfy equation (6.44). Combining these two equations yields the following expression for the tolls:

$$\delta_j = \lambda(-U''(\lambda)) + \lambda_j G'_j(\lambda_j) \, , \; j \in J \, .$$

Recall that the socially optimal toll at facility j is just the external effect, $\lambda_j G'_j(\lambda_j)$. Just as in the single-facility model of Chapter 2, we see that in the facility-optimal case the extra nonnegative term,

$$\lambda(-U''(\lambda)) \, ,$$

is added to each of the tolls.

What is the effect of this addition to the tolls on the optimal total arrival rate? In the case of social optimality, the full price associated with arrival rate λ is given by $\alpha(\lambda)$, the value of α that makes $\sum_{j \in J} \lambda_j = \lambda$, where each λ_j is the solution of equation (6.43). In the case of facility optimality, the full price associated with arrival rate λ is given by $\pi(\lambda) = \alpha(\lambda) + \lambda(-U''(\lambda)) \geq \alpha(\lambda)$. Hence the supply curve for facility optimality lies uniformly above the supply curve for social optimality, from which it follows that it intersects the demand curve (the graph of $U'(\lambda)$) at a smaller value of λ. (See Figure 6.1.) That is,

$$\lambda^f \leq \lambda^s \, .$$

6.1.3.4 Heavy-Tailed Rewards

What happens in the multi-facility setting if $\lambda U'(\lambda) \to \infty$ as $\lambda \to 0$? Recall that this is the case if the reward distribution, $F(x) = P\{\boldsymbol{R} \leq x\}$, has a heavy tail and $x\bar{F}(x) \to \infty$ as $x \to \infty$ (cf. Section 2.3.3 of Chapter 2). In the single-facility model of Chapter 2 we saw that in this case the facility operator can earn an arbitrarily large profit by choosing an arbitrarily small arrival rate (equivalently, charging an arbitrarily large toll).

Not surprisingly, the same is true in the multi-facility setting with cooperating facilities. It remains to examine how the arrival rate λ is allocated to the various facilities as $\lambda \to 0$. We shall find it convenient to work with the formulation of the facility optimization problem in terms of the arrival rates, λ_j, $j \in J$, as the decision variables, namely Problem (**F**), which we rewrite here for convenience:

$$\max_{\{\pi, \lambda, \lambda_j, j \in J\}} \quad \lambda\pi - \sum_{j \in J} \lambda_j G_j(\lambda_j)$$

(**F**)   s.t. $$\sum_{j \in J} \lambda_j = \lambda \, ,$$

$$U'(\lambda) = \pi \, ,$$

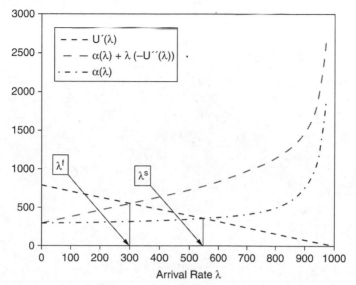

Figure 6.1 *Comparison of S.O. and F.O. Supply-Demand Curves for Variable* λ

$$\lambda_j \geq 0 \, , \, j \in J \, .$$

Now consider a fixed, arbitrary λ. Let $\pi = U'(\lambda)$. Since $\lambda\pi$ is fixed, the optimal $\lambda_j, j \in J$, corresponding to this value of λ will solve the following minimization problem:

$$\min_{\{\lambda_j, j \in J\}} \quad \sum_{j \in J} \lambda_j G_j(\lambda_j)$$

$$\text{s.t.} \quad \sum_{j \in J} \lambda_j = \lambda \, ,$$

$$\lambda_j \geq 0 \, , \, j \in J \, .$$

The necessary and sufficient *KKT* conditions for this problem are

$$\alpha \leq G_j(\lambda_j) + \lambda_j G'(\lambda_j) \, , \, j \in J \, ,$$
$$\alpha = G_j(\lambda_j) + \lambda_j G'(\lambda_j) \, , \quad , \text{if } \lambda_j > 0 \, , \, j \in J \, ,$$
$$\sum_{j \in J} \lambda_j = \lambda \, ,$$
$$\lambda_j \geq 0 \, , \, j \in J \, .$$

Assume the facilities are ordered so that

$$G_1(0) < G_2(0) < \cdots < G_n(0) \, .$$

(For simplicity we assume strict inequalities. The modifications required to allow for equalities are straightforward and we leave them to the reader.)

Theorem 6.9 *For sufficiently small $\lambda > 0$, the cooperative facility optimal allocation of arrival rates is*

$$\lambda_1 = \lambda \; ; \; \lambda_j = 0 \; , \; j = 2, 3, \ldots, n \; .$$

Proof It suffices to show that this allocation satisfies the above KKT conditions for sufficiently small λ. Since $G_1(\lambda_1) + \lambda_1 G'(\lambda_1)$ is a continuous function of λ_1, for sufficiently small λ we have

$$\alpha = G_1(\lambda) + \lambda G_1'(\lambda) < G_j(0) \; , \; j \neq 1 \; ,$$

so that

$$\begin{aligned} \alpha &= G_1(\lambda_1) + \lambda_1 G_1'(\lambda_1) \; , \\ \alpha &< G_j(\lambda_j) + \lambda_j G_j'(\lambda_j) \; , \; j \neq 1 \; , \\ \sum_{j \in J} \lambda_j &= \lambda \; , \end{aligned}$$

with $\lambda_1 = \lambda$ and $\lambda_j = 0$, $j \neq 1$. Thus the KKT conditions are satisfied.   ∎

As a corollary of this theorem, we have:

Corollary 6.10 *Suppose $\lambda U'(\lambda) \to \infty$ as $\lambda \to 0$. Then the cooperative facility optimal solution has an unbounded objective function as $\lambda \to 0$. The optimal solution consists in allocating the total arrival rate, λ, to facility 1 (i.e., the facility with the minimal no-flow waiting cost, $G_j(0)$) and then letting $\lambda \to 0$.*

6.1.3.5 Competing Facilities

Suppose now that each facility has a separate operator, who sets the toll at that facility and whose goal is to maximize the total profit at that facility. Again we assume that the costs of operating the facilities are fixed, so that maximizing the profit at facility j is equivalent to maximizing the revenue,

$$\lambda_j \delta_j \; ,$$

received from the tolls paid by the customers who use that facility.

Assuming (as usual) that the arriving customers are individual optimizers, any particular set of values δ_j for the tolls will result in arrival rates λ_j that satisfy the equilibrium conditions, (6.2)-(6.7). As in the model with cooperating facilities, we shall focus on the case where the total arrival rate, λ, is positive. Then the equilibrium conditions may be written in the following form:

$$\begin{aligned} U'(\lambda) &= \pi \; , \\ \sum_{j \in J} \lambda_j &= \lambda \; , \\ G_j(\lambda_j) + \delta_j &\geq \pi \; , \; j \in J \; , \\ \lambda_j(G_j(\lambda_j) + \delta_j - \pi) &= 0 \; , \; j \in J \; , \\ \lambda_j &\geq 0 \; , \; j \in J \; . \end{aligned}$$

Each of the (competing) revenue-maximizing facility operators seeks to find a value for the toll at its facility that maximizes its revenue, with the above equilibrium conditions as constraints. We seek a Nash equilibrium solution for the tolls, that is, a set of values for δ_j, $j \in J$, such that, at each facility j, δ_j maximizes the revenue, $\lambda_j \delta_j$, at that facility, assuming that δ_k, $k \neq j$, do not change. In other words, no facility operator can benefit (i.e., achieve a larger revenue at its facility) by unilaterally changing the value of its toll.

Focusing on a particular facility j, the set of tolls, $\{\delta_k, k \in J\}$, satisfies the Nash-equilibrium characterization with respect to that facility, if δ_j maximizes $\lambda_j \delta_j$, with the above i.o. equilibrium conditions as constraints (in which δ_k, $k \neq j$, are fixed). That is, δ_j achieves the maximum in the following problem:

$$\max_{\{\pi,\lambda,\delta_j;\lambda_k,k\in J\}} \quad \lambda_j \delta_j$$

$$(\mathbf{F}_j) \qquad \text{s.t.} \qquad \pi \leq \delta_k + G_k(\lambda_k) \, , \ k \in J \, ,$$

$$\pi = \delta_k + G_k(\lambda_k) \, , \text{ if } \lambda_k > 0 \, , \ k \in J \, ,$$

$$\sum_{k \in J} \lambda_k = \lambda \, ,$$

$$U'(\lambda) = \pi \, ,$$

$$\lambda_k \geq 0 \, , \ k \in J \, ,$$

with δ_k, $k \neq j$, fixed.

In the case where all arrival rates are positive, the formulation of each Problem $(\mathbf{F}_j)$ simplifies. As with the single-facility model of Chapter 2 and the cooperating-facilities model of the previous section, we can use the constraint, $\pi = \delta_j + G_j(\lambda_j)$, to substitute for δ_j in the objective function. Thus, in this case an equivalent formulation of the optimization problem for facility j is the following:

$$\max_{\{\pi,\delta_j,\lambda_k,k\in J\}} \quad \lambda_j(\pi - G_j(\lambda_j))$$

$$(\mathbf{F}_j) \qquad \text{s.t.} \qquad U'(\lambda) = \pi \, ,$$

$$\pi = \delta_k + G_k(\lambda_k) \, , \ k \in J \, ,$$

$$\sum_{k \in J} \lambda_k = \lambda \, ,$$

with δ_k, $k \neq j$, fixed.

6.1.3.6 Competing Facilities: Fixed Total Arrival Rate

In this subsection we consider the competitive facility-optimal solution in the case of a fixed total arrival rate λ.

We shall begin our analysis with the special case $n = 2$. To keep the exposition simple, we shall assume that the arrival rates at both facilities are positive in an optimal solution. In this case the optimization problem for facility 1 (with δ_2 fixed) takes the form:

$$\max_{\{\pi,\delta_1,\lambda_1,\lambda_2\}} \quad \lambda_1\delta_1$$

$$\text{s.t.} \quad \pi = \delta_1 + G_1(\lambda_1) \,,$$
$$\pi = \delta_2 + G_2(\lambda_2) \,,$$
$$\lambda_1 + \lambda_2 = \lambda \,,$$
$$\lambda_1 \geq 0 \,, \ \lambda_2 \geq 0 \,.$$

Using the equality constraints to eliminate the variables π and λ_2, we obtain the following equivalent formulation:

$$\max_{\{\lambda_1,\delta_1\}} \quad \lambda_1\delta_1$$

$$\text{s.t.} \quad \delta_1 = \delta_2 + G_2(\lambda - \lambda_1) - G_1(\lambda_1) \,,$$
$$0 \leq \lambda_1 \leq \lambda \,.$$

Using the equality constraint to substitute for δ_1 in the objective function yields

$$\max_{\{\lambda_1\}} \quad \lambda_1(\delta_2 + G_2(\lambda - \lambda_1) - G_1(\lambda_1))$$

$$\text{s.t.} \quad \delta_1 = \delta_2 + G_2(\lambda - \lambda_1) - G_1(\lambda_1) \,,$$
$$0 \leq \lambda_1 \leq \lambda \,.$$

The equality constraint now serves simply to define δ_1 and can be disregarded in the optimization.

Differentiating the objective function with respect to λ_1 yields the following (necessary) first-order condition for a value of $\lambda_1 \in (0, \lambda)$ to be optimal for facility 1, with δ_2 fixed:

$$\delta_1 = \delta_2 + G_2(\lambda - \lambda_1) - G_1(\lambda_1) = \lambda_1(G_1'(\lambda_1) + G_2'(\lambda - \lambda_1)) \,. \tag{6.46}$$

Symmetrically, the (necessary) first-order condition for a value of $\lambda_2 \in (0, \lambda)$ to be optimal for facility 2, with δ_1 fixed, is:

$$\delta_2 = \delta_1 + G_1(\lambda - \lambda_2) - G_2(\lambda_2) = \lambda_2(G_1'(\lambda - \lambda_2) + G_2'(\lambda_2)) \,. \tag{6.47}$$

Rewriting (6.47) in terms of λ_1 yields

$$\delta_2 = (\lambda - \lambda_1)(G_1'(\lambda_1) + G_2'(\lambda - \lambda_1)) \,,$$

and subtracting this equation from equation (6.46) leads to

$$\delta_1 - \delta_2 = (2\lambda_1 - \lambda)(G_1'(\lambda_1) + G_2'(\lambda - \lambda_1)) \,.$$

But the equilibrium condition for individual optimality on the part of the customers implies that

$$\delta_1 - \delta_2 = G_2(\lambda - \lambda_1) - G_1(\lambda_1) \,.$$

Combining these two equations yields

$$G_2(\lambda - \lambda_1) - G_1(\lambda_1) = (2\lambda_1 - \lambda)(G_1'(\lambda_1) + G_2'(\lambda - \lambda_1)) \,. \tag{6.48}$$

The solutions to this equation (note that the solution need not be unique) are the candidates for an interior facility-optimal value of λ_1. The facility-optimal value of λ_2 can then be found using the equation, $\lambda_1 + \lambda_2 = \lambda$.

Remark 11 Note that equations (6.46) and (6.47) can also be combined to yield the following interesting relationship between the facility-optimal tolls and arrival rates at the two facilities:

$$\frac{\delta_1}{\lambda_1} = \frac{\delta_2}{\lambda_2} \, .$$

Note also that the full price π for this problem satisfies the equations

$$\pi = G_1(\lambda_1) + \lambda_1 G_1'(\lambda_1) + \lambda_1 G_2'(\lambda_2) \, ,$$
$$\pi = G_2(\lambda_2) + \lambda_2 G_2'(\lambda_2) + \lambda_2 G_1'(\lambda_1) \, .$$

The toll at facility 1 has two components: the external effect, $\lambda_1 G_1'(\lambda_1)$, of a marginal increase in λ_1 on waiting costs at that facility, and a second component, $\lambda_1 G_2'(\lambda_2)$, which measures the increase in waiting costs at facility 2 caused by a marginal increase in λ_1.

Remark 12 Let us look again at the optimization problem for facility 1 (given a fixed value of δ_2) expressed in terms of the decision variable, λ_1:

$$\max_{\{\lambda_1\}} \quad \lambda_1(\delta_2 + G_2(\lambda - \lambda_1) - G_1(\lambda_1))$$
$$\text{s.t.} \quad 0 \le \lambda_1 \le \lambda \, . \tag{6.49}$$

Consider the following "surrogate" utility function for facility 1:

$$U_1(\lambda_1) := \int_0^{\lambda_1} (\delta_2 + G_2(\lambda - \nu_1)d\nu_1 \, , \ 0 \le \lambda_1 \le \lambda \, .$$

Then

$$U_1'(\lambda_1) = \delta_2 + G_2(\lambda - \lambda_1) \, , \ 0 \le \lambda_1 \le \lambda \, ,$$

and the optimization problem (6.49) can be rewritten as:

$$\max_{\{\lambda_1\}} \quad \lambda_1 U_1'(\lambda_1) - \lambda G_1(\lambda_1)$$
$$\text{s.t.} \quad 0 \le \lambda_1 \le \lambda \, .$$

Since $G_2(\cdot)$ is a nonnegative, nondecreasing, and continuous function, the function, $U_1(\lambda_1)$, is nondecreasing, concave, and differentiable and thus satisfies the default assumptions for a utility function. We can therefore interpret the optimization problem for facility 1 as that of finding a facility optimal arrival rate for a single facility with utility function, $U_1(\lambda_1)$, waiting-cost function, $G_1(\lambda_1)$, and a maximal arrival rate, λ. As we have remarked many times already, this problem is in turn equivalent to a social optimization problem with transformed utility function, $\tilde{U}_1(\lambda_1) := \lambda_1 U_1'(\lambda_1)$. In general, however, $\tilde{U}_1(\lambda_1)$ may be neither nondecreasing nor concave. If it is concave, then the solution to the optimization problem for facility 1, with δ_2 fixed, is unique. Of course, the same observations hold for the optimization problem for facility 2, with δ_1 fixed, *mutatis mutandi*. (Note, however, that even if each of these optimization problems has a unique solution, there may still be more than one Nash equilibrium.)

It is instructive to view the competitive facility optimization problem in dynamic terms. Given that facility 2 is charging a toll δ_2, the "best reply" of facility 1 is to choose a toll, δ_1, and corresponding arrival rate, λ_1, to maximize its revenue. This can be done by solving problem (6.49) for an optimal λ_1 and then setting δ_1 equal to $\lambda_1(G_1'(\lambda_1) + G_2'(\lambda - \lambda_1))$. The best reply of facility 2 to this value of δ_1 is then found by solving the symmetric optimization problem for λ_2 and setting δ_2 equal to $\lambda_2(G_2'(\lambda_2) + G_1'(\lambda - \lambda_2))$. If the resulting value of δ_2 equals the starting value, then the pair, (δ_1, δ_2), constitute a Nash equilibrium for the competitive facility optimization problem. If they are not equal, then one could in principle make this procedure the basis for a discrete-time dynamical-system algorithm, along the lines of those discussed in Chapter 3, and analyze its convergence properties. We shall leave this as an exercise for the reader.

Example: M/M/1 *Facilities with Linear Waiting Costs.*

Suppose each of the two facilities is an *M/M/1* queue operating in steady state with a linear waiting cost function:

$$G_j(\lambda_j) = \frac{h_j}{\mu_j - \lambda_j} , \ j = 1, 2 .$$

We assume that $\mu_j > \lambda$, $j = 1, 2$, so that it is feasible to assign all the traffic to either facility. First we show that in this case the transformed surrogate utility function, $\tilde{U}_j(\lambda_j) = \lambda_j U_j'(\lambda_j)$ is nondecreasing and concave, so that there is a unique solution to the optimality condition for the "best reply" for each facility. It suffices to consider facility 1. We have

$$\begin{aligned} U_1'(\lambda_1) &= \delta_2 + G_2(\lambda - \lambda_1) \\ &= \delta_2 + \frac{h_2}{\mu_2 - \lambda + \lambda_1} , \end{aligned}$$

and hence

$$\begin{aligned} \tilde{U}_1'(\lambda_1) &= U_1'(\lambda_1) + \lambda_1 U_1''(\lambda_1) \\ &= \delta_2 + \frac{h_2}{\mu_2 - \lambda + \lambda_1} - \frac{\lambda_1 h_2}{(\mu_2 - \lambda + \lambda_1)^2} \\ &= \delta_2 + \frac{h_2(\mu_2 - \lambda)}{(\mu_2 - \lambda + \lambda_1)^2} \geq 0 . \end{aligned}$$

Hence $\tilde{U}_1(\cdot)$ is nondecreasing. Moreover,

$$\begin{aligned} \tilde{U}_1''(\lambda_1) &= 2U_1''(\lambda_1) + \lambda_1 U_1'''(\lambda_1) \\ &= \frac{-2h_2}{(\mu_2 - \lambda + \lambda_1)^2} + \frac{2h_2\lambda_1}{(\mu_2 - \lambda + \lambda_1)^3} \end{aligned}$$

$$= \frac{-2h_2(\mu_2 - \lambda)}{(\mu_2 - \lambda + \lambda_1)^3} \le 0 .$$

Hence $\tilde{U}_1(\cdot)$ is concave. Therefore, the best reply for facility 1 to the toll δ_2 at facility 2 is the unique solution to the first-order optimality condition, which in this case reduces to:

$$\frac{h_1}{(\mu_1 - \lambda_1)^2} - \frac{h_2}{(\mu_2 - \lambda + \lambda_1)^2} = \delta_2 .$$

By symmetry, the same is true for the the best reply for facility 2 to the toll δ_1 at facility 1, *mutatis mutandi*.

Now consider the condition (6.48) which must be satisfied in order for a pair (λ_1, λ_2) to be a Nash equilibrium for the competitive facility optimization problem. This condition may be rewritten in symmetric form (subject to $\lambda_1 + \lambda_2 = \lambda$) as:

$$G_1(\lambda_1) + (\lambda_1 - \lambda_2)G_1'(\lambda_1) = G_2(\lambda_2) + (\lambda_2 - \lambda_1)G_2'(\lambda_2)$$

For the present example, this equation becomes:

$$\frac{h_1}{\mu_1 - \lambda_1} + \frac{(\lambda_1 - \lambda_2)h_1}{(\mu_1 - \lambda_1)^2} = \frac{h_2}{\mu_2 - \lambda_2} + \frac{(\lambda_2 - \lambda_1)h_2}{(\mu_2 - \lambda_2)^2}$$

Let $x_1 := \mu_1 - \lambda_1$, $x_2 := \mu_2 - \lambda_2$, $\mu := \mu_1 + \mu_2$, $c := \mu - \lambda$, and $b := \mu_2 - \mu_1$. Without loss of generality, assume $b \ge 0$ ($\mu_2 \ge \mu_1$). After substitution and algebraic simplification, we obtain the following equivalent equations:

$$h_1 \cdot x_2^2(x_2 - b) = h_2 \cdot x_1^2(x_1 + b)$$
$$x_1 + x_2 = c$$

Substituting $c - x_1$ for x_2 in the first equation leads to:

$$h_1 \cdot (c - x_1)^2(c - x_1 - b) = h_2 \cdot x_1^2(x_1 + b) .$$

The r.h.s. of this equation is positive and strictly increasing for $x_1 \in (0, c)$. The l.h.s is positive for $x_1 \in (0, c - b)$ and negative for $x_1 \in (c - b, c)$; strictly decreasing for $x_1 \in (0, c - 2b/3)$ and strictly increasing for $x_1 \in (c - 2b/3, c)$. It follows that this equation has a unique solution in $(0, c)$ and it lies in the interval $(0, c - b)$.

Figure 6.2 illustrates this result. The parameters are $\mu_1 = 400$, $\mu_2 = 600$, $\lambda = 500$, $h_1 = 0.00002$, $h_2 = 0.0001$. (Hence, $c = 500$, $b = 200$.)

Now let us turn our attention to the problem with more than two facilities ($n > 2$), in which we have the following formulation for the optimization problem at facility j (in which δ_k, $k \ne j$, are fixed):

$$\max_{\{\delta_j; \lambda_k, k \in J; \pi\}} \quad \lambda_j \delta_j$$

($\mathbf{F}_j$)

$$\text{s.t.} \quad \pi = \delta_k + G_k(\lambda_k) , \quad k \in J ,$$

$$\sum_{k \in J} \lambda_k = \lambda .$$

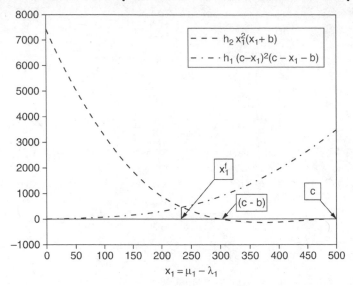

Figure 6.2 *Nash Equilibrium for Two Competitive* M/M/1 *Facilities*

Now, attach Lagrange multipliers, β_k, $k \in J$, to the constraints involving π, and Lagrange multiplier, α, to the last constraint, and form the Lagrangean:

$$L(\delta_j; \lambda_k, k \in J; \pi) = \lambda_j \delta_j - \sum_{k \in J} \beta_k(\pi - \delta_k - G_k(\lambda_k)) - \alpha \left(\sum_{j \in J} \lambda_j - \lambda \right)$$

Differentiating the Lagrangean with respect to the decision variables, δ_j, λ_k, $k \in J$, and π, leads to the following first-order optimality conditions:

$$\frac{\partial L}{\partial \delta_j} = \lambda_j + \beta_j = 0 \, ,$$

$$\frac{\partial L}{\partial \lambda_j} = \delta_j + \beta_j G_j'(\lambda_j) - \alpha = 0 \, ,$$

$$\frac{\partial L}{\partial \lambda_k} = \beta_k G_k'(\lambda_k) - \alpha = 0 \, , \ k \neq j \, ,$$

$$\frac{\partial L}{\partial \pi} = -\sum_{k \in J} \beta_k = 0 \, .$$

Combining and simplifying, we obtain

$$\delta_j = \lambda_j G_j'(\lambda_j) + \alpha \, , \tag{6.50}$$

$$\alpha = \beta_k G_k'(\lambda_k) \, , \ k \neq j \, , \tag{6.51}$$

$$\lambda_j = \sum_{k \neq j} \beta_k \, . \tag{6.52}$$

Together with the constraints

$$\lambda = \sum_{k \in J} \lambda_k ,$$

$$\delta_k = \pi - G_k(\lambda_k) , \ k \in J ,$$

these equations constitute the *KKT* conditions, which are necessary for an optimal solution to the original problem.

Note that, as a consequence of these equations,

$$\delta_j = \lambda_j G'_j(\lambda_j) + \beta_k G'_k(\lambda_k) , \ k \neq j .$$

This equation generalizes the equation for δ_j in the case $n = 2$ and the fact that $\beta_k = \lambda_j$ when $n = 2$ (with $j = 1$, $k = 2$, and with $j = 2$, $k = 1$).

6.1.3.7 Competing Facilities: Variable Total Arrival Rate

Recall that the optimization problem for each facility j in the case of a variable total arrival rate λ with utility function $U(\lambda)$ may be written as:

$$(\mathbf{F}_j) \quad \begin{aligned} &\max_{\{\delta_j ; \lambda_k, k \subset J; \pi\}} && \lambda_j(\pi - G_j(\lambda_j)) \\ &\text{s.t.} && \pi = \delta_k + G_k(\lambda_k) , \ k \in J , \\ & && \sum_{k \in J} \lambda_k = \lambda , \\ & && U'(\lambda) = \pi , \end{aligned}$$

with δ_k, $k \neq j$, fixed. (Again we are assuming, for simplicity, that all arrival rates are positive.) The set of tolls, $\{\delta_k, k \in J\}$, constitutes a Nash equilibrium for the competing facility operators if, together with a set of arrival rates λ_j, $j \in J$, total arrival rate, λ, and full price, π, they simultaneously satisfy the optimization problems $(\mathbf{F}_j)$ for all $j \in J$.

Now, attach Lagrange multipliers, β_k, $k \in J$, to the constraints involving π, and Lagrange multipliers, α and β, to the last two constraints, and form the Lagrangean:

$$L(\delta_j, \lambda_k, k \in J; \pi) \ := \ \lambda_j \delta_j - \sum_{k \in J} \beta_k(\pi - \delta_k - G_k(\lambda_k))$$

$$-\alpha \left(\sum_{j \in J} \lambda_j - \lambda \right) - \beta(U'(\lambda) - \pi) .$$

Differentiating the Lagrangean with respect to the decision variables, δ_j, λ_k, $k \in J$, λ, and π, leads to the following first-order optimality conditions

$$\frac{\partial L}{\partial \delta_j} = \lambda_j + \beta_j = 0 ,$$

$$\frac{\partial L}{\partial \lambda_j} = \delta_j + \beta_j G'_j(\lambda_j) - \alpha = 0 ,$$

$$\frac{\partial L}{\partial \lambda_k} = \beta_k G'_k(\lambda_k) - \alpha = 0 \ , \ k \neq j \ ,$$

$$\frac{\partial L}{\partial \lambda} = \alpha - \beta U''(\lambda) = 0 \ ,$$

$$\frac{\partial L}{\partial \pi} = -\sum_{k \in J} \beta_k + \beta = 0 \ .$$

Combining and simplifying, we obtain

$$\delta_j = \lambda_j G'_j(\lambda_j) + \alpha \ , \tag{6.53}$$

$$\alpha = \beta_k G'_k(\lambda_k) \ , \ k \neq j \ , \tag{6.54}$$

$$\alpha = (-\lambda_j + \sum_{k \neq j} \beta_k) U''(\lambda) \ . \tag{6.55}$$

Together with the constraints

$$\lambda = \sum_{k \in J} \lambda_k \ ,$$

$$\delta_k = \pi - G_k(\lambda_k) \ , \ k \in J \ ,$$

these equations constitute the *KKT* conditions, which are necessary for an optimal solution to the original problem.

Note that, as a consequence of these equations,

$$\delta_j = \lambda_j G'_j(\lambda_j) + \beta_k G'_k(\lambda_k) \ , \ k \neq j \ .$$

This is formally the same equation as in the case of a fixed arrival rate λ. The difference is that now α – the common value of $\beta_k G'_k(\lambda_k)$, $k \neq j$ – must also satisfy the equation

$$\alpha = (-\lambda_j + \sum_{k \neq j} \beta_k) U''(\lambda) \ ,$$

whereas in the case of a fixed arrival rate α was a parameter whose value was chosen so that the constraint, $\sum_{k \in J} \lambda_k = \lambda$, was satisfied.

6.1.3.8 Linear Utility Function

Consider now the special case of a linear utility function, $U(\lambda) = r \cdot \lambda$. In this case $U''(\lambda) \equiv 0$. Thus $\alpha = 0$ and the optimality conditions imply $\delta_j = \lambda_j G_j(\lambda_j)$. Hence the facility-optimal solution in the case of competing servers with a variable total arrival rate coincides with the socially optimal solution, just as we saw was the case in the single-facility model studied in Chapter 2. Moreover, the linear utility function renders the problem completely separable, so that each server can maximize its profit with no regard for what the other servers are doing.

6.1.3.9 Heavy-Tailed Rewards

In this section we again consider what happens if $\lambda U'(\lambda) \to \infty$ as $\lambda \to 0$, but now in the setting of competitive facilities.

In the case of cooperating facilities, we saw that the facility operators acting together can earn an arbitrarily large total revenue by charging an arbitrarily large toll (and, incidentally, directing all traffic to the facility with the smallest zero-flow waiting cost). The result is that a vanishingly small fraction of the customer demand is satisfied. *A priori* it is not clear if the same phenomenon can occur when the facilities compete. In fact, economic intuition suggests that competition among facilities should lead to a significantly larger fraction of the demand being satisfied, perhaps even to a solution that is close to being socially optimal. We shall see, however, that this not necessarily the case.

Recall that a set of tolls, $\{\delta_j, j \in J\}$, is a Nash equilibrium for competitive facility operators, if for each $j \in J$, δ_j achieves the maximum in the problem,

$$\max_{\{\pi, \lambda, \delta_j; \lambda_k, k \in J\}} \quad \lambda_j \delta_j$$

$$(\mathbf{F}_j) \qquad \text{s.t.} \qquad \pi \leq \delta_k + G_k(\lambda_k) \,, \ k \in J \,,$$

$$\pi = \delta_k + G_k(\lambda_k) \,, \ \text{if } \lambda_k > 0 \,, \ k \in J \,,$$

$$\sum_{k \in J} \lambda_k = \lambda \,,$$

$$U'(\lambda) = \pi \,,$$

$$\lambda_k \geq 0 \,, \ k \in J \,,$$

with δ_k, $k \neq j$, fixed.

Let us focus first on the case of two facilities. The problem for facility 1 then takes the form,

$$\max_{\{\lambda, \delta_1; \lambda_k, k=1,2\}} \quad \lambda_1 \delta_1$$

$$\text{s.t.} \qquad U'(\lambda) \leq \delta_k + G_k(\lambda_k) \,, \ k = 1, 2 \,,$$

$$U'(\lambda) = \delta_k + G_k(\lambda_k) \,, \ \text{if } \lambda_k > 0 \,, \ k = 1, 2 \,,$$

$$\lambda_1 + \lambda_2 = \lambda \,,$$

$$\lambda_k \geq 0 \,, \ k = 1, 2 \,,$$

with δ_2 fixed. Since a solution with $\lambda_1 = 0$ is clearly suboptimal, we can use the equality, $\delta_1 = U'(\lambda) - G_1(\lambda_1)$, to substitute for δ_1 in the objective function, leading to the equivalent problem,

$$\max_{\{\lambda; \lambda_k, k=1,2\}} \quad \lambda_1 (U'(\lambda) - G_1(\lambda_1))$$

$$\text{s.t.} \qquad U'(\lambda) \leq \delta_2 + G_2(\lambda_2) \,,$$

$$U'(\lambda) = \delta_2 + G_2(\lambda_2) \,, \ \text{if } \lambda_2 > 0 \,,$$

$$\lambda_1 + \lambda_2 = \lambda \,,$$

$$\lambda_k \geq 0 \,, \ k = 1, 2 \,,$$

with δ_2 fixed and $\delta_1 = U'(\lambda) - G_1(\lambda_1)$. The problem for facility 2 is symmetric.

We begin by considering a special case which, at first glance, might seem of little practical interest: the case of constant waiting costs.

Theorem 6.11 *Assume* $\lambda U'(\lambda) \uparrow \infty$ *as* $\lambda \downarrow 0$. *Suppose the waiting cost at each facility is constant:* $G_k(\lambda_k) = G_k(0)$, *for all* $\lambda_k \geq 0$, $k = 1, 2$. *Then* every *pair of tolls,* (δ_1, δ_2), *such that* $\delta_1 - \delta_2 = G_2(0) - G_1(0)$ *generates a Nash equilibrium for the competing facilities. As a consequence, there exist Nash equilibria with arbitrarily small total arrival rate in which at least one of the facilities earns an arbitrarily large profit.*

Proof   Consider an arbitrary pair of tolls, (δ_1, δ_2), such that $\delta_1 - \delta_2 = G_2(0) - G_1(0)$. Let λ, λ_1, and λ_2 be a solution to the conditions,

$$U'(\lambda) = \delta_1 + G_1(0), \tag{6.56}$$

$$U'(\lambda) = \delta_2 + G_2(0), \tag{6.57}$$

$$\lambda = \lambda_1 + \lambda_2, \tag{6.58}$$

$$\lambda_k \geq 0, \; k = 1, 2. \tag{6.59}$$

Then we have an individually optimal equilibrium. Note that λ is unique but λ_1 and λ_2 may be any nonnegative numbers summing to λ.

We need to show that neither facility has an incentive to deviate unilaterally from its choice of toll, assuming the other facility does not change its toll. Obviously, there is no incentive for either facility to increase its toll, as this would result in all the traffic going to the other facility. What about a decrease in the toll? Suppose facility 1 decreases its toll from δ_1 to $\delta_1 - \epsilon_1$, for some $\epsilon_1 > 0$. As a result the total arrival rate will change to $\lambda + \epsilon$, for some $\epsilon > 0$, and all of the traffic will go to facility 1. The new individually optimal equilibrium will satisfy the conditions

$$U'(\lambda + \epsilon) = \delta_1 + \epsilon_1 + G_1(0) < \delta_2 + G_2(0) = U'(\lambda),$$

with $\lambda_1 = \lambda + \epsilon$ and $\lambda_2 = 0$. The additional profit earned by facility 1 will be

$$(\lambda + \epsilon)(U'(\lambda + \epsilon) - G_1(0)) - \lambda_1(U'(\lambda) - G_1(0)) =$$
$$(\lambda + \epsilon)U'(\lambda + \epsilon) - \lambda_1 U'(\lambda) - \epsilon G_1(0).$$

But, since $(\lambda + \epsilon)(U'(\lambda + \epsilon)$ is a decreasing function of ϵ, this additional profit is maximized by letting $\epsilon \to 0$. Thus, facility 1 will want to choose a vanishingly small ϵ to accomplish the goal of capturing all of the traffic while maximizing its profit. The result is that in the limit the toll at facility 1 remains equal to δ_1, but now facility 1 has arrival rate $\lambda_1 = \lambda$. The associated individually optimal equilibrium solution still satisfies the conditions (6.56)–(6.59), with the same values of δ_1, δ_2, and λ, but now with $\lambda_1 = \lambda$ and $\lambda_2 = 0$.

Of course, the same argument applies to facility 2: given a fixed value of δ_1, it will maximize its profit by choosing a toll $\delta_2 - \epsilon_2$ and letting $\epsilon_2 \to 0$. The associated individually optimal equilibrium solution will have the same values of δ_1, δ_2, and λ, but now with $\lambda_1 = 0$ and $\lambda_2 = \lambda$.   ∎

Perhaps this behavior is best understood in the context of a best-reply discrete-time dynamic algorithm (cf. Chapter 3) in which the facilities alternate their choices of toll. The result will be that the tolls and the total arrival

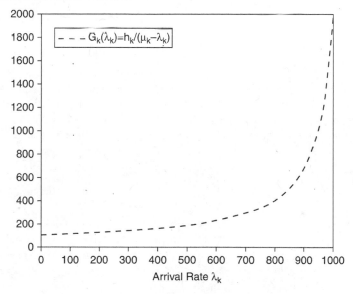

Figure 6.3 *Waiting-Cost Function for* $M/M/1$ *Queue*

rate do not change but the arrival rate pair, (λ_1, λ_2), will alternate between $(\lambda, 0)$ and $(0, \lambda)$.

Of course, in the real world the concept of a "vanishingly small" decrease in the toll does not have any meaning. But the incentive will still be there for each facility to decrease its toll by the smallest amount possible. The result may be that the total arrival rate "creeps" up over time, but the rate of increase could be quite small and negligible over the time scale of interest. Our idealized model captures the limiting case of this behavior.

Also, in a realistic congestion model it seems beside the point to assume that the waiting costs are constant. One can imagine, however, a situation in which the waiting costs are constant for sufficiently small λ and do not begin to increase until the arrival rate reaches a certain threshold. (Indeed, there are classical queueing models in which this property can occur, for example, a *GI/GI/1* queue in which the interarrival-time and service-time distributions are such that $P\{T \geq S\} = 1$, for sufficiently small values of the parameter λ, where S and T are, respectively, a generic service time and a generic inter-arrival time). Moreover, for some queueing systems the waiting-cost function increases very slowly at first (and much faster later) and may be well approximated by a constant for relatively small values of λ. An example is a steady state *M/M/1* queue with linear waiting cost, in which

$$G_k(\lambda_k) = \frac{h_k}{\mu_k - \lambda_k} \,,\ 0 \leq \lambda_k < \mu_k \,.$$

(See Figure 6.3.)

The result may be that, even with nonconstant waiting-cost functions, the rate of increase of the facility arrival rates in the discrete-time algorithm may be negligible over the time scale of interest. We quantify this assertion in the following theorem.

Theorem 6.12 *Assume* $\lambda U'(\lambda) \to \infty$ *as* $\lambda \to 0$. *Assume also that there exists a* $\bar{\lambda} > 0$ *such that* $\lambda U'(\lambda)$ *is strictly decreasing in* $\lambda \in (0, \bar{\lambda}]$. *Let* n *be an arbitrary positive integer. Then for all* $\lambda > 0$, *there exists a* $\lambda^0 \in (0, \lambda)$ *such that the sequential discrete-time algorithm takes at least* n *iterations to reach* λ, *starting from* λ^0.

Proof Let $\lambda^0 \in (0, \bar{\lambda}]$ and let $\delta_2^0 := U'(\lambda^0) - G_2(\lambda^0)$. Let $\delta_1 = U'(\lambda^0) - G_1(0)$. Then

$$U'(\lambda^0) = \delta_2^0 + G_2(\lambda^0) = \delta_1 + G_1(0)) ,$$

and therefore $\lambda_1 = 0$ and $\lambda_2 = \lambda^0$ (with $\lambda = \lambda_1 + \lambda_2$) constitute an individually optimal equilibrium allocation corresponding to the given pair of tolls, (δ_1, δ_2^0).

We shall apply the sequential discrete-time algorithm to this starting point, with facility 1 choosing a new toll, δ_1, at iteration k=1, facility 2 choosing a new toll, δ_2, at iteration k=2, facility 1 choosing a new toll, δ_1, at iteration k=3, facility 2 choosing a new toll, δ_2, at iteration k=4, and so forth.

At iteration $k = 1$ the problem for facility 1 takes the form

$$\max_{\{\lambda_1, \lambda_2, \lambda\}} \quad \lambda_1(U'(\lambda) - G_1(\lambda_1))$$

$$\text{s.t.} \quad U'(\lambda) \leq \delta_2^0 + G_2(\lambda_2) ,$$
$$U'(\lambda) = \delta_2^0 + G_2(\lambda_2) , \text{ if } \lambda_2 > 0 ,$$
$$\lambda_1 + \lambda_2 = \lambda ,$$
$$\lambda_1 \geq 0 , \ \lambda_2 \geq 0 ,$$

with $\delta_1 = U'(\lambda) - G_1(\lambda_1)$, where δ_2^0 is fixed at $\delta_2^0 = U'(\lambda^0) - G_2(\lambda^0)$. (Note that it is never optimal for facility 1 to increase its toll, since $\lambda_1 = 0$ already at the current solution. Therefore we have restricted attention to decreases in the toll, or equivalently, to positive values of λ_1.) Let λ^1 be the (unique) solution to

$$U'(\lambda) = \delta_2^0 + G_2(0) ,$$

and note that $\lambda^1 > \lambda^0$. Suppose $\lambda^1 < \bar{\lambda}$. Then it is suboptimal for facility 1 to choose $\lambda_1 = \lambda > \lambda^1$, since $\lambda_2 = 0$ for any such choice and $\lambda(U'(\lambda) - G_1(\lambda))$ is strictly decreasing in $\lambda \in (0, \bar{\lambda}]$. Therefore, without loss of generality, assume that $\lambda_1 = \lambda^1$ (with $\lambda_2 = 0$) maximizes the objective function in the above problem. (If not, then λ^1 provides an upper bound on the new value of λ and the subsequent proof still works, *mutatis mutandi*.) We have

$$U'(\lambda^1) = \delta_1^1 + G_1(\lambda^1) = \delta_2^0 + G_2(0) ,$$

and therefore $\lambda_1 = \lambda^1$ and $\lambda_2 = 0$ (with $\lambda = \lambda_1 + \lambda_2 = \lambda^1$) constitute an individually optimal equilibrium allocation corresponding to the given pair of tolls, (δ_1^1, δ_2^0). Let $\eta_1 := G_1(\lambda^1) - G_1(0)$ and $\epsilon_1 := \lambda^1 - \lambda^0$. This completes iteration $k = 1$.

For iteration $k = 2$ the roles of facilities 1 and 2 are reversed. The problem for facility 2 takes the form

$$\max_{\{\lambda_1,\lambda_2,\lambda\}} \quad \lambda_2(U'(\lambda) - G_2(\lambda_2))$$

$$\text{s.t.} \quad U'(\lambda) \leq \delta_1^1 + G_1(\lambda_1) \, ,$$
$$U'(\lambda) = \delta_1^1 + G_1(\lambda_1) \, , \text{ if } \lambda_1 > 0 \, ,$$
$$\lambda_1 + \lambda_2 = \lambda \, ,$$
$$\lambda_1 \geq 0 \, , \ \lambda_2 \geq 0 \, ,$$

with $\delta_2 = U'(\lambda) - G_1(\lambda_2)$, where δ_1^1 is fixed at $\delta_1^1 = U'(\lambda^1) - G_1(\lambda^1)$. Let λ^2 be the (unique) solution to

$$U'(\lambda) = \delta_1^1 + G_1(0) \, ,$$

and note that $\lambda^2 > \lambda^1$. Suppose $\lambda^2 < \bar{\lambda}$. Again, it is suboptimal for facility 2 to choose $\lambda_2 = \lambda > \lambda^2$, since $\lambda_1 = 0$ for any such choice and $\lambda(U'(\lambda) - G_2(\lambda))$ is strictly decreasing in $\lambda \in (0, \bar{\lambda}]$. Therefore, without loss of generality, assume that $\lambda_2 = \lambda^2$ (with $\lambda_1 = 0$) maximizes the objective function in the above problem. We have

$$U'(\lambda^2) = \delta_1^1 + G_1(0) = \delta_2^2 + G_2(\lambda^2) \, ,$$

and therefore $\lambda_1 = 0$ and $\lambda_2 = \lambda^2$ (with $\lambda = \lambda_1 + \lambda_2 = \lambda^2$) constitute an individually optimal equilibrium allocation corresponding to the given pair of tolls, (δ_1^1, δ_2^2). Let $\eta_2 := G_2(\lambda^2) - G_2(0)$ and $\epsilon_2 := \lambda^2 - \lambda^1$. This completes iteration $k = 2$.

This process is repeated until we reach iteration n. Figure 6.4 illustrates the iterations of the algorithm for the case $n = 4$.

Now suppose n is a given positive integer. The above analysis shows that, if $\lambda^n \in (0, \bar{\lambda}]$, then the sequential discrete-time dynamic algorithm reaches λ^n in n steps, starting from λ^0, provided $\lambda^0 = \lambda^n - \sum_{k=1}^n \epsilon_k > 0$. Hence we need to check whether $\sum_{k=1}^n \epsilon_k < \lambda^n$ for the given value of n. To this end, let

$$s_k := \frac{U'(\lambda^{k-1}) - U'(\lambda^k)}{\lambda^k - \lambda^{k-1}} \, ,$$

and note that

$$\epsilon_k = \frac{\eta_k}{s_k} \leq \frac{G(\lambda^{k-1}) - G(0)}{s_k} \, ,$$

$k = 1, \ldots, n$, where $G(\lambda) := \max\{G_1(\lambda), G_2(\lambda)\}$, $\lambda \geq 0$. Consider a particular k, $1 \leq k \leq n$. From the mean-value theorem it follows that

$$s_k = -U''(\tilde{\lambda}) \, ,$$

for some $\tilde{\lambda} \in [\lambda^{k-1}, \lambda^k]$. But, from the assumption that $\lambda U'(\lambda)$ is strictly decreasing in $\lambda \in (0, \bar{\lambda}]$, it follows that

$$-U''(\tilde{\lambda}) > U'(\tilde{\lambda})/\tilde{\lambda} \geq U'(\lambda^k)/\lambda^k \, .$$

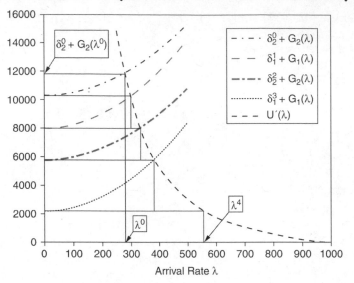

Figure 6.4 *Illustration of Sequential Discrete-Time Algorithm*

Combining these inequalities leads to

$$\epsilon_k \ \leq \ \frac{G(\lambda^{k-1}) - G(0)}{s_k}$$

$$< \ \frac{G(\lambda^{k-1}) - G(0)}{U'(\lambda^k)/\lambda^k}$$

$$\leq \ \frac{G(\lambda^n) - G(0)}{U'(\lambda^n)/\lambda^n}$$

so that

$$\sum_{k=1}^{n} \epsilon_k < n \left(\frac{\lambda^n (G(\lambda^n) - G(0))}{U'(\lambda^n)} \right) .$$

Hence, $\sum_{k=1}^{n} \epsilon_k < \lambda^n$ if

$$\frac{U'(\lambda^n)}{G(\lambda^n) - G(0)} > n .$$

Since $U'(\lambda)$ increases and $G(\lambda) - G(0)$ decreases as $\lambda \to 0$, we can ensure that this inequality holds if we choose λ^n sufficiently small.

Now let $\lambda > 0$ be arbitrary. In order for the algorithm to reach λ from λ^0, it must first reach $\lambda^n = \lambda_0 + \sum_{k=1}^{n} \epsilon_k$. We have just shown that by choosing λ^n sufficiently small (which is equivalent to choosing λ_0 sufficiently small), this will take (at least) n iterations. Therefore, it will take at least n iterations to reach λ from λ^0. ∎

6.2 Optimal Service Rates

In this section we consider parallel facilities in which the service rates at the facilities are decision variables, instead of the arrival rates. Our model is similar to the model introduced at the beginning of Section 6.1, except that now the waiting cost at each facility depends on the variable service rate and there is a service-cost function associated with the service rate at each facility. Our model can also be viewed as a generalization of the model of Section 5.2.2 of Chapter 5 to parallel facilities. For the reader's convenience, we provide a self-contained description of the complete model here.

We consider a system with n independent parallel facilities with variable service rates, μ_j, and fixed arrival rates, λ_j, $j \in J := \{1, \ldots, n\}$. The variable μ_j measures the average number of services completed per unit time while facility j is busy. For example, when the service facility is a $GI/GI/1$ queue, μ_j is the service rate of the single server (the reciprocal of the expected service time). Throughout this section, μ_j will be a decision variable, chosen from a feasible region $B_j \subseteq [0, \infty)$, $j \in J$. Our default assumption is that $B_j = [0, \infty)$.

We assume that at each facility the average waiting cost per customer is a function of the service rate at that facility. For a given value of the service rate, μ_j, let $G_j(\mu_j)$ denote the average waiting cost of a job at facility j, averaged over all customers who arrive during the period in question. We make the same assumptions about each $G_j(\mu_j)$ as we made for the single-facility model in Chapter 5; we repeat them here for convenient reference.

For all $j \in J$, we assume that $G_j(\mu_j)$ is well defined and finite and differentiable in μ_j, $\mu_j > \lambda_j$, with $G_j(\mu_j) = \infty$ for $0 \le \mu_j \le \lambda_j$. We also assume that $G_j(\mu_j)$ is strictly decreasing in μ_j, $\mu_j > \lambda_j$, and $G_j(\mu_j) \to \infty$ as $\mu_j \downarrow \lambda_j$.

Finally, for each facility $j \in J$ there is a service-cost rate, $c_j(\mu_j)$, which measures the average cost per unit time associated with operating the facility at rate μ_j. Our default assumption is that $c_j(\cdot)$ is a nondecreasing, convex, and differentiable function of $\mu_j \in B_j$, with $c_j(0) = 0$ and $c_j(\mu_j) \to \infty$ as $\mu_j \to \infty$, $j \in J$.

As in the case of a single facility, since the arrival rates are fixed, the criteria of individual and facility optimality are not relevant in the present context.

6.2.1 Social Optimality

Let $\boldsymbol{\mu} := (\mu_1, \ldots, \mu_n)$. When the decision criterion is social optimality, the objective function, to be minimized, is given by

$$C(\boldsymbol{\mu}) := \sum_{j \in J} (c_j(\mu_j) + \lambda_j G_j(\mu_j)) .$$

Since $C(\boldsymbol{\mu})$ is separable, we can minimize each term by itself. The problem for facility j is therefore to minimize $c_j(\mu_j) + \lambda_j G_j(\mu_j)$. This problem was solved in Chapter 5. In particular, it follows that the socially optimal service rate,

μ_j^s, at facility j is the unique solution in (λ_j, ∞), to the equation

$$c_j'(\mu_j) + \lambda_j \frac{\partial}{\partial \mu_j} G_j(\mu_j) = 0 \,. \qquad (6.60)$$

The reader is referred to Section 5.2.2 of Chapter 5 for further results.

6.2.1.1 Fixed Total Service Rate

Consider a variant of the above problem, in which there is no service cost $(c_j(\cdot) \equiv 0, \ j \in J)$ but the total service rate, $\mu = \sum_{j \in J} \mu_j$, is fixed. Then the social optimization problem takes the form:

$$\min_{\{\mu_j, j \in J\}} \quad \sum_{j \in J} \lambda_j G_j(\mu_j)$$

$$\text{s.t.} \quad \sum_{j \in J} \mu_j = \mu \,, \qquad (6.61)$$

$$\mu_j > \lambda_j \,, \ j \in J \,.$$

We can solve this problem using a Lagrange multiplier, α, to incorporate the equality constraint into the objective function. The Lagrangean minimization problem is:

$$\min_{\{\mu_j, j \in J\}} \quad \sum_{j \in J} \lambda_j G_j(\mu_j) + \alpha \Big(\sum_{j \in J} \mu_j - \mu \Big) \qquad (6.62)$$

$$\text{s.t.} \quad \mu_j > \lambda_j \,, \ j \in J \,.$$

The solution to (6.62) is parameterized by α, which can be interpreted as an imputed cost per unit time per unit of service rate. Let $\mu_j^s(\alpha)$ denote the optimal service rate at queue j as a function of α. If α is chosen so that $\sum_{j \in J} \mu_j^*(\alpha) = \mu$, then we have solved the original social optimization problem.

Problem (6.62) is separable, so we can minimize the terms in the objective function separately for each facility. This results in n separate minimization problems, each in the same form as the single-facility example in Section 5.2.2 of Chapter 5, namely,

$$\min_{\{\mu_j : \mu_j > \lambda_j\}} \{\lambda_j G_j(\mu_j) + \alpha \mu_j\} \,.$$

It follows that $\mu_j^s(\alpha)$ is the unique solution to the following optimality condition:

$$-\lambda_j \frac{\partial}{\partial \mu_j} G_j(\mu_j) = \alpha \,. \qquad (6.63)$$

6.2.1.2 Example: M/M/1 Queue With Linear Waiting Costs

We now consider the special case where each facility consists of an *M/M/1* queue operating in steady state with a linear waiting-cost function. Specifically, we assume that the arrival process at facility j is Poisson with parameter $\lambda_j, \ j \in J$. Successive service times at each facility are i.i.d. with an exponential distribution with parameter μ_j, $j \in J$. Let h_j denote the cost per customer per unit of time waiting in the system at facility j, $j \in J$.

The expected steady-state waiting cost per customer at facility j is

$$G_j(\mu_j) = \frac{h_j}{\mu_j - \lambda_j} \, , \, \mu > \lambda_j \, , \tag{6.64}$$

with $G_j(\mu_j) = \infty$ for $0 \le \mu_j \le \lambda_j$. In this case the Lagrangean minimization problem for facility j takes the form

$$\min_{\{\mu_j : \mu_j > \lambda_j\}} \left\{ \frac{\lambda_j h_j}{\mu_j - \lambda_j} + \alpha \mu_j \right\} \, ,$$

the solution to which is

$$\mu_j^s = \mu_j^s(\alpha) = \lambda_j + \sqrt{\lambda_j h_j / \alpha} \, . \tag{6.65}$$

(See Section 1.1 of Chapter 1.) Now that we have an explicit expression for $\mu_j^s(\alpha)$ as a function of α, we can choose α to satisfy the constraint, $\sum_{j \in J} \mu_j = \mu$:

$$\mu = \sum_{j \in J} \mu_j^s(\alpha) = \sum_{j \in J} \lambda_j + \sqrt{\alpha^{-1}} \sum_{j \in J} \sqrt{\lambda_j h_j} \, .$$

Therefore,

$$\sqrt{\alpha^{-1}} = (\mu - \sum_{j \in J} \lambda_j) / \sum_{j \in J} \sqrt{\lambda_j h_j} \, .$$

Substituting in (6.65), we obtain

$$\mu_j^s = \lambda_j + \left(\frac{\sqrt{\lambda_j h_j}}{\sum_{k \in J} \sqrt{\lambda_k h_k}} \right) \left(\mu - \sum_{k \in J} \lambda_k \right) \, . \tag{6.66}$$

We see that an optimal choice of service rates allocates the total excess service capacity, $\mu - \sum_{k \in J} \lambda_k$, to each of the individual facilities in proportion to the square root of the arrival rate times the waiting-cost rate at that facility.

6.2.1.3   Generalization: Open Jackson Network

Consider an open Jackson network, consisting of n nodes, $j \in J$. Each node j consists of a single exponential server, with mean service rate μ_j. Customers arrive at the nodes j from outside the system according to independent Poisson processes, where the mean external arrival rate at node j is ν_j. A customer who completes service at node j goes next to node k with probability r_{jk}, independently of the state of the system and previous nodes visited (Markovian routing of customers). For $j \in J$, let

$$r_{j0} := 1 - \sum_{k=1}^{n} r_{jk} \, ,$$

so that r_{j0} is the probability that a customer leaves the system upon completing service at node j. Assume that the ν_j and the r_{jk} are such that a customer can reach any node from outside the system and can leave the system from any node, either directly or indirectly via a path through other nodes. Under these conditions, the *traffic equations*

$$\lambda_j = \nu_j + \sum_{i=1}^{n} \lambda_i r_{ij} \, , \, j \in J \, , \tag{6.67}$$

have a unique positive solution, $\{\lambda_j, j \in J\}$, and that each node j behaves in steady state like an $M/M/1$ queue with mean arrival rate λ_j and mean service rate μ_j, provided $\lambda_j < \mu_j$, $j \in J$.

It follows that the steady-state expected number of customers at node j is given by:

$$L_j(\mu_j) := \frac{\lambda_j}{\mu_j - \lambda_j} \, .$$

Thus all the above results about optimal service rates μ_j^* for n parallel queues extend to an open Jackson network.

6.3 Optimal Arrival Rates and Service Rates

In this section we consider an optimal design model in which both the arrival rates (or tolls) and the service rates at the facilities $j \in J$ are design variables. Together the tolls and the service rates determine the arrival rate, through the equilibrium condition for individually optimizing customers.

Our model is essentially the same as the model introduced at the beginning of Section 6.1, with the additional features that the waiting cost at each facility depends on the variable service rate as well as the arrival rate and there is a service-cost function associated with the service rate at each facility. Our model can also be viewed as a generalization of the model of Section 5.5 of Chapter 5 to parallel facilities. For the reader's convenience, we provide a self-contained description of the complete model here.

We consider a system with n independent parallel facilities with variable service rates, μ_j, and variable arrival rates, λ_j, $j \in J := \{1, \dots, n\}$. The variable μ_j measures the average number of services completed per unit time while facility j is busy. For example, when the service facility is a $GI/GI/1$ queue, μ_j is the service rate of the single server (the reciprocal of the expected service time). Throughout this section, μ_j will be a decision variable, chosen from a feasible region $B_j \subseteq [0, \infty)$, $j \in J$. Our default assumption is that $B_j = [0, \infty)$. The set of feasible values for λ_j is denoted by A_j. Our default assumption will be that $A_j = [0, \mu_j)$.

The system earns a utility, $U(\lambda)$, per unit time when the total arrival rate is $\lambda = \sum_{j \in J} \lambda_j$. As usual, we assume that $U(\lambda)$ is nondecreasing, differentiable, and concave in $\lambda \geq 0$.

We assume that at each facility the average waiting cost per customer is a function of the arrival rate and service rate at that facility. For given values of the arrival rate, λ_j, and service rate, μ_j, let $G_j(\lambda_j, \mu_j)$ denote the average waiting cost of a job at facility j, averaged over all customers who arrive during the period in question. We make the same assumptions about each $G_j(\lambda_j, \mu_j)$ as we made for the single-facility model in Chapter 5; we repeat them here for convenient reference.

For all $j \in J$, we assume that $G_j(\lambda_j, \mu_j)$ is well defined and finite and differentiable in (λ_j, μ_j), $0 \le \lambda_j < \mu_j$, with $G_j(\lambda_j, \mu_j) = \infty$ for $\lambda_j \ge \mu_j \ge 0$. We also assume that:

(i) for each fixed $\mu_j > 0$, $G_j(\lambda_j, \mu_j)$ is strictly increasing in λ_j, $0 \le \lambda_j < \mu_j$, and $G(\lambda_j, \mu_j) \to \infty$ as $\lambda_j \uparrow \mu_j$;

(ii) for each fixed $\lambda_j \ge 0$, $G_j(\lambda_j, \mu_j)$ is strictly decreasing in μ_j, $\mu_j > \lambda_j$, and $G_j(\lambda_j, \mu_j) \to \infty$ as $\mu_j \downarrow \lambda_j$.

There may also be an admission fee (or toll) δ_j which is charged to each customer who joins facility j, $j \in J$.

Finally, for each facility $j \in J$ there is a service-cost rate, $c_j(\mu_j)$, which measures the average cost per unit time associated with operating the facility at rate μ_j. Our default assumption is that $c_j(\cdot)$ is a nondecreasing, convex, and differentiable function of $\mu_j \in B_j$, with $c_j(0) = 0$ and $c_j(\mu_j) \to \infty$ as $\mu_j \to \infty$, $j \in J$. (In Section 6.3.2.1, however, we also consider the case of *concave* service-cost functions.)

As usual, the solution to the decision problem depends on who is making the decision and what criterion is being used. The decision may be made by the individual customers, each concerned only with its own net utility (*individual optimality*), by an agent for the customers as a whole, who might be interested in maximizing the aggregate net utility to all customers (*social optimality*), or by a system operator at each facility j interested in maximizing profit at that facility (*facility optimality*).

As in the case of a single facility, the tolls may be either fixed or variable. The analysis of the model with fixed tolls is a direct extension of the analysis of the corresponding single-facility model and is left to the reader. For the remainder of this section we shall assume that the tolls (as well as the arrival rates and service rates) are decision variables.

6.3.1 Individual Optimality

For given values of the tolls, δ_j, and the service rates, μ_j, we can apply the analysis of Section 6.1.1. Recall that the individually optimal allocation of flows, $\{\lambda_j^e, j \in J\}$, uniquely satisfies conditions (6.2)-(6.7). We rewrite these conditions below, with $G_j(\lambda_j)$ now written as $G_j(\lambda_j, \mu_j)$, to indicate the dependence on the service rate:

$$U'(\lambda) \le \pi, \qquad (6.68)$$

$$\lambda(\pi - U'(\lambda)) = 0, \qquad (6.69)$$

$$\sum_{j \in J} \lambda_j = \lambda, \qquad (6.70)$$

$$G_j(\lambda_j, \mu_j) + \delta_j \ge \pi, \ j \in J, \qquad (6.71)$$

$$\lambda_j(G_j(\lambda_j, \mu_j) + \delta_j - \pi) = 0, \ j \in J, \tag{6.72}$$

$$\lambda_j \geq 0, \ j \in J. \tag{6.73}$$

6.3.2 Social Optimality

Now consider the problem from the point of view of social optimization. The objective is to choose the arrival rates, λ_j, and service rates, μ_j, at each facility $j \in J$, to maximize the steady-state expected net benefit (utility minus sum of waiting costs and service costs) per unit time. Thus the problem takes the form:

$$\max_{\{\lambda, \lambda_j, \mu_j\}} \quad U(\lambda) - \sum_{j \in J} (\lambda_j G_j(\lambda_j, \mu_j) + c_j(\mu_j))$$

$$\text{s.t.} \quad \sum_{j \in J} \lambda_j = \lambda, \tag{6.74}$$

$$0 \leq \lambda_j < \mu_j, \ j \in J.$$

We use a Lagrange multiplier to eliminate the equality constraint on the total arrival rate, leading to the following Lagrangian maximization problem:

$$\max_{\{\lambda, \lambda_j, \mu_j\}} \quad U(\lambda) - \sum_{j \in J} (\lambda_j G_j(\lambda_j, \mu_j) + c_j(\mu_j)) + \alpha(\sum_{j \in J} \lambda_j - \lambda) \tag{6.75}$$

$$\text{s.t.} \quad 0 \leq \lambda_j \leq \mu, \ j \in J.$$

Rewriting the Lagrangean objective function in equivalent form,

$$U(\lambda) - \alpha\lambda + \sum_{j \in J} [\lambda_j(\alpha - G_j(\lambda_j, \mu_j)) - c_j(\mu_j)],$$

reveals that it is separable in λ and (λ_j, μ_j), $j \in J$, for any fixed value of α.

For the system as a whole the objective function, $U(\lambda) - \lambda\alpha$, is that of a single-facility arrival-rate-optimization problem (cf. Chapter 2), with utility function $U(\lambda)$ and constant waiting-cost function, $G(\lambda) \equiv \alpha$. The necessary (and sufficient) optimality conditions for this problem are

$$U'(\lambda) \leq \alpha, \ \lambda \geq 0,$$

$$U'(\lambda) = \alpha, \ \text{if } \lambda > 0.$$

The objective function for each facility $j \in J$ is that of a single-facility social optimization problem with variable arrival and service rates with a linear utility function, $\alpha\lambda_j$. The necessary optimality conditions for this problem are

$$\alpha \leq G_j(\lambda_j, \mu_j) + \lambda_j \frac{\partial}{\partial \lambda_j} G_j(\lambda_j, \mu_j), \ \lambda_j \geq 0, \ j \in J,$$

$$\alpha = G_j(\lambda_j, \mu_j) + \lambda_j \frac{\partial}{\partial \lambda_j} G_j(\lambda_j, \mu_j), \ \text{if } \lambda_j > 0, \ j \in J,$$

$$c'(\mu_j) = -\lambda_j \frac{\partial}{\partial \mu_j} G_j(\lambda_j, \mu_j), \ j \in J.$$

In Section 5.5.2 of Chapter 5 we saw that these conditions are not in general sufficient for a global maximum pair, (λ_j, μ_j), because the objective function is not in general jointly concave, even when $H_j(\lambda_j, \mu_j) = \lambda_j G_j(\lambda_j, \mu_j)$ is convex in λ_j and convex in μ_j, and $c_j(\mu_j)$ is convex in μ_j (our default assumptions).

Combining the optimality conditions for the system with those for each facility and adding the constraint, $\sum_{j \in J} \lambda_j = \lambda$, leads to the following *KKT* conditions, which are necessary (but not in general sufficient) conditions for the original problem:

$$U'(\lambda) \leq \alpha, \lambda \geq 0, \tag{6.76}$$

$$U'(\lambda) = \alpha, \text{ if } \lambda > 0, \tag{6.77}$$

$$\lambda = \sum_{j \in J} \lambda_j, \tag{6.78}$$

$$\alpha \leq G_j(\lambda_j, \mu_j) + \lambda_j \frac{\partial}{\partial \lambda_j} G_j(\lambda_j, \mu_j), \lambda_j \geq 0, j \in J, \tag{6.79}$$

$$\alpha = G_j(\lambda_j, \mu_j) + \lambda_j \frac{\partial}{\partial \lambda_j} G_j(\lambda_j, \mu_j), \text{ if } \lambda_j > 0, j \in J, \tag{6.80}$$

$$c'(\mu_j) = -\lambda_j \frac{\partial}{\partial \mu_j} G_j(\lambda_j, \mu_j), j \in J. \tag{6.81}$$

As in the single-facility problem, because of the lack of joint concavity, it is not enough simply to look for a solution to these conditions. A more careful analysis is called for. The following example illustrates the complications which may arise.

6.3.2.1 Example: Linear Utility Function

Suppose the utility function is linear: $U(\lambda) = r \cdot \lambda$. Then the original social optimization problem (6.74) is separable and the objective function can be maximized separately for each facility $j \in J$.

We shall follow the approach for a single-facility problem introduced in Section 5.5.2.3 of Chapter 5. The social optimization problem for each facility j takes the form

$$\max_{\lambda_j, \mu_j} B_j(\lambda_j, \mu_j) := r \cdot \lambda_j - \lambda_j G(\lambda_j, \mu_j) - c_j(\mu_j)$$

$$\text{s.t.} \quad 0 \leq \lambda_j < \mu_j \leq \bar{\mu}_j,$$

where $\bar{\mu}_j \leq \infty$. We make the following assumptions (cf. Assumptions 2 and 3 in Section 5.5.2.3 of Chapter 5).

Assumption 1 At each facility j, the service cost, $c_j(\mu_j)$, is a *concave*, strictly increasing, and differentiable function of $\mu_j \in [0, \bar{\mu}_j]$.

Assumption 2 The waiting cost per unit time, $H_j(\lambda_j, \mu_j) = \lambda_j G(\lambda_j, \mu_j)$, takes the form

$$H_j(\lambda_j, \mu_j) = f_j(\lambda_j/\mu_j),$$

where $f_j(\rho)$ is a convex, strictly increasing, and differentiable function of $\rho \in [0, 1)$, with $f_j(0) = 0$ and $\lim_{\rho \to 1} f_j(\rho) = \infty$ and $f_j(\rho) = \infty$ for $\rho \geq 1$.

For a given value of μ_j, let $\lambda_j(\mu_j)$ denote the value of λ_j that maximizes the objective function, $B_j(\lambda_j, \mu_j)$, for facility j, subject to $0 \leq \lambda_j < \mu_j$. Let μ_{j0} be the unique solution to $r = G_j(0, \mu_j)$. Then for $0 \leq \mu_j \leq \mu_{j0}$, we have $\lambda_j(\mu_j) = 0$. For $\mu_j > \mu_{j0}$, $\lambda_j(\mu_j)$ maximizes the concave function, $r \cdot \lambda_j - f_j(\lambda_j/\mu_j)$, and is therefore a solution to the optimality condition,

$$r = \left(\frac{1}{\mu_j}\right) f'_j(\lambda_j/\mu_j) . \tag{6.82}$$

Let $\psi_j(\mu_j) := B_j(\lambda_j(\mu_j), \mu_j)$, $\mu_j \geq 0$. Then

$$\psi_j(\mu_j) = r \cdot \lambda_j(\mu_j) - f_j(\lambda_j(\mu_j)/\mu_j) - c_j(\mu_j) .$$

It follows from Theorem 5.8 that $\psi_j(\mu_j)$ is not concave. Rather, it is a *convex* function of μ_j and therefore attains its maximum at a boundary point of the feasible region for μ_j. Hence, the socially optimal pair, (λ_j^s, μ_j^s) for facility j is either $(0, 0)$ or $(\lambda_j(\bar{\mu}_j), \bar{\mu}_j)$.

Thus we see that each facility will either be closed or operate at the maximum possible service rate.

Of course, a linear utility function is an extreme case. When the utility function is strictly concave, the optimal solution, μ_j^s, for a particular facility j may lie at an interior point of the feasible region, $[0, \bar{\mu}_j]$, for that facility. However, it may still be the case that there are multiple solutions to the necessary *KKT* conditions, some of which are relative *minima*, rather than maxima, of the objective function for social optimization.

6.3.2.2 Fixed Total Arrival Rate: Facility Dominance

Now consider the social optimization problem in which the total arrival rate λ is fixed. The problem now takes the form:

$$\min_{\{\lambda_j, \mu_j\}} \quad \sum_{j \in J} (\lambda_j G_j(\lambda_j, \mu_j) + c_j(\mu_j))$$

$$\text{s.t.} \quad \sum_{j \in J} \lambda_j = \lambda , \tag{6.83}$$

$$0 \leq \lambda_j \leq \lambda , \ j \in J .$$

(The addition of the redundant constraints, $\lambda_j \leq \lambda$, is without loss of optimality.)

Once more we use a Lagrange multiplier to eliminate the equality constraint on the total arrival rate. Consider the following Lagrangean minimization problem:

$$\min_{\{\lambda_j, \mu_j\}} \quad \sum_{j \in J} (\lambda_j G_j(\lambda_j, \mu_j) + c_j(\mu_j)) - \alpha(\sum_{j \in J} \lambda_j - \lambda) \tag{6.84}$$

$$\text{s.t.} \quad 0 \leq \lambda_j \leq \lambda , \ \mu_j > \lambda_j , \ j \in J .$$

One can easily verify that, if $\{\lambda_j, \mu_j, j \in J\}$ achieve the minimum in the Lagrangean problem for given α and $\sum_{j \in J} \lambda_j = \lambda$, then $\{\lambda_j, \mu_j, j \in J\}$ are optimal for the original problem. Rewriting the Lagrangean objective function in equivalent form (omitting the constant term, $\alpha\lambda$),

$$\sum_{j \in J} [\lambda_j (G_j(\lambda_j, \mu_j) - \alpha) + c_j(\mu_j)] ,$$

reveals that it is separable in (λ_j, μ_j), $j \in J$, for any fixed value of α. Thus, we can minimize the objective function separately with respect to the arrival and service rate, (λ_j, μ_j), at each facility $j \in J$, subject to $0 \leq \lambda_j \leq \lambda$.

The Lagrangean minimization problem for each facility j is equivalent to a single-facility social optimization problem with variable arrival and service rates, with a linear utility function, $\alpha\lambda_j$, in which the objective function (to be maximized) is

$$B_j(\lambda_j, \mu_j; \alpha) := \alpha\lambda_j - \lambda_j G_j(\lambda_j, \mu_j) + c_j(\mu_j) .$$

Even under our default assumptions that $H_j(\lambda_j, \mu_j) = \lambda_j G_j(\lambda_j, \mu_j)$ is convex in λ_j and convex in μ_j and $c_j(\mu_j)$ is convex in μ_j, we know that $B_j(\lambda_j, \mu_j; \alpha)$ need not be jointly convex in λ_j and μ_j. Indeed, when $c_j(\mu_j)$ is linear (or, more generally, concave) the optimal solution may lie at an extreme point of the feasible region for (λ_j, μ_j). We analyzed this problem in the previous section, with an upper bound on μ_j rather than on λ_j. Essentially the same analysis applies here, except that the upper bound on μ_j is now implicit rather than explicit.

As in the previous section, we shall make Assumptions 1 and 2. Let μ_{j0} be the unique solution to $\alpha = G_j(0, \mu_j)$. Let μ_{j1} be the unique solution to

$$\alpha = \left(\frac{1}{\mu_j} \right) f_j'(\lambda/\mu_j) .$$

For $\mu_{j0} < \mu_j < \mu_{j1}$, let $\tilde{\lambda}_j(\mu_j)$ be the unique solution to

$$\alpha = \left(\frac{1}{\mu_j} \right) f_j'(\lambda_j/\mu_j) .$$

Let $\lambda_j(\mu_j)$ be the value of λ_j that maximizes $B_j(\lambda_j, \mu_j; \alpha)$ over $0 \leq \lambda_j \leq \lambda$. Then

$$
\begin{aligned}
\lambda_j(\mu_j) &:= 0 , \; 0 \leq \mu_j \leq \mu_{j0} \\
&:= \tilde{\lambda}_j(\mu_j) , \; \mu_{j0} < \mu_j < \mu_{j1} \\
&:= \lambda , \; \mu_{j1} \leq \mu_j .
\end{aligned}
$$

Now let $\mu_j(\lambda)$ denote the value of μ_j that maximizes $B_j(\lambda, \mu_j; \alpha)$ over $\mu_j > \lambda$. Let $\psi_j(\mu_j) := B_j(\lambda_j(\mu_j), \mu_j; \alpha)$, $0 \leq \mu_j < \infty$. It follows from the analysis of the case of a linear utility function in the previous section that $\psi_j(\mu_j)$ is convex in $\mu_j \in [0, \mu_{j1}]$ and therefore the maximum of $\psi_j(\mu_j)$ over this range occurs either at $\mu_j = 0$ or $\mu_j = \mu_j(\lambda)$.

Thus we have proved the following lemma.

Lemma 6.13 *Under Assumptions 1 and 2, for any $\alpha \geq 0$ the optimal pair,* $(\lambda_j(\alpha), \mu_j(\alpha))$ *for the Lagrangean maximization problem for facility j,*

$$\max_{\{\lambda_j, \mu_j\}} \quad B_j(\lambda_j, \mu_j; \alpha) := \alpha\lambda_j - \lambda_j G_j(\lambda_j, \mu_j) - c_j(\mu_j)$$

$$\text{s.t.} \quad 0 \leq \lambda_j \leq \lambda \, , \, \mu_j > \lambda_j \, ,$$

is either $(0,0)$ *or* $(\lambda, \mu_j(\lambda))$.

It remains to find a value of α for which the solutions to these Lagrangean problems together satisfy the constraint, $\sum_{j \in J} \lambda_j = \lambda$. From Lemma 6.13 we see that this will be the case if and only if, for some $j \in J$, we have $\lambda_j = \lambda$, with $\lambda_k = 0$, for $k \neq j$.

Thus we have the following theorem.

Theorem 6.14 *Under Assumptions 1 and 2, the socially optimal allocation of arrival rates and service rates for the problem with a fixed total arrival rate assigns all traffic to one facility (facility dominance).*

Which facility j dominates depends, of course, on the value of λ. Because of the continuity of the functions involved, for each facility j there is a (possibly empty) interval of values of λ over which facility j dominates. The set of all such intervals constitutes a partition of the interval $[0, \infty)$ of possible values for λ.

The following example illustrates this phenomenon.

Example: M/M/1 Facilities with Linear Waiting and Service Costs.

Consider a system of n parallel facilities, in which facility $j \in J = \{1, \ldots, n\}$ operates as an $M/M/1$ queue in steady state, with variable arrival rate, λ_j, and variable service rate, μ_j. The waiting cost per customer per unit time at facility j is h_j, so that

$$G_j(\lambda_j, \mu_j) = \frac{h_j}{\mu_j - \lambda_j} \, , \, 0 \leq \lambda_j < \mu_j \, .$$

The service-cost function at each facility $j \in J$ is linear:

$$c_j(\mu_j) = c_j \cdot \mu_j \, , \, \mu_j \geq 0 \, .$$

We analyzed the single-facility version of this model in Section 1.3 of Chapter 1.

For this model, the Lagrangean maximization problem for facility j takes the form

$$\max_{\{\lambda_j, \mu_j\}} \quad B_j(\lambda_j, \mu_j; \alpha) := \alpha \cdot \lambda_j - \frac{\lambda_j h_j}{\mu_j - \lambda_j} - c_j \cdot \mu_j$$

$$\text{s.t.} \quad 0 \leq \lambda_j \leq \lambda \, , \, \mu_j > \lambda_j \, .$$

For a given α, we have

$$B_j(\lambda, \mu_j(\lambda); \alpha) = (\alpha - c_j)\lambda - 2\sqrt{\lambda h_j c_j} \, ,$$

from which it follows that

$$\lambda_j(\alpha) = \begin{cases} 0 & \text{, if } \alpha \le c_j + 2\sqrt{h_j c_j}/\lambda \\ \lambda & \text{, if } \alpha > c_j + 2\sqrt{h_j c_j}/\lambda \end{cases} .$$

As a function of λ, define

$$\alpha_k(\lambda) := c_k + 2\sqrt{h_k c_k/\lambda} , \ k \in J ,$$

and

$$j(\lambda) := \arg \min_{\{k \in J\}} \alpha_k(\lambda) .$$

Then among all the facilities, $k \in J$, facility $j = j(\lambda)$ attains the minimum total cost when assigned the total arrival rate, λ. That is, $C_j(\lambda) = \min_{k \in J} C_k(\lambda)$, where

$$C_k(\lambda) := c_k \lambda + 2\sqrt{\lambda h_k c_k} = \lambda \alpha_k(\lambda) , \ k \in J .$$

For a given λ, let $j = j(\lambda)$. The socially optimal arrival rates λ_k^s, $k \in J$ are given by

$$\lambda_j^s = \lambda ; \ \lambda_k^s = 0 , \ k \ne j .$$

The socially optimal service rates are

$$\mu_j^s = \mu_j(\lambda) = \lambda + \sqrt{\lambda h_j/c_j} ; \ \mu_k^s = 0 , \ k \ne j .$$

How does the value of λ affect which facility is dominant? For a particular value of λ, facility j dominates facility k if and only if $\alpha_j < \alpha_k$. (Note that we are construing dominance in a strict sense here.) It follows that

1. If $c_j < c_k$ and $h_j c_j < h_k c_k$, then facility j dominates facility k for all $\lambda > 0$.

2. If $c_j > c_k$ and $h_j c_j < h_k c_k$, then

facility j dominates facility k for $\quad 0 < \lambda < 4 \left(\dfrac{\sqrt{h_k c_k} - \sqrt{h_j c_j}}{c_j - c_k} \right)^2 ,$

facility k dominates facility j for $\quad \lambda > 4 \left(\dfrac{\sqrt{h_k c_k} - \sqrt{h_j c_j}}{c_j - c_k} \right)^2 .$

Figure 6.5 illustrates the dominance relations for three parallel facilities, with

$$\begin{aligned} c_1 &= 10 &,& \quad h_1 = 90 ; \\ c_2 &= 20 &,& \quad h_2 = 20 ; \\ c_3 &= 30 &,& \quad h_3 = 30 . \end{aligned}$$

6.3.2.3 Variable Total Arrival Rate: Facility Dominance

Now let us return to the social optimization problem in which the total arrival rate λ is variable. Throughout this section we again make Assumptions 1 and 2.

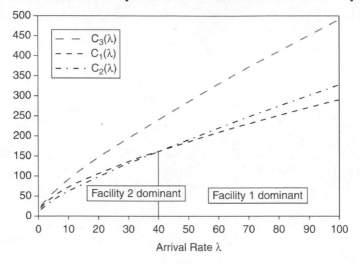

Figure 6.5 *Facility Dominance as a Function of* λ

Recall that the general form for this problem is:

$$\max_{\{\lambda,\lambda_j,\mu_j\}} \quad U(\lambda) - \sum_{j \in J}(\lambda_j G_j(\lambda_j,\mu_j) + c_j(\mu_j))$$

$$\text{s.t.} \quad \sum_{j \in J}\lambda_j = \lambda\,, \tag{6.85}$$

$$0 \le \lambda_j < \mu_j\,,\ j \in J\,.$$

Now for any particular value of λ (including, of course, the socially optimal value), the facility arrival rates, $\{\lambda_j, j \in J\}$, must maximize the above objective function with λ fixed at that value. That is, they must be optimal for the problem discussed in the previous section, which takes the form:

$$\min_{\{\lambda_j,\mu_j\}} \quad \sum_{j \in J}(\lambda_j G_j(\lambda_j,\mu_j) + c_j(\mu_j))$$

$$\text{s.t.} \quad \sum_{j \in J}\lambda_j = \lambda\,, \tag{6.86}$$

$$0 \le \lambda_j \le \lambda\,,\ j \in J\,.$$

But we have shown (cf. Theorem 6.14) that an optimal solution to this problem exhibits *facility dominance*; that is, $\lambda_j = \lambda$, $\lambda_k = 0$, $k \ne j$, for some $j \in J$. It follows from the above reasoning that the same will be true for the problem with a variable total arrival rate, and therefore we have the following theorem.

Theorem 6.15 *Under Assumptions 1 and 2, the socially optimal allocation of arrival rates and service rates for the problem with a variable total arrival rate assigns all traffic to one facility* (facility dominance).

Consider a particular facility j. As we observed in the previous section, facility j will be dominant over a (possibly empty) interval of values of λ, say $[\underline{\lambda}_j, \bar{\lambda}_j]$, where $\underline{\lambda}_j < \bar{\lambda}_j$. Over this interval, the objective for the social optimization problem is to maximize $U(\lambda) - C_j(\lambda)$, where $(j \in J)$

$$C_j(\lambda) := \min_{\mu_j > \lambda} \{\lambda G(\lambda, \mu_j) + c_j(\mu_j)\}, \ \lambda \geq 0. \qquad (6.87)$$

Since $C_j(\lambda)$ is concave (cf. Section 6.3.2.1), the function $U(\lambda) - C_j(\lambda)$ may not be concave. (Indeed, it may be convex, as is the case when the utility function is linear: $U(\lambda) = r \cdot \lambda$.) Thus the maximum of $U(\lambda) - C_j(\lambda)$ over the interval, $[\underline{\lambda}_j, \bar{\lambda}_j]$, may occur at one of the endpoints or at a point in the interior of the interval at which the optimality condition,

$$U'(\lambda) = C_j'(\lambda),$$

holds. (There may be more than one such point.)

Following the approach suggested by the above observations, there is no easy way to solve the social optimization problem without additional conditions on the functions involved. One must identify the intervals, $[\underline{\lambda}_j, \bar{\lambda}_j]$, over which each facility j dominates, and then find the optimal value of λ over each such interval, which, as we have noted, may require comparing several candidate solutions.

An alternative approach is first to solve a separate optimization problem for each facility j, in which one seeks the pair (λ_j, μ_j) that achieves the maximum in the following single-facility problem:

$$\max_{\lambda_j, \mu_j} \ \mathcal{U}_j(\lambda_j, \mu_j) \ := \ U(\lambda_j) - \lambda_j G_j(\lambda_j, \mu_j) - c_j(\mu_j)$$

$$\text{s.t.} \qquad 0 \leq \lambda_j < \mu_j.$$

We discussed problems of this form in Section 5.5.2 of Chapter 5. There we saw that there may in general be more than one solution to the necessary KKT conditions, and hence we must compare the value of the objective function, $\mathcal{U}_j(\lambda_j, \mu_j)$, at the various solutions to identify which one is globally optimal.

Let (λ_j^*, μ_j^*) denote the globally optimal pair for the above problem. Then, in order to find the socially optimal allocation, $(\lambda^s, \lambda_k^s, \mu_k^s, k \in J)$, for the multi-facility problem, we must identify the facility j that achieves

$$\max_{\{j \in J\}} \mathcal{U}_j(\lambda_j^*, \mu_j^*).$$

This is the dominant facility and the socially optimal solution is given by

$$\lambda_j^s \ = \ \lambda_j^*, \ \mu_j^s = \mu_j^*;$$
$$\lambda_k^s \ = \ 0, \ \mu_k^s = 0, \ k \neq j.$$

Again we use a simple example to illustrate these points.

Example: M/M/1 Facilities with Linear Waiting and Service Costs.

Consider (again) a system of n parallel facilities, in which facility $j \in J = \{1, \ldots, n\}$ operates as an *M/M/1* queue in steady state, with variable arrival rate, λ_j, and variable service rate, μ_j. The waiting cost per customer per unit time at facility j is h_j, so that

$$G_j(\lambda_j, \mu_j) = \frac{h_j}{\mu_j - \lambda_j} \, , \ 0 \le \lambda_j < \mu_j \, .$$

The service-cost function at each facility $j \in J$ is linear:

$$c_j(\mu_j) = c_j \cdot \mu_j \, , \ \mu_j \ge 0 \, .$$

For any given utility function, $U(\lambda)$, because of facility dominance, the social optimization problem may be written as

$$\max_{\{\lambda\}} \quad U(\lambda) - C(\lambda)$$

$$\text{s.t.} \quad \lambda \ge 0$$

where $C(\lambda) := \min_{j \in J} C_j(\lambda)$ and (cf. equation (6.87) above)

$$C_j(\lambda) = \min_{\{\mu_j > \lambda\}} \{\lambda G_j(\lambda, \mu_j) + c_j(\mu_j)\} \, , \ \lambda \ge 0 \, , \ j \in J \, .$$

For the present example, we have

$$C_j(\lambda) = c_j \cdot \lambda + 2\sqrt{\lambda h_j c_j} \, , \ \lambda \ge 0 \, , \ j \in J \, .$$

Suppose the utility function is given by

$$U(\lambda) = a\lambda \left(1 - \frac{\lambda}{2\Lambda}\right) \, , \ 0 \le \lambda \le \Lambda \, .$$

Note that $U(\lambda)$ is (strictly) increasing and (strictly) concave, with

$$U'(\lambda) = a(1 - \lambda/\Lambda) \, ,$$

over the interval $[0, \Lambda]$. (Recall that this utility function arises in the case where potential customers arrive at rate Λ and follow a probabilistic joining rule, with a uniform reward distribution, $F(r) = r/a, 0 \le r \le a$, in which case we have $U'(\lambda) = \bar{F}(\lambda/\Lambda)$. See Section 2.4 of Chapter 2 and Section 5.5.2.1 of Chapter 5.)

For the same three-facility example as in the previous section, with the above utility function, with $a = 50$, $\Lambda = 100$, we have graphed the functions $U'(\lambda)$ and $C'(\lambda)$ in Figure 6.6. Note that $C'(\lambda)$ has a discontinuity at $\lambda = 40$, where dominance switches from facility 2 to facility 1. The points where the graphs of $U'(\lambda)$ and $C'(\lambda)$ intersect are the candidates for an interior optimum.[*] In this case, there are two such points, λ^1 and λ^2. The former is a relative minimum of the objective function and the latter a relative maximum. The value of the objective function at λ^2 must be compared to the value at $\lambda = 0$ to determine the global maximum.

---

[*] More generally, any point at which $U'(\lambda-) - C'(\lambda-) \ge 0 \ge U'(\lambda) - C'(\lambda)$ is a relative maximum and hence a candidate for the global maximum.

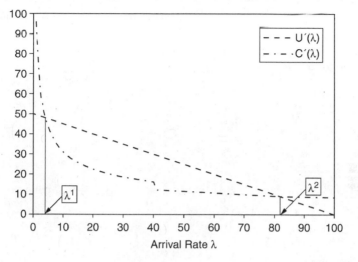

Figure 6.6 *Graphs of $U'(\lambda)$ and $C'(\lambda)$ for Parallel-Facility Example*

6.3.3 Facility Optimality

Now we consider the system from the point of view of a facility operator whose goal is to set service rates and tolls that will maximize its revenue. As was the case with fixed service rates, each of the facilities may have its own operator (*competing* facilities), or there may be a single operator who manages all the facilities (*cooperating* facilities). We shall consider the case of cooperating facilities first.

6.3.3.1 Cooperating Facilities

Suppose a single operator (or a team of cooperating operators) sets the tolls, δ_j, and service rates, μ_j, for all facilities $j \in J$. The goal is to maximize the total profit,

$$\sum_{j\in J}[\lambda_j\delta_j - c_j(\mu_j)] \,,$$

that is, the revenue received from the tolls paid by the entering customers minus the sum of the service costs.

Assuming (as usual) that the arriving customers are individual optimizers, any particular set of service rates, μ_j, and tolls, δ_j, will result in arrival rates λ_j that satisfy the equilibrium conditions, (6.68)-(6.73), which we rewrite here in equivalent form:

$$
\begin{aligned}
U'(\lambda) &\leq \delta_j + G_j(\lambda_j,\mu_j) \,, \ \lambda_j \geq 0 \,, \ j \in J \,, \\
U'(\lambda) &= \delta_j + G_j(\lambda_j,\mu_j) \,, \quad \text{, if } \lambda_j > 0 \,, \ j \in J \,, \\
\sum_{j\in J}\lambda_j &= \lambda \,.
\end{aligned}
$$

(Here, for simplicity, we are again assuming that $\lambda > 0$.)

Using the same arguments as in Section 6.1.3.1 we can show that finding the facility optimal arrival rates, λ_j^f, and service rates, μ_j^f, $j \in J$, reduces to solving the following problem:

$$\max_{\{\lambda, \lambda_j, \mu_j, j \in J\}} \quad \lambda U'(\lambda) - \sum_{j \in J} (\lambda_j G_j(\lambda_j, \mu_j) + c_j(\mu_j))$$

$$\text{s.t.} \quad \sum_{j \in J} \lambda_j = \lambda \,,$$

$$0 \le \lambda_j < \mu_j \,, \ j \in J \,.$$

A set of facility-optimal tolls, δ^f, then must satisfy the following conditions:

$$\delta_j \ \ge \ U'(\lambda^f) - G_j(\lambda_j^f, \mu_j^f) \,, \ j \in J$$

$$= \ U'(\lambda^f) - G_j(\lambda_j^f, \mu_j^f) \,, \text{ if } \lambda_j^f > 0 \,, \ j \in J \,,$$

where $\lambda^f := \sum_{j \in J} \lambda_j^f$.

The above optimization problem is equivalent to a social optimization problem in which the utility function $U(\lambda)$ is replaced by $\lambda U'(\lambda)$ (cf. the analysis of the facility optimization problem in the case of a single facility in Section 2.1.3 of Chapter 2). Hence we can apply the analysis of the previous section. Now, however, in addition to the possible non-joint-convexity of the waiting-cost functions, $\lambda_j G_j(\lambda_j, \mu_j)$, we must contend with the fact that the function, $\lambda U'(\lambda)$, may not be concave. (It may also fail to be nondecreasing.)

6.3.3.2 Cooperating Facilities: Fixed Total Arrival Rate

In this subsection we consider the cooperative facility optimal solution in the case of a fixed total arrival rate λ. The problem takes the following form:

$$\max_{\{\pi, \lambda_j, \mu_j, \delta_j, j \in J\}} \quad \sum_{j \in J} [\lambda_j \delta_j - c_j(\mu_j)]$$

$$\text{s.t.} \quad \pi \le \delta_j + G_j(\lambda_j, \mu_j) \,, \ j \in J \,,$$

$$\pi = \delta_j + G_j(\lambda_j, \mu_j) \,, \ \ \text{, if } \lambda_j > 0 \,, \ j \in J \,,$$

$$\sum_{j \in J} \lambda_j = \lambda \,,$$

$$0 \le \lambda_j < \mu_j \,, \ j \in J \,.$$

Again, we can transform this problem into one in which the only decision variables are π and λ_j, μ_j, $j \in J$, namely:

$$\max_{\{\pi, \lambda_j, \mu_j, j \in J\}} \quad \lambda \pi - \sum_{j \in J} [\lambda_j G_j(\lambda_j, \mu_j) - c_j(\mu_j)] \tag{6.88}$$

$$\text{s.t.} \quad \sum_{j \in J} \lambda_j = \lambda \,,$$

$$0 \le \lambda_j < \mu_j \,, \ j \in J \,.$$

A set of facility-optimal tolls, δ^f, then must satisfy the following conditions:

$$\delta_j \geq \pi - G_j(\lambda_j^f, \mu_j^f), \; j \in J, \tag{6.89}$$

$$= \pi - G_j(\lambda_j^f, \mu_j^f), \text{ if } \lambda_j^f > 0, \; j \in J, \tag{6.90}$$

where $\lambda^f := \sum_{j \in J} \lambda_j^f$.

For any fixed value of π, since λ is fixed, the problem takes the following form:

$$\min_{\{\lambda_j, \mu_j, j \in J\}} \quad \sum_{j \in J} [\lambda_j G_j(\lambda_j, \mu_j) + c_j(\mu_j)]$$

$$\text{s.t.} \quad \sum_{j \in J} \lambda_j = \lambda,$$

$$0 \leq \lambda_j < \mu_j, \; j \in J.$$

We refer the reader to Section 6.3.2.2, where we studied this problem in detail. In particular we showed (cf. Theorem 6.14) that, if Assumptions 1 and 2 hold, then an optimal allocation of arrival rates and service rates assigns all traffic to one facility (*facility dominance*). Note that which facility receives all the traffic depends on the parameter λ but not on π.

Now let π once again be a variable and consider the facility-maximization problem (6.88). It follows that an optimal solution to this problem will assign all traffic to a single facility, which does not depend on the value of π, and that the objective function can be made arbitrarily large by allowing π to become arbitrarily large, while choosing the tolls, δ_j, to satisfy the conditions (6.89) and (6.90).

Thus we see that the facility-optimization problem with a fixed total arrival rate and variable arrival and service rates at each facility has the unbounded-profit property that we already observed (in Section 6.1.3.2) in the corresponding problem with fixed service rates. Now, however, the optimal allocation of arrival rates and service rates for each fixed value of π exhibits facility dominance.

6.3.3.3 Cooperating Facilities: Variable Total Arrival Rate

We remarked earlier that the facility optimization problem with a variable total arrival rate λ is equivalent to a social optimization problem in which the utility function, $U(\lambda)$, is replaced by $\tilde{U}(\lambda) := \lambda U'(\lambda)$. The analysis of this problem thus combines the techniques used in the analysis of the single-facility model (cf. Section 2.1.3 of Chapter 2) to deal with the fact that $\tilde{U}(\lambda)$ may fail to be concave (or even nondecreasing) with the techniques used in Section 6.3.2 of this chapter to study the social optimization problem with variable arrival and service rates, $\lambda_j, \mu_j, j \in J$. We leave it to the reader to carry out this analysis in detail, but note here that (once again) if Assumptions 1 and 2 hold, then an optimal allocation of arrival rates and service rates assigns all traffic to one facility (*facility dominance*). Since this property holds for any fixed valuwe of the total arrival rate, λ, it holds regardless of whether $\tilde{U}(\lambda)$ is concave or nondecreasing.

6.3.3.4 Cooperating Facilities: Heavy-Tailed Rewards

As in the model with fixed service rates, we are interested in what happens in the cooperative multi-facility setting if $\lambda U'(\lambda) \to \infty$ as $\lambda \to 0$. Once again, we expect that the facility operator will be able to earn an arbitrarily large profit by choosing an arbitrarily small arrival rate (equivalently, charging an arbitrarily large toll). But we also would like to know how the arrival and service rates at each facility vary as $\lambda \to 0$.

We shall find it convenient to work with the formulation of the facility optimization problem in terms of the arrival rates, λ_j, and service rates, μ_j, $j \in J$, as the decision variables:

$$\max_{\{\lambda, \lambda_j, \mu_j, j \in J\}} \quad \lambda U'(\lambda) - \sum_{j \in J}(\lambda_j G_j(\lambda_j, \mu_j) + c_j(\mu_j))$$

$$\text{s.t.} \quad \sum_{j \in J} \lambda_j = \lambda \,,$$

$$0 \leq \lambda_j < \mu_j \,, \ j \in J \,.$$

Now consider a fixed, arbitrary λ. Since $\lambda U'(\lambda)$ is fixed, the optimal λ_j, $j \in J$, corresponding to this value of λ will solve the following minimization problem:

$$\min_{\{\lambda_j, \mu_j, j \in J\}} \quad \sum_{j \in J}(\lambda_j G_j(\lambda_j, \mu_j) + c_j(\mu_j))$$

$$\text{s.t.} \quad \sum_{j \in J} \lambda_j = \lambda \,, \tag{6.91}$$

$$0 \leq \lambda_j < \mu_j \,, \ j \in J \,.$$

Again, we refer the reader to Section 6.3.2.2, where we showed (cf. Theorem 6.14) that, if Assumptions 1 and 2 hold, then an optimal allocation of arrival rates and service rates assigns all traffic to one facility (*facility dominance*). In particular, for all sufficiently small λ, the same facility will receive all the traffic. Without loss of generality, suppose it is facility 1. It follows that, when $\lambda U'(\lambda) \to \infty$, the cooperating facility operators can earn an arbitrarily large profit by directing all the traffic to facility 1 and charging arbitrarily large tolls, in accordance with the above results.

The facility-dominance property used in this argument holds for all values of the arrival rate, λ, but it requires Assumptions 1 and 2. If all we are interested in is facility dominance for sufficiently small λ, it turns out that we can establish this result under a weak assumption, following an approach similar to that used in Section 5.5.2.3 of Chapter 5.

Assumption 3   For all $j \in J$, the service cost $c_j(\mu_j)$ satisfies the following asymptotic condition:

$$\liminf_{\mu_j \to \infty} \frac{c_j(\mu_j)}{\mu_j} > 0 . \tag{6.92}$$

As we observed in Section 5.5.2.3 of Chapter 5, Assumption 3 says that $c_j(\mu_j)$ grows asymptotically at least linearly.

Theorem 6.16 *Suppose Assumption 3 holds. Then there exists $\lambda_0 > 0$ such that for $0 \le \lambda \le \lambda_0$, the optimal arrival and service rates for the problem with a fixed total arrival rate, λ, given by (6.91), exhibit facility dominance. That is, for some facility $j \in J$, $\lambda_j = \lambda$, $\mu_j = \mu_j(\lambda)$, with $\lambda_k = \mu_k = 0$, $k \ne j$.*

Proof To solve problem (6.91), it suffices to solve the following Lagrangean problem, where the Lagrange multiplier, α, is chosen so that $\sum_{j \in J} \lambda_j(\alpha) = \lambda$:

$$\min_{\{\lambda_j, \mu_j, j \in J\}} \quad \sum_{j \in J}(\lambda_j G_j(\lambda_j, \mu_j) + c_j(\mu_j) - \alpha\lambda_j)$$

$$\text{s.t.} \quad 0 \le \lambda_j < \mu_j , \ j \in J . \tag{6.93}$$

(Here, $\lambda_j(\alpha)$, $\mu_j(\alpha)$, $j \in J$, denote the optimal arrival and service rates for the Lagrangean problem (6.93) as a function of the parameter, α.) The objective function for this problem is separable and therefore can be solved separately for each facility $j \in J$. The problem for a particular facility j can be written as

$$\max_{\{\lambda_j, \mu_j\}} \quad \alpha\lambda_j - \lambda_j G_j(\lambda_j, \mu_j) - c_j(\mu_j)$$

$$\text{s.t.} \quad 0 \le \lambda_j < \mu_j .$$

Note that the optimal arrival rate for this problem, $\lambda_j(\alpha)$, is nondecreasing in α.

We analyzed this problem in Section 5.5.2.3 of Chapter 5. From Assumption 3 and Theorem 5.9, it follows that there exists a $\alpha_{0j} > 0$ such that, for $0 \le \alpha \le \alpha_{0j}$, the socially optimal arrival and service rates for this problem are $\lambda_j = 0$ and $\mu_j = 0$. Now let us order the facilities so that

$$\alpha_{01} < \alpha_{02} < \ldots < \alpha_{0n} .$$

(For simplicity we assume strict inequalities. The modifications required to allow for equalities are straightforward and we leave them to the reader.) Then, for $\alpha_{01} < \alpha \le \alpha_{02}$, the optimal solution to the Lagrangean problem (6.93) is

$$\lambda_1 = \lambda_1(\alpha) > 0 \quad , \quad \mu_1 = \mu_1(\alpha) > 0 ;$$

$$\lambda_k = 0 \quad , \quad \mu_k = 0 , \ k \ne 1 .$$

Since $\lambda_1(\alpha)$ is nondecreasing in α, it follows that there exists a $\lambda_0 > 0$ such that for $0 \le \lambda \le \lambda_0$, the value of α for which $\sum_{j \in J} \lambda_j(\alpha) = \lambda$ lies in $(\alpha_{01}, \alpha_{02}]$. Hence, for $0 \le \lambda \le \lambda_0$, the optimal solution to the problem has $\lambda_1 = \lambda$, $\mu_1 = \mu_1(\lambda)$, with $\lambda_k = \mu_k = 0$, $k \ne 1$. $\blacksquare$

As a corollary of this theorem, we have

Corollary 6.17 *Suppose Assumption 3 holds and $\lambda U'(\lambda) \to \infty$ as $\lambda \to 0$. Then the cooperative facility optimal solution has an unbounded objective function as $\lambda \to 0$. The optimal solution consists in allocating the total arrival rate, λ, to facility 1 (the facility j with the minimal α_{0j}), that is, setting $\lambda_1 = \lambda$ and $\mu_1 = \mu_1(\lambda)$, and then letting $\lambda \to 0$.*

6.3.3.5 Competing Facilities

Suppose now that each facility has an operator who sets the service rate and the toll at that facility and whose goal is to maximize the total profit at that facility,

$$\lambda_j \delta_j - c_j(\mu_j) \,,$$

that is, the revenue received from the tolls paid by the customers who use that facility minus the service cost.

Assuming (as usual) that the arriving customers are individual optimizers, any particular set of values for the service rates, μ_j, and the tolls, δ_j, $j \in J$, will result in arrival rates λ_j that satisfy the equilibrium conditions,

$$
\begin{aligned}
U'(\lambda) &\leq \delta_j + G_j(\lambda_j, \mu_j) \,, \ \lambda_j \geq 0 \,, \ j \in J \,, \\
U'(\lambda) &= \delta_j + G_j(\lambda_j, \mu_j) \,, \quad , \text{if } \lambda_j > 0 \,, \ j \in J \,, \\
\sum_{j \in J} \lambda_j &= \lambda \,.
\end{aligned}
$$

Each of the (competing) revenue-maximizing facility operators seeks to find values for the toll and service rate at its facility that maximize its profit, with the above equilibrium conditions as constraints. We seek a Nash equilibrium solution for the tolls, that is, a set of values for (μ_j, δ_j), $j \in J$, such that, at each facility j, the pair (μ_j, δ_j) maximizes the profit, $\lambda_j \delta_j - c_j(\mu_j)$, at that facility, assuming that (μ_k, δ_k), $k \neq j$, do not change. In other words, no facility operator can benefit (i.e., achieve a larger revenue at its facility) by unilaterally changing the value of its service rate or toll.

Thus, the set of service rates and tolls (μ_k, δ_k), $k \in J$, satisfies the Nash-equilibrium characterization with respect to that facility, if the pair (μ_j, δ_j), achieves the maximum in the following problem:

$$
\begin{aligned}
&\max_{\{\pi, \lambda, \mu_j, \delta_j; \lambda_k, k \in J\}} && \lambda_j \delta_j - c_j(\mu_j) \\
&(\mathbf{F}_j) \qquad \text{s.t.} && \pi \leq \delta_k + G_k(\lambda_k, \mu_k) \,, \ k \in J \,, \\
& && \pi = \delta_k + G_k(\lambda_k, \mu_k) \,, \ \text{if } \lambda_k > 0 \,, \ k \in J \,, \\
& && \sum_{k \in J} \lambda_k = \lambda \,, \\
& && U'(\lambda) = \pi \,, \\
& && 0 \leq \lambda_k < \mu_k \,, \ k \in J \,,
\end{aligned}
$$

.with μ_k, δ_k, $k \neq j$, fixed.

When all arrival rates are positive, we can use the constraint, $\pi - \delta_j$ $G_j(\lambda_j)$, to substitute for δ_j in the objective function. Then an equivalent formulation of the optimization problem for facility j is the following:

$$\max_{\{\pi,\lambda,\mu_j,\delta_j;\lambda_k,k\in J\}} \quad \lambda_j(\pi - G_j(\lambda_j,\mu_j))$$

$$(\mathbf{F}_j) \qquad \text{s.t.} \qquad U'(\lambda) = \pi\,,$$

$$\pi = \delta_k + G_k(\lambda_k)\,,\ k \in J\,,$$

$$\sum_{k\in J} \lambda_k = \lambda\,,$$

with δ_k, μ_k, $k \neq j$, fixed. In this case, the set (μ_j, δ_j), $j \in J$, constitutes a Nash equilibrium for the competing facility operators if, together with a set of arrival rates λ_j, $j \in J$, total arrival rate λ, and full price π, they simultaneously solve the optimization problems $(\mathbf{F}_j)$ for all $j \in J$.

We shall consider only the case of a fixed total arrival rate. The analysis of the case of a variable total arrival rate can be carried out by a combination of the techniques used for a fixed total arrival rate with those used in Section 6.1.3.7 for the case of fixed service rates and a variable total arrival rate. We leave this to the reader.

6.3.3.6 Competing Facilities: Fixed Total Arrival Rate

In this subsection we consider the competitive facility-optimal solution in the case of a fixed total arrival rate λ. We focus on the case in which all facility arrival rates are positive.

We shall begin our analysis with the special case $n = 2$. In this case the optimization problem for facility 1 takes the form,

$$\max_{\{\pi,\mu_1,\delta_1,\lambda_j,j=1,2\}} \quad \lambda_1\delta_1 - c_1(\lambda_1)$$

$$\text{s.t.} \qquad \pi = \delta_1 + G_1(\lambda_1,\mu_1)\,,$$

$$\pi = \delta_2 + G_2(\lambda_2,\mu_2)\,,$$

$$\lambda_1 + \lambda_2 = \lambda\,,$$

with μ_2 and δ_2 fixed. Using the equality constraints to eliminate the variables π and λ_2, we obtain the following equivalent formulation:

$$\max_{\{\lambda_1,\mu_1,\delta_1\}} \quad \lambda_1\delta_1 - c_1(\mu_1)$$

$$\text{s.t.} \qquad \delta_1 = \delta_2 + G_2(\lambda - \lambda_1,\mu_2) - G_1(\lambda_1,\mu_1)\,,$$

$$0 \leq \lambda_1 \leq \lambda\,.$$

Using the equality constraint to substitute for δ_1 in the objective function yields

$$\max_{\{\lambda_1,\mu_1\}} \quad \lambda_1(\delta_2 + G_2(\lambda - \lambda_1,\mu_2) - G_1(\lambda_1,\mu_1)) - c_1(\mu_1)$$

$$\text{s.t.} \qquad \delta_1 = \delta_2 + G_2(\lambda - \lambda_1,\mu_2) - G_1(\lambda_1,\mu_1)\,,$$

$$0 \le \lambda_1 \le \lambda .$$

As in the case of fixed service rates (cf. Remark 12 in Section 6.1.3.6), for fixed δ_2 and μ_2 we can set

$$U_1'(\lambda_1) = \delta_2 + G_2(\lambda - \lambda_1, \mu_2) ,$$

and write the above optimization problem equivalently as

$$\max_{\{\lambda_1, \mu_1\}} \quad \lambda_1 U'(\lambda_1) - \lambda_1 G_1(\lambda_1, \mu_1)) - c_1(\mu_1)$$

$$\text{s.t.} \quad 0 \le \lambda_1 \le \lambda , \tag{6.94}$$

and then use the constraint,

$$\delta_1 = \delta_2 + G_2(\lambda - \lambda_1, \mu_2) - G_1(\lambda_1, \mu_1) ,$$

to define the toll, δ_1. Problem (6.94) takes the form of a facility optimization problem for a single facility with variable arrival and service rates, which we considered in Section 5.5.3 of Chapter 5. Recall that it may or may not have a unique solution, depending on whether

1. the transformed utility function,

$$\tilde{U}_1(\lambda_1) := \lambda_1 U'(\lambda_1) ,$$

   is concave; and

2. the cost function,

$$C_1(\lambda_1) := \lambda_1 G_1(\lambda_1, \mu_1(\lambda_1)) - c_1(\mu_1(\lambda_1)) ,$$

   is convex.

Here, as usual, we are defining $\mu_1(\lambda_1)$ as the value of the service rate, μ_1, that solves the minimization problem:

$$\min_{\{\mu_1\}} \quad \lambda_1 G_1(\lambda_1, \mu_1) + c_1(\mu_1)$$

$$\text{s.t.} \quad \mu_1 > \lambda_1 .$$

We sketch the analysis of this problem here, leaving the details (and the extension to more than two facilities) to the reader.

First, suppose $\tilde{U}_1(\lambda_1)$ is concave and $C_1(\lambda_1)$ is convex. Then the objective function in Problem (6.94) is concave in λ_1 and has a unique solution. If the corresponding concave/convex conditions hold for the corresponding optimization problem for facility 2, with δ_1 and μ_1 fixed, then this problem also has a unique solution. This problem is essentially equivalent to the problem discussed in Section 6.1.3.6, with $\lambda_1 G_1(\lambda_1)$ replaced by $C_1(\lambda_1)$, and the analysis carried out there may also be applied here.

Now suppose $C_(\lambda_1)$ is concave. Recall that this is the case if $c_1(\lambda_1)$ satisfies Assumption 1 and $H_1(\lambda_1, \mu_1) = \lambda_1 G_1(\lambda_1, \mu_1)$ satisfies Assumption 2 (cf. Section 6.3.2.1). Then, depending on the properties of the transformed utility function, $\tilde{U}_1(\lambda_1)$, the objective function in Problem (6.94) may have several local maxima. In particular, the global maximum may occur at one of the two extreme points, $\lambda_1 = 0$ (with $\mu_1 = 0$) or $\lambda_1 = \lambda$ (with $\mu_1 = \mu_1(\lambda)$).

The Nash equilibrium for the competing two facilities must simultaneously solve Problem (6.94) and the corresponding problem for facility 2, with δ_1 and μ_1 fixed. If one or both of these problems has multiple local maxima, then in general there may also be multiple Nash equilibria.

6.4 Endnotes

Section 6.1.2.2

The material in this section is largely based on Bell and Stidham [18].

Section 6.2.1.3

For a reference on networks in which the traffic equations (6.67) hold and have a unique solution, see, e.g., Kelly [103]).

CHAPTER 7

Single-Class Networks of Queues

In this chapter we show how some of the models and techniques presented in the previous chapters can be extended to networks of queues. A network of queues is a system consisting of several service facilities connected by logical or physical links along which customers move from one facility to another, in some cases choosing among alternative paths through the system. In the previous chapter we studied systems of parallel queueing facilities, which are the simplest examples of networks of queues: each path through the system contains exactly one facility. As we shall see, many of the results for such systems carry over without too much additional effort to more general networks of queues. But there are also some important differences and new results.

This chapter focuses on a network of queues with a single class of customers. Multi-class networks of queues are the subject of the next chapter.

7.1 Basic Model

In Section 6.1 of Chapter 6 we studied a model for selection of the arrival rate λ_j to each facility j among a set of parallel facilities. The jobs were assumed to belong to a single class, with a given utility function, but different facilities could have different delay-cost functions. The general model of this chapter is a generalization of this model; there is still a single class of jobs, but the set of parallel facilities is replaced by a network of queues.

The total arrival rate to the system, λ, is a decision variable. The set of feasible values for λ is denoted by A. Our default assumption is that $A = [0, \infty)$. The system earns a utility, $U(\lambda)$, per unit time when the total arrival rate is λ. As usual, we assume that $U(\lambda)$ is nondecreasing, differentiable, and concave in $\lambda \geq 0$. The system is now a network consisting of a set J of facilities and a set R of routes. Each route $r \in R$ consists of a subset of facilities, and we use the notation $j \in r$ to indicate that facility j is on route r.*

---

* This abstract characterization of a network is sufficiently general to include both classical models of networks of queues and road traffic networks, as well as more recent models of communication networks. In queueing-network models (e.g., a Jackson network), each queue (service facility) is modeled as a node, with a directed arc from node j to node k if service at queue j may be followed immediately by service at queue k. In communication-network models it is more common (and more natural) to consider each transmission link as a service facility, with a queue of jobs (messages or packets) at the node (router/server) at the head of the link, waiting to be transmitted. In road traffic networks, both nodes (intersections) and links (road segments between intersections) are service facilities in the sense that they are potential sources of congestion and waiting.

Each job that enters the system must be assigned to one of the routes, $r \in R$. Let λ_r denote the flow assigned to route r, $r \in R$. The flows, λ_r, $r \in R$, are decision variables, subject to the constraint that the total flow must equal λ:

$$\sum_{r \in R} \lambda_r = \lambda \ .$$

As in many of our previous models we assume that at each facility the average waiting cost per job is a function of the flow (arrival) rate at that facility. The flow rate ν_j at each facility j is the sum of the flow rates on all the routes that use that facility:

$$\nu_j = \sum_{r : j \in r} \lambda_r \ , \ j \in J \ .$$

Let $G_j(\nu_j)$ denote the average waiting cost of a job at facility j. We make the same assumptions about each function $G_j(\nu_j)$ as we made for $G(\lambda)$ in the single-facility model: that $G_j(\nu_j)$ takes values in $[0, \infty]$ and is strictly increasing and differentiable in ν_j, $0 \le \nu_j < \infty$.

Let $H_j(\nu_j) = \nu_j G_j(\nu_j)$: the average waiting cost per unit time at facility j, $j \in J$. We assume that $H_j(\nu_j)$ is convex in $\nu_j \ge 0$, $j \in J$.

Given the flow rates, ν_j, at the various facilities, the total waiting cost incurred by a job that follows a particular route is the sum of the resulting waiting costs at the facilities on that route:

$$\sum_{j \in r} G_j(\nu_j) \ , \ r \in R \ .$$

There may also be a toll δ_j which is charged to each customer who uses facility j, $j \in J$. In this case the total cost (full price) for a job assigned to route r is given by

$$\sum_{j \in r} (\delta_j + G_j(\nu_j)) \ .$$

As in the previous models for selection of arrival rates, the solution to the decision problem depends on who is making the decision and what criteria are being used. The decision may be made by the individual customers, each concerned only with its own net utility (*individual optimality*), by an agent for the customers as a whole who might be interested in maximizing the aggregate net utility to all customers (*social optimality*), or by a system operator interested in maximizing profit (*facility optimality*).

7.2 Individually Optimal Arrival Rates and Routes

As usual, a natural starting point for all the optimality criteria is the notion of individually optimal (equilibrium) arrival rates. In this case (roughly speaking) the arrival rates are chosen so as to balance the marginal utility per unit time with the full price on each route that is used.

Once again it is useful to separate the decision about the total arrival rate

λ from the decision about the flow rates λ_r on each route r. As in the single-facility model (Chapter 2) and the parallel-facility model (Chapter 6), the total arrival rate is found by equating the marginal utility to the price per unit of flow. Let π denote the price. Then the equilibrium arrival rate, λ^e, for the system as a whole is the unique solution of

$$U'(\lambda) = \pi , \qquad (7.1)$$

in the interval $[0, \infty)$, provided such a solution exists. This is the case if and only if $U'(0) \geq \pi$. If $U'(0) < \pi$, then $\lambda^e := 0$.

The next step is to calculate the full price π and, in the process, to determine the equilibrium arrival rate, λ^e_r, on each route $r \in R$. Of course, if $\lambda^e = 0$, then $\lambda^e_r = 0$ for all $r \in R$. Suppose $\lambda^e > 0$, in which case $U'(\lambda^e) = \pi$. Consider the behavior of a marginal user who joins the system. At equilibrium, such a user will choose a route that offers the minimum full price, which implies that

$$\pi = \min_{r \in R} \sum_{j : j \in r} (\delta_j + G_j(\nu_j))$$

If, to the contrary, a route with a larger full price receives positive flow, then such a solution cannot be an equilibrium, since there is an incentive to divert some of this flow to a route that achieves the minimum price. Thus $\lambda^e_r > 0$ only if $\sum_{j : j \in r}(\delta_j + G_j(\nu_j)) = \pi$.

We can summarize these observations as follows. An allocation of flows, $(\lambda_r, r \in R)$, is individually optimal (denoted $(\lambda^e_r, r \in R)$) if and only if it satisfies the following system of equations and inequalities:

$$U'(\lambda) \quad \leq \quad \pi , \qquad (7.2)$$

$$\lambda(\pi - U'(\lambda)) \quad = \quad 0 , \qquad (7.3)$$

$$\sum_{r \in R} \lambda_r \quad = \quad \lambda , \qquad (7.4)$$

$$\sum_{j : j \in r} (\delta_j + G_j(\nu_j)) \quad \geq \quad \pi , \, r \in R , \qquad (7.5)$$

$$\lambda_r \left(\sum_{j : j \in r} (\delta_j + G_j(\nu_j)) - \pi \right) \quad = \quad 0 , \, r \in R , \qquad (7.6)$$

$$\sum_{r : j \in r} \lambda_r \quad = \quad \nu_j , \, j \in J , \qquad (7.7)$$

$$\lambda_r \quad \geq \quad 0 , \, j \in J . \qquad (7.8)$$

The formulation of this problem is essentially the same as that for the parallel-facility model, except that now each facility is replaced by a route, which may consist of several facilities. Moreover, the fact that different routes may share the same facility leads to interactions which make the solution of the problem more difficult and the properties of that solution more complex.

Using an argument similar to that used for parallel facilities, we can establish that the equilibrium conditions for an individually optimal allocation have a unique solution. As in Section 6.1.1 of Chapter 6, we can show that the equilibrium conditions are the (necessary and sufficient) optimality conditions for a related maximization problem with a concave objective function and linear constraints. The maximization problem is the following:

$$\max_{\{\lambda; \lambda_r, r \in R; \nu_j \in J\}} \quad U(\lambda) - \sum_{j \in J} \int_0^{\nu_j} (\delta_j + G_j(\eta)) d\eta$$

$$\text{s.t.} \quad \sum_{r \in R} \lambda_r = \lambda \, ,$$

$$\sum_{r: j \in r} \lambda_r = \nu_j \, ,$$

$$\lambda_r \geq 0 \, , \ r \in R \, .$$

Since the objective function is jointly concave in $(\lambda, \nu_j, \ j \in J)$ and the constraints are linear, the *KKT* conditions are necessary and sufficient for a global maximum to this problem. These conditions have a unique solution and it is easily verified that they are identical to the equilibrium conditions, (6.2)-(6.7), for an individually optimal solution.

7.3 Socially Optimal Arrival Rates and Routes

Now let us consider the problem from the point of view of social optimization. The objective is to choose the system arrival rate, λ, and flow rates, λ_r, for each route $r \in R$, to maximize the steady-state expected net benefit (utility minus waiting costs) per unit time. Thus the problem takes the form:

$$\max_{\{\lambda; \lambda_r, r \in R; \nu_j, j \in J\}} \quad U(\lambda) - \sum_{r \in R} \lambda_r \sum_{j: j \in r} G_j(\nu_j)$$

$$\text{(S)} \qquad \text{s.t.} \quad \sum_{r \in R} \lambda_r = \lambda \, ,$$

$$\sum_{r: j \in r} \lambda_r = \nu_j \, ,$$

$$\lambda_r \geq 0 \, , \ r \in R \, .$$

We shall use a Lagrange multiplier to eliminate the constraint on the total arrival rate. The Lagrangean problem is:

$$\max \quad U(\lambda) - \sum_{r \in R} \lambda_r \sum_{j: j \in r} G_j(\nu_j) - \alpha \left(\sum_{r \in R} \lambda_r - \lambda \right)$$

$$\text{s.t.} \quad \sum_{r: j \in r} \lambda_r = \nu_j \, , \ j \in J \, ,$$

$$\lambda_r \geq 0 \, , \ r \in R \, ,$$

$$\lambda \geq 0 \, .$$

The solution is parameterized by α, which can be interpreted as an imputed cost per unit time per unit of arrival rate. Rewriting the objective function in equivalent form,

$$U(\lambda) - \alpha\lambda + \sum_{r \in R} \lambda_r(\alpha - \sum_{j:j \in r} G_j(\nu_j)) \, ,$$

reveals that it is separable in λ and λ_j, $j = 1, \ldots, m$, for any fixed value of α. Thus we can maximize the objective function separately with respect to the arrival rate λ for the system as a whole and with respect to the arrival rate λ_j for each facility j. If the resulting solution satisfies the linking constraint

$$\sum_{r \in R} \lambda_r = \lambda \, ,$$

then it is optimal for the original problem.

For the system as a whole the problem takes the form of a single-facility arrival-rate-optimization problem (cf. Chapter 2), with utility function $U(\lambda)$ and constant waiting-cost function, $G(\lambda) \equiv \alpha$. That is,

$$\max \quad U(\lambda) - \alpha\lambda$$
$$\text{s.t.} \quad \lambda \geq 0 \, .$$

We know how to solve this problem and the reader may consult Chapter 2 for the details. For example, if $U(\cdot)$ is differentiable, then we know that λ maximizes $U(\lambda) - \alpha\lambda$ if (and only if) either

(i) $\lambda > 0$ and $U'(\lambda) = \alpha$, or

(ii) $\lambda = 0$ and $U'(0) \leq \alpha$.

For each $r \in R$, the optimal value of λ_r for the Lagrangean problem has the following characterization (together with the defining constraints, $\nu_j = \sum_{s:j \in s} \lambda_s$, $j \in J$):

$$\alpha \leq \sum_{j:j \in r}(G_j(\nu_j) + \nu_j\frac{\partial}{\partial\nu_j}G_j(\nu_j)) \, , \ \lambda_r \geq 0 \, ,$$

$$\alpha = \sum_{j:j \in r}(G_j(\nu_j) + \nu_j\frac{\partial}{\partial\nu_j}G_j(\nu_j)) \, , \ \text{if } \lambda_r > 0 \, .$$

As mentioned above, a solution λ, λ_r, $r \in R$, to these conditions will be optimal for the original problem if α is chosen so that the constraint, $\sum_{r \in R} \lambda_r = \lambda$, is satisfied.

Thus we see that a socially optimal allocation, denoted $(\lambda^s, \lambda_r^s, r \in R)$, is characterized by the following necessary and sufficient KKT optimality conditions:

$$U'(\lambda) \leq \alpha \, , \tag{7.9}$$
$$\lambda(\alpha - U'(\lambda)) = 0 \, , \tag{7.10}$$
$$\sum_{r \in R} \lambda_r = \lambda \, , \tag{7.11}$$

$$\sum_{j:j\in r} (G_j(\nu_j) + \nu_j \frac{\partial}{\partial \nu_j} G_j(\nu_j)) \geq \alpha \, , \, r \in R \, , \qquad (7.12)$$

$$\lambda_r \left(\sum_{j:j\in r} (G_j(\nu_j) + \nu_j \frac{\partial}{\partial \nu_j} G_j(\nu_j)) - \alpha \right) = 0 \, , \, r \in R \, , \qquad (7.13)$$

$$\sum_{r:j\in r} \lambda_r = \nu_j \, , \, j \in J \, , \qquad (7.14)$$

$$\lambda_r \geq 0 \, , \, r \in R \, . \qquad (7.15)$$

As in the single-facility model, we can interpret the term $G_j(\nu_j)$ as the *internal effect* of a marginal increase in the flow (arrival rate) ν_j at facility j. It is the portion of the marginal increase in aggregate waiting cost that is borne by the "marginal" arriving customer when the arrival rate is ν_j. Similarly, we can interpret the term $\nu_j \frac{\partial}{\partial \nu_j} G_j(\nu_j)$ as the *external effect*: the rate of increase in waiting cost borne by all users as a result of a marginal increase in the arrival rate ν_j. By charging a toll at each facility j equal to the external effect – that is, $\delta_j = \nu_j \frac{\partial}{\partial \nu_j} G_j(\nu_j)$ – one can render the individually optimal allocation socially optimal.

In a similar fashion, the social optimization problem with a fixed arrival rate, λ, may be formulated as a constrained minimization problem:

$$\min_{\{\lambda_r, r\in R; \nu_j, j\in J\}} \sum_{r\in R} \lambda_r \sum_{j:j\in r} G_j(\nu_j)$$

$$\text{s.t.} \quad \sum_{r\in R} \lambda_r = \lambda \, ,$$

$$\sum_{r:j\in r} \lambda_r = \nu_j \, , \, j \in J \, ,$$

$$\lambda_r \geq 0 \, , \, r \in R \, .$$

The necessary and sufficient *KKT* conditions for this problem are (7.11), (7.12), (7.13), (7.14), and (7.15).

7.4 Comparison of S.O. and Toll-Free I.O. Solutions

In the single-facility and parallel-facility models of Chapters 2 and 6, we were able to make strong statements about the relationship between the socially optimal and the individually optimal flow allocations, when no tolls are charged in the individually optimal case. In particular we could show that the full price of admission is always higher in the socially optimal than in the individually optimal allocation. As a consequence of this result, we saw that for the model with variable total arrival rate and a concave utility function, the total arrival rate is always smaller in the socially optimal allocation than in the individually optimal allocation without tolls, for both the single-facility and the parallel-facility models.

In fact we showed a more general monotonicity result: that the full price of admission to the system under an individually optimal allocation of flows is monotonically increasing in the waiting-cost function at each facility. Since a socially optimal allocation may be implemented by charging appropriate nonnegative tolls at the various facilities and then assuming an individually optimal allocation, this result implies that the full price of admission is always higher in the socially optimal than in the individually optimal allocation.

As we mentioned in Chapter 6, this is an intuitively plausible property. After all, the users are all trying to find a path through the system that has least cost. It seems to make sense that increasing any particular cost should lead to an increase in the average cost, that is, the cost of the path(s) they follow in an equilibrium solution. As we shall see below, however, this monotonicity property may not hold in more general networks, in which users choose between routes, each of which may consist of a sequence of facilities. In such cases the topology of the network may lead to complicated interactions between different customers, whose chosen routes may or may not share particular facilities.

If we were concerned with only socially optimal allocations, then the monotonicity property would hold trivially. The total (and therefore the average) cost of a socially optimal allocation is monotonic in the waiting-cost functions at the various facilities. Moreover, for any *fixed* flow allocation (in particular, for the socially optimal allocation) the average cost is always smaller than the marginal cost, since the waiting-cost functions at the various facilities are convex.

The fact that customers are self-optimizing, however, can lead to surprising and nonintuitive behavior.

To compare the total arrival rates under the two criteria, it is convenient once again to consider separately the problem of setting the total arrival rate and the problem of choosing flow rates on the various routes. From the optimality conditions for the two problems, we see that in both cases the total arrival rate is found by equating the marginal utility to a price parameter. In the case of social optimization, the equation is

$$U'(\lambda) = \alpha \, , \qquad (7.16)$$

whereas in the case of social optimization, the equation is

$$U'(\lambda) = \pi \, . \qquad (7.17)$$

The difference of course lies in how the two parameters, α and π, are calculated.

Recall that in the case of a simple network consisting of parallel facilities (Chapter 6), we showed that for any given λ, $\alpha(\lambda) \geq \pi(\lambda)$, where $\alpha(\lambda)$ and $\pi(\lambda)$ are, respectively, the marginal cost of a socially optimal allocation and the average cost (full price) of an individually optimal allocation, with λ fixed. From this inequality and equations (7.16) and (7.17) it follows that $\lambda^s \leq \lambda^e$.

It turns out that neither the general monotonicity result nor this relation between arrival rates need hold in more general networks. We shall demon-

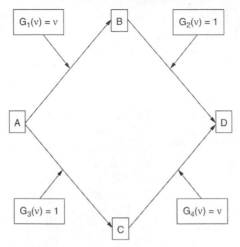

Figure 7.1 *First Example Network for Braess's Paradox*

strate these negative results by means of counterexamples, each of which uses a relatively simple network of four nodes. The first counterexample is one in which an increase in the waiting cost at one facility results in a decrease in the average total cost along all the routes used by customers in the individually optimal flow allocation. It is an example of what has become known in the literature as *Braess's Paradox*, named after the traffic engineer who first discovered and exhibited it in the context of vehicular traffic flow. In Braess's example (and in ours, which is slightly different) the result is particularly dramatic, in that the cost at the facility in question is changed from zero to infinity. [†]

7.4.1 Braess's Paradox

Consider the four-node network in Figure 7.1. All traffic enters the network (at rate λ) at node 1 and leaves the network at node 4. There are two routes. Route 1 consists of links (facilities) 1 and 2 and route 2 consists of links (facilities) 3 and 4. The cost functions on the links (facilities) are as follows:

$$G_1(\nu) = G_4(\nu) = \nu \,, \; \nu \geq 0 \,,$$
$$G_2(\nu) = G_3(\nu) = 1 \,, \; \nu \geq 0 \,.$$

An individually optimal flow allocation allocates flows λ_1 and λ_2 to routes 1 and 2, respectively, so that the total flow is allocated,

$$\lambda_1 + \lambda_2 = \lambda \,,$$

[†] Actually, Braess's Paradox is usually stated in the following equivalent form: adding a zero-cost link to an existing network can result in an increase in the total cost incurred by each user. Adding such a link, of course, is equivalent to having the link there all the time, but changing its cost from infinity to zero.

and the average total cost per customer on each is the same,

$$G_1(\lambda_1) + G_2(\lambda_1) = G_3(\lambda_2) + G_4(\lambda_2) ,$$

that is,

$$\lambda_1 + 1 = 1 + \lambda_2 .$$

The individually optimal solution is therefore $\lambda_1 = \lambda_2 = \lambda/2$, with average cost $\pi(\lambda) = \lambda/2 + 1$. It is easy to see that the socially optimal allocation is also $\lambda_1 = \lambda_2 = \lambda/2$, since this allocation equates the total marginal cost on each route. That is, with $\hat{G}_j(\nu) := G_j(\nu) + \nu G'_j(\nu)$, we have

$$\hat{G}_1(\lambda_1) + \hat{G}_2(\lambda_1) = \hat{G}_3(\lambda_2) + \hat{G}_4(\lambda_2) ,$$

or, equivalently,

$$\lambda_1 + \lambda_1 + 1 = 1 + \lambda_2 + \lambda_2 .$$

Thus the marginal cost of the socially optimal allocation is $\alpha(\lambda) = \lambda + 1$.

Remark 1   Note that for this network we have $\alpha(\lambda) > \pi(\lambda)$, which agrees with our intuition that the full price of the socially optimal allocation should be larger than that of the individually optimal allocation. (So this is not the promised counterexample of this intuition: that will come later.) Indeed, since the full price of the socially optimal allocation is a marginal cost and the full price of the individually optimal allocation is an average cost (per customer), it is tempting to conclude that this inequality will always hold, since "it is well known that" the marginal cost always exceeds the average cost when the total cost is a convex function (as it is in the present case). But this is a specious argument, since we are comparing the marginal cost of one allocation (namely, the socially optimal allocation) with the average cost of another allocation (namely, the individually optimal allocation), and, of course, these two allocations are not in general the same. (They are in this particular example, however.)

Now consider what happens when we add a link (facility) leading from node 2 to node 3. Call this link 5. The resulting network appears in Figure 7.2. Suppose the cost on this link is zero, regardless of the flow:

$$G_5(\nu) = 0 , \ \nu \geq 0 .$$

There is now an additional route from the origin to the destination, consisting of links 1, 5, and 4, in that order. Call it route 3. The individually optimal allocation now depends on λ. We consider three cases.

Case 1: $0 \leq \lambda \leq 1$

The individually optimal allocation for this case assigns all λ units of flow to route 3; that is, $\lambda_1 = \lambda_2 = 0$, $\lambda_3 = \lambda$. The average cost is $\tilde{\pi}(\lambda) = \lambda + 0 + \lambda = 2\lambda$. To see this, note that the necessary and sufficient conditions for this allocation to be individually optimal are:

$$\pi \ \leq \ G_1(\nu_1) + G_2(\nu_2) = \nu_1 + 1 = \lambda + 1 ,$$

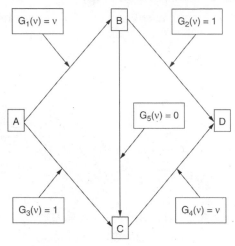

Figure 7.2 *Second Example Network for Braess's Paradox*

$$\pi \leq G_3(\nu_3) + G_4(\nu_4) = 1 + \nu_4 = 1 + \lambda \,,$$
$$\pi = G_1(\nu_1) + G_5(\nu_5) + G_4(\nu_4) = \nu_1 + 0 + \nu_4 = \lambda + 0 + \lambda = 2\lambda \,.$$

These conditions are satisfied if (and only if) $0 \leq \lambda \leq 1$.

Case 2: $1 \leq \lambda \leq 2$

The individually optimal allocation for this case is $\lambda_1 = \lambda_2 = \lambda - 1$, $\lambda_3 = 2 - \lambda$, with average cost $\tilde{\pi}(\lambda) = 2$. To see this, note that

$$\nu_1 = \lambda_1 + \lambda_3 = 1 \,,$$
$$\nu_2 = \lambda_1 = \lambda - 1 \,,$$
$$\nu_3 = \lambda_2 = \lambda - 1 \,,$$
$$\nu_4 = \lambda_2 + \lambda_3 = 1 \,,$$
$$\nu_5 = \lambda_3 = 2 - \lambda \,.$$

The necessary and sufficient conditions for this allocation to be individually optimal are satisfied, as seen below:

$$\pi = G_1(\nu_1) + G_2(\nu_2) = \nu_1 + 1 = 1 + 1 = 2 \,,$$
$$\pi = G_3(\nu_3) + G_4(\nu_4) = 1 + \nu_4 = 1 + 1 = 2 \,,$$
$$\pi = G_1(\nu_1) + G_5(\nu_5) + G_4(\nu_4) = \nu_1 + 0 + \nu_4 = 1 + 0 + 1 = 2 \,.$$

Case 3: $\lambda \geq 2$

The individually optimal allocation for this case is $\lambda_1 = \lambda_2 = \lambda/2$, $\lambda_3 = 0$. The average cost is $\tilde{\pi}(\lambda) = \lambda/2 + 1$. (Note that this allocation coincides with the individually optimal allocation for the network without link 5.) To see this, note that the necessary and sufficient conditions for this allocation to be individually optimal are:

$$\pi = G_1(\nu_1) + G_2(\nu_2) = \nu_1 + 1 = \lambda/2 + 1 \,,$$
$$\pi = G_3(\nu_3) + G_4(\nu_4) = 1 + \nu_1 = \lambda/2 + 1 \,,$$
$$\pi \leq G_1(\nu_1) + G_5(\nu_5) + G_4(\nu_4) = \nu_1 + 0 + \nu_4 = \lambda/2 + 0 + \lambda/2 = \lambda \,.$$

These conditions are satisfied if (and only if) $\lambda \geq 2$.

Now let us compare the average cost, $\tilde{\pi}(\lambda)$, of the individually optimal allocation with link 5 included in the network to the average cost, $\pi(\lambda)$, of the individually optimal allocation for the original network. We see that

$$\tilde{\pi}(\lambda) < \pi(\lambda) \,, \text{ for } 0 \leq \lambda < 2/3 \,,$$
$$\tilde{\pi}(\lambda) = \pi(\lambda) \,, \text{ for } \lambda = 2/3 \,,$$
$$\tilde{\pi}(\lambda) > \pi(\lambda) \,, \text{ for } 2/3 < \lambda < 2 \,,$$
$$\tilde{\pi}(\lambda) = \pi(\lambda) \,, \text{ for } \lambda \geq 2 \,.$$

In particular, the average cost with link 5 included exceeds that for the original network for $2/3 < \lambda < 2$, with the maximum difference (as well as the maximum percentage difference) occurring at $\lambda = 1$, where $\pi(\lambda) = 3/2$ and $\tilde{\pi}(\lambda) = 2$ (a 33 1/3 % difference).

Thus, as promised, we have given an example where adding a zero-cost link to a network results in each user incurring a strictly larger total cost to traverse the network under an individually optimal allocation.[‡] As noted above, this example is therefore a counterexample to the (intuitive) assertion that increasing the cost at any link (facility) cannot decrease the average cost for each user to traverse the network under an individually optimal allocation. (Recall that this assertion *is* true in the special case of a system consisting of a single facility or parallel facilities.)

What about the socially optimal allocation for the new network? In particular, does its marginal cost exceed the average cost of the individually optimal allocation?

The socially optimal allocation now also depends on λ, and we consider three cases.

Case 1: $0 \leq \lambda \leq 1/2$

The socially optimal allocation for this case assigns all λ units of flow to route 3; that is, $\lambda_1 = \lambda_2 = 0$, $\lambda_3 = \lambda$. The marginal cost is $\tilde{\alpha}(\lambda) = 2\lambda + 0 + 2\lambda = 4\lambda$. To see this, note that the necessary and sufficient conditions for this allocation to be a social optimum are:

$$\alpha \leq 2\nu_1 + 1 = 2\lambda + 1 \,,$$
$$\alpha \leq 1 + 2\nu_4 = 1 + 2\lambda \,,$$
$$\alpha = 2\nu_1 + 0 + 2\nu_4 = 2\lambda + 0 + 2\lambda = 4\lambda \,.$$

[‡] Braess's original example was slightly different, and more complicated. Our example is taken from Correa et al. [43].

These conditions are satisfied if (and only if) $0 \leq \lambda \leq 1/2$.

Case 2: $1/2 \leq \lambda \leq 1$

The socially optimal allocation for this case is $\lambda_1 = \lambda_2 = \lambda - 1/2$, $\lambda_3 = 1 - \lambda$, with marginal cost $\tilde{\alpha}(\lambda) = 2$. To see this, note that

$$\begin{aligned}
\nu_1 &= \lambda_1 + \lambda_3 = 1/2 \,, \\
\nu_2 &= \lambda_1 = \lambda - 1/2 \,, \\
\nu_3 &= \lambda_2 = \lambda - 1/2 \,, \\
\nu_4 &= \lambda_2 + \lambda_3 = 1/2 \,, \\
\nu_5 &= \lambda_3 = 1 - \lambda \,.
\end{aligned}$$

The necessary and sufficient conditions for this allocation to be individually optimal are satisfied, as seen below:

$$\begin{aligned}
\alpha &= 2\nu_1 + 1 = 1 + 1 = 2 \,, \\
\alpha &= 1 + 2\nu_4 = 1 + 1 = 2 \,, \\
\alpha &= 2\nu_1 + 0 + 2\nu_4 = 1 + 0 + 1 = 2 \,.
\end{aligned}$$

Case 3: $\lambda \geq 1$

The socially optimal allocation for this case is: $\lambda_1 = \lambda_2 = \lambda/2$, $\lambda_3 = 0$. The marginal cost is $\tilde{\alpha}(\lambda) = \lambda + 1$. (Note that this allocation coincides with the individually optimal allocation for the network without link 5.) To see this, note that the necessary and sufficient conditions for this allocation to be individually optimal are:

$$\begin{aligned}
\alpha &= 2\nu_1 + 1 = \lambda + 1 \,, \\
\alpha &= 1 + 2\nu_4 = 1 + \lambda \,, \\
\alpha &\leq 2\nu_1 + 0 + 2\nu_4 = \lambda + 0 + \lambda = 2\lambda \,.
\end{aligned}$$

These conditions are satisfied if (and only if) $\lambda \geq 1$.

Now let us compare the marginal cost, $\tilde{\alpha}(\lambda)$, of the socially optimal allocation with link 5 included in the network to the marginal cost, $\alpha(\lambda)$, of the socially optimal allocation for the original network. We see that

$$\begin{aligned}
\tilde{\alpha}(\lambda) &< \alpha(\lambda) \,, \text{ for } 0 \leq \lambda < 1/3 \,, \\
\tilde{\alpha}(\lambda) &= \alpha(\lambda) \,, \text{ for } \lambda = 1/3 \,, \\
\tilde{\alpha}(\lambda) &> \alpha(\lambda) \,, \text{ for } 1/3 < \lambda < 1 \,, \\
\tilde{\alpha}(\lambda) &= \alpha(\lambda) \,, \text{ for } \lambda \geq 1 \,.
\end{aligned}$$

In particular, the marginal cost with link 5 included exceeds that for the original network for $1/3 < \lambda < 1$, with the maximum difference (as well as the maximum percentage difference) occurring at $\lambda = 1/2$, where $\alpha(\lambda) = 3/2$ and $\tilde{\alpha}(\lambda) = 2$ (a 33 1/3 % difference).

Thus we have given an example where adding a zero-cost link to a network results in each user incurring a strictly larger marginal cost associated with traversing the network under a socially optimal allocation. This result parallels that for the individually optimal flow allocation and is equally counterintuitive. That is, we have a counterexample to the (intuitive) assertion that increasing the marginal cost at any link (facility) cannot decrease the marginal cost for each user to traverse the network under a socially optimal allocation. (Recall that this assertion *is* true in the special case of a system consisting of a single facility or parallel facilities, just as it is for individually optimal allocations. See Chapters 2 and 6.)

7.4.2 Effect on Optimal Arrival Rates for the Variable-Rate Model

This surprising behavior has consequences for the problem with a variable total arrival rate, λ, and a concave utility function $U(\lambda)$. Recall that for this problem the individually optimal arrival rate, λ^e, satisfies the equation

$$U'(\lambda) = \pi ,$$

where π is the full price of admission. For the original network, $\pi = \pi(\lambda)$, whereas for the network with link 5 added, $\pi = \tilde{\pi}(\lambda)$. Now suppose the solutions to both equations,

$$U'(\lambda) = \pi(\lambda) ,$$

and

$$U'(\lambda) = \tilde{\pi}(\lambda) ,$$

occur in the interval $(2/3, 2)$, in which $\tilde{\pi}(\lambda) > \pi(\lambda)$. Call these solutions λ^e and $\tilde{\lambda}^e$, respectively. Then, since $U'(\cdot)$ is nonincreasing and both $\pi(\lambda)$ and $\tilde{\pi}(\lambda)$ are increasing, it follows that

$$\lambda^e > \tilde{\lambda}^e .$$

That is, the individually optimal arrival rate decreases when the zero-cost link 5 is added to the network: adding capacity to the network actually results in fewer customers receiving service from the system under an individually optimal flow allocation.

Now consider the socially optimal arrival rate, λ^s, which satisfies the equation

$$U'(\lambda) = \alpha ,$$

where α is the marginal cost of admission. For the original network, $\alpha = \alpha(\lambda)$, whereas for the network with link 5 added, $\alpha = \tilde{\alpha}(\lambda)$. Now suppose the solutions to both equations,

$$U'(\lambda) = \alpha(\lambda) ,$$

and

$$U'(\lambda) = \tilde{\alpha}(\lambda) ,$$

occur in the interval $(1/3, 1)$, in which $\tilde{\alpha}(\lambda) > \alpha(\lambda)$. Call these solutions λ^s

and $\tilde{\lambda}^s$, respectively. Then, since $U'(\cdot)$ is nonincreasing and both $\alpha(\lambda)$ and $\tilde{\alpha}(\lambda)$ are increasing, it follows that

$$\lambda^s > \tilde{\lambda}^s .$$

That is, the socially optimal arrival rate decreases when the zero-cost link 5 is added to the network: adding capacity to the network actually results in fewer customers receiving service from the system under a socially optimal flow allocation.

Because of the special nature of the waiting-cost functions for this example, the marginal cost of the socially optimal allocation is simply a rescaled replica of the average cost of the individually optimal allocation: $\tilde{\alpha}(\lambda) = \tilde{\pi}(2\lambda)$. Note that $\tilde{\alpha}(\lambda) \geq \tilde{\pi}(\lambda)$ for all $\lambda \geq 0$, with strict inequality for all λ except $\lambda = 0$ and $\lambda = 1$, where $\tilde{\alpha}(\lambda) = \tilde{\pi}(\lambda)$.

Thus the present example is *not* a counterexample to the "folk theorem" that the marginal cost of the socially optimal solution is always at least as great as the average cost of the individually optimal solution, although it comes close, in that the two are equal at $\lambda = 1$.

The next subsection uses a variant of the network which we used to illustrate Braess's Paradox. In this example, the marginal cost of the socially optimal solution is actually strictly smaller than the average cost of the individually optimal solution. For the case of a variable total arrival rate, λ, and a concave utility function, $U(\lambda)$, this result leads to a counterexample to the folk theorem that the socially optimal arrival rate is never larger than the individually optimal arrival rate.

7.4.3 Example with $\alpha(\lambda) < \pi(\lambda)$

We use the same four-node graph as used in the illustration of Braess's Paradox, with link 5 present. The cost functions on some of the links are different, however. (See Figure 7.3.) Specifically, we continue to assume that links 2 and 3 have the same, constant waiting-cost function,

$$G_2(\nu) = G_3(\nu) = a , \; \nu \geq 0 ,$$

where $a > 0$. The cost function on link 5 is now a linear function,

$$G_5(\nu) = d \cdot \nu , \; \nu \geq 0 ,$$

where $d > 0$. Links 1 and 4 again have identical cost functions, but now

$$G_1(\nu) = G_4(\nu) = f(\nu) ,$$

where $f(\cdot)$ is an increasing, nonnegative function. For the moment we make no further assumptions about $f(\cdot)$. We denote the marginal cost on these two links by $\hat{f}(\nu)$. That is,

$$\hat{f}(\nu) = f(\nu) + \nu f'(\nu) .$$

The marginal costs on the five links are therefore given by

$$\hat{G}_1(\nu) \;\; = \;\; \hat{f}(\nu) ,$$

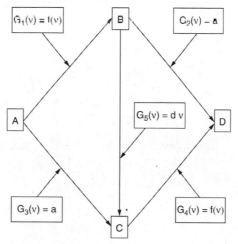

Figure 7.3 *Example Network with* $\alpha(\lambda) < \pi(\lambda)$

$$\hat{G}_2(\nu) = a\,,$$
$$\hat{G}_3(\nu) = a\,,$$
$$\hat{G}_4(\nu) = \hat{f}(\nu)\,,$$
$$\hat{G}_5(\nu) = 2d\nu\,.$$

In our counterexample we want the individually optimal allocation to divide its flow among the three routes as follows: $\lambda_1 = \lambda_2 = \epsilon$, $\lambda_3 = \lambda - 2\epsilon$, where $0 < \epsilon < \lambda/2$. It follows that

$$\nu_1 = \lambda_1 + \lambda_3 = \lambda - \epsilon\,,$$
$$\nu_2 = \epsilon\,,$$
$$\nu_3 = \epsilon\,,$$
$$\nu_4 = \lambda_2 + \lambda_3 = \lambda - \epsilon\,,$$
$$\nu_5 = \lambda_3 = \lambda - 2\epsilon\,.$$

The necessary and sufficient conditions for this allocation to be individually optimal are:

$$\pi = f(\lambda - \epsilon) + a\,,$$
$$\pi = a + f(\lambda - \epsilon)\,,$$
$$\pi = 2f(\lambda - \epsilon) + d(\lambda - 2\epsilon)\,.$$

These conditions hold if and only if

$$f(\lambda - \epsilon) = a - d\lambda + 2d\epsilon\,. \tag{7.18}$$

We also want the socially optimal allocation to divide its flow among the three routes, with $\lambda_1 = \lambda_2 = \delta$, $\lambda_3 = \lambda - 2\delta$, where $0 < \delta < \lambda/2$. It follows that

$$\begin{aligned}
\nu_1 &= \lambda_1 + \lambda_3 = \lambda - \delta , \\
\nu_2 &= \delta , \\
\nu_3 &= \delta , \\
\nu_4 &= \lambda_2 + \lambda_3 = \lambda - \delta , \\
\nu_5 &= \lambda_3 = \lambda - 2\delta .
\end{aligned}$$

The necessary and sufficient conditions for this allocation to be a social optimum are:

$$\begin{aligned}
\alpha &= \hat{f}(\lambda - \delta) + a , \\
\alpha &= a + \hat{f}(\lambda - \delta) , \\
\alpha &= 2\hat{f}(\lambda - \delta) + 2d(\lambda - 2\delta) .
\end{aligned}$$

These conditions hold if and only if

$$\hat{f}(\lambda - \delta) = a - d\lambda - d(\lambda - 4\delta) . \tag{7.19}$$

In our example we want $\alpha < \pi$, which holds if and only if

$$\hat{f}(\lambda - \delta) < f(\lambda - \epsilon) . \tag{7.20}$$

Since $\hat{f}(\lambda) \geq f(\lambda)$ for all $\lambda \geq 0$, we see that a necessary condition for this inequality is that $\delta > \epsilon$. That is, the socially optimal allocation must send more of its flow on routes 1 and 2 than the individually optimal allocation.

So it suffices to find a function, $f(\cdot)$, and positive values of a, d, λ, ϵ, and δ satisfying (7.18), (7.19), and (7.20). Clearly a necessary condition is that $4\delta - 2\epsilon < \lambda$.

We now show that the two necessary conditions,

$$4\delta - 2\epsilon < \lambda , \tag{7.21}$$

and

$$\epsilon < \delta , \tag{7.22}$$

are also sufficient for the existence of positive values of a, d, λ, ϵ, and δ satisfying (7.18) and (7.19), for any given function $f(\lambda)$ that satisfies (7.20). To this end, suppose we are given values of λ, δ, and ϵ such that (7.21), (7.22), and (7.20) hold. Then it suffices to show that there exist positive values of a and d satisfying the following two linear equations:

$$\begin{aligned}
a - (\lambda - 2\epsilon)d &= f(\lambda - \epsilon) , \\
a - 2(\lambda - 2\delta)d &= \hat{f}(\lambda - \delta) .
\end{aligned}$$

Solving these equations in closed form is easy, and in doing so we demonstrate constructively the existence of a (unique) solution:

$$\begin{aligned}
d &= (\lambda - 4\delta + 2\epsilon)^{-1}(f(\lambda - \epsilon) - \hat{f}(\lambda - \delta)) > 0 , \\
a &= (\lambda - 4\delta + 2\epsilon)^{-1}(2(\lambda - 2\delta)f(\lambda - \epsilon) - \lambda\hat{f}(\lambda - \delta)) > 0 .
\end{aligned}$$

We are left with a final question: under what conditions on the function $f(\cdot)$ and the arrival rate λ do there exist values of δ and ϵ satisfying (7.21), (7.22), and (7.20), which we have just shown to be necessary and sufficient for our network to exhibit the sought-for example in which $\alpha(\lambda) < \pi(\lambda)$? A little reflection will show that this will be the case if and only if

$$\text{there exists a } \lambda > 0 \text{ such that } \hat{f}(3\lambda/4) < f(\lambda) . \tag{7.23}$$

This condition puts definite restrictions on the shape of the waiting-cost function $f(\cdot)$. In particular, note that it is not satisfied by a linear function. There are many realistic waiting-cost functions, however, that do satisfy it. An example is the waiting-cost function for an $M/M/1$ queue.

Example 1. Suppose $f(\cdot)$ is given by

$$
\begin{aligned}
f(\lambda) \quad &= \quad \frac{1}{\mu - \lambda} \, , \; 0 \le \lambda < \mu \, , \\
&= \quad \infty , \qquad \lambda \ge \mu \, ,
\end{aligned}
$$

where μ is a positive constant. Then

$$
\begin{aligned}
\hat{f}(\lambda) \quad &= \quad \frac{\mu}{(\mu - \lambda)^2} \, , \; 0 \le \lambda < \mu \, , \\
&= \quad \infty , \qquad \lambda \ge \mu \, ,
\end{aligned}
$$

and therefore

$$\hat{f}(3\lambda/4) < f(\lambda) \quad \Leftrightarrow \quad \mu(\mu - \lambda) < (\mu - 3\lambda/4)^2$$
$$\Leftrightarrow \quad \mu^2 - \lambda\mu < \mu^2 - 3\lambda\mu/2 - 9(\lambda)^2/16 \quad \Leftrightarrow \quad \lambda > 8/9 \, .$$

Thus for this example Condition (7.23) holds. Indeed, it is satisfied by *all* $\lambda > 8/9$.

Example 2. Suppose there exists a constant μ, $0 < \mu < \infty$, such that

$$
\begin{aligned}
f(\lambda) \quad &< \quad \infty , \, 0 \le \lambda < \mu \\
&\to \quad \infty , \text{ as } \lambda \uparrow \mu \\
&= \quad \infty , \, \lambda \ge \mu \, .
\end{aligned}
$$

Let δ and $\epsilon < \delta$ be positive constants such that $4\delta - 2\epsilon < \mu$. Then there exists a $\lambda_0 < \mu$ such that for all $\lambda > \lambda_0$, both inequalities (7.21) and (7.22) hold. To be specific, let $\lambda = \lambda_1$ be the solution to

$$f(\lambda - \epsilon) = \hat{f}(\mu - \delta) \, ,$$

and take

$$\lambda_0 := \max\{4\delta - 2\epsilon, \lambda_1\} \, . \tag{7.24}$$

Then $\lambda > \lambda_0$ implies $\lambda > 4\delta - 2\epsilon$, so that (7.21) holds, and

$$f(\lambda - \epsilon) > \hat{f}(\mu - \delta) > \hat{f}(\lambda - \delta) \, ,$$

so that (7.22) holds.

Now let us turn to the question of the worst-case behavior of the ratio

$$\frac{\pi(\lambda)}{\alpha(\lambda)} = \frac{f(\lambda - \epsilon) + a}{\hat{f}(\lambda - \delta) + a} \,,$$

and examine this question in the context of this example. Since $\hat{f}(\lambda - \delta) < \hat{f}(\mu - \delta)$ for all $\lambda < \mu$, we conclude that

$$\frac{\pi(\lambda)}{\alpha(\lambda)} > \frac{f(\lambda - \epsilon) + a}{\hat{f}(\mu - \delta) + a} \,,$$

and, since $f(\lambda - \epsilon) \uparrow \infty$ as $\lambda \uparrow \mu$, it follows that

$$\frac{\pi(\lambda)}{\alpha(\lambda)} \to \infty \,, \text{ as } \lambda \uparrow \mu \,.$$

In other words, in this example there is no finite upper bound on the ratio of the average cost of the individually optimal allocation to the marginal cost of the socially optimal allocation.

7.4.4 Comparison of Individually Optimal and Socially Optimal Arrival Rates

Now we turn once again to the problem with a variable total arrival rate, λ, and a concave utility function $U(\lambda)$ and explore the implications of the example we have just given. Recall that for this problem the individually optimal arrival rate, λ^e, satisfies the equation

$$U'(\lambda) = \pi(\lambda) \,,$$

where $\pi(\lambda)$ is the full price of admission, that is, the average total waiting cost incurred by a customer who enters the system when the arrival rate is λ. The socially optimal arrival rate, λ^s, satisfies the equation

$$U'(\lambda) = \alpha(\lambda) \,,$$

where $\alpha(\lambda)$ is the marginal cost of admission, that is, the rate of increase of the total cost borne by all entering customers when the arrival rate is λ.

For the case in which $f(\lambda) \uparrow \infty$ as $\lambda \uparrow \mu$ (see Example 2 above), we have shown that, for any given δ and ϵ such that $4\delta - 2\epsilon < \mu$, $\alpha(\lambda) < \pi(\lambda)$ for all $\lambda > \lambda_0$, where λ_0 is given by (7.24). Moreover, $\pi(\lambda)/\alpha(\lambda) \to \infty$ as $\lambda \uparrow \mu$. Suppose, then, that both λ^e and λ^s lie in the interval, (λ_0, μ). Since $U'(\cdot)$ is nonincreasing and both $\pi(\lambda)$ and $\alpha(\lambda)$ are increasing, it follows that

$$\lambda^s > \lambda^e \,.$$

That is, the socially optimal arrival rate is actually *larger* than the individually optimal arrival rate. This is a surprising contradiction to the "folk theorem" that the socially optimal arrival rate is always smaller than (or equal to) the individually optimal arrival rate.

7.4.5 The Price of Anarchy

So far our comparison of the socially optimal and toll-free individually optimal flow allocations has focused on the marginal cost, $\alpha(\lambda)$, of a socially optimal allocation and the average cost (or full price), $\pi(\lambda)$, of an individually optimal allocation, as functions of the arrival rate, λ. We have shown how, contrary to intuition, each of these measures may *increase* as the waiting cost at one or more of the facilities *decreases* (Braess's Paradox). We have also exhibited an example network for which $\alpha(\lambda) < \pi(\lambda)$ for a range of values of λ. As a result, the socially optimal arrival rate may be *larger* than the individually optimal arrival rate in a problem with a variable arrival rate and a concave utility function (again, contrary to intuition).

For the problem with a fixed arrival rate, a question we have not yet addressed is the following: how bad (relative to the socially optimal allocation) can an individually optimal allocation be? More precisely, what is the worst-case behavior of the ratio of the total cost of an individually optimal allocation to the total cost of the socially optimal allocation? Using more colorful language: what is the "price of anarchy"? In this setting, "anarchy" means letting customers make their own route choices.

Let us begin with the formulation of the social optimization problem with a fixed arrival rate, λ, as a constrained minimization problem (introduced at the beginning of Section 7.3):

$$\min_{\{\lambda_r, r \in R; \nu_j, j \in J\}} \quad \sum_{r \in R} \lambda_r \sum_{j:j \in r} G_j(\nu_j)$$

$$\text{s.t.} \quad \sum_{r \in R} \lambda_r = \lambda \,, \tag{7.25}$$

$$\sum_{r:j \in r} \lambda_r = \nu_j \,, \ j \in J \,, \tag{7.26}$$

$$\lambda_r \geq 0 \,, \ r \in R \,. \tag{7.27}$$

Using the equality constraints, (7.26), we may rewrite the objective function as follows:

$$\sum_{r \in R} \lambda_r \sum_{j:j \in r} G_j(\nu_j) = \sum_{j \in J} \sum_{r:j \in r} \lambda_r G_j(\nu_j)$$

$$= \sum_{j \in J} \nu_j G_j(\nu_j) \,.$$

Let $\boldsymbol{\nu} := (\nu_j, j \in J)$ and define the feasible set $\mathcal{N}$ as the set of all $\boldsymbol{\nu}$ for which there exists a $\boldsymbol{\lambda} = (\lambda_r, r \in R)$ such that $\boldsymbol{\nu}$ and $\boldsymbol{\lambda}$ satisfy the constraints, (7.25), (7.26), and (7.27). Then the social optimization problem may be rewritten with decision variables, $\boldsymbol{\nu} = (\nu_j, j \in J)$, as follows:

$$\min_{\{\boldsymbol{\nu} \in \mathcal{N}\}} \quad C(\boldsymbol{\nu}) := \sum_{j \in J} \nu_j G_j(\nu_j) \,.$$

Now let $\boldsymbol{\nu}^e = (\nu_j^e, j \in J)$ denote the vector of facility flow rates corresponding to an individually optimal allocation, $\boldsymbol{\lambda}^e = (\lambda_r^e, r \in R)$, of flow rates on
the various routes. We wish to establish an upper bound on the total cost,
$C(\boldsymbol{\nu}^e)$, of $\boldsymbol{\nu}^e$, which will make it possible to bound the percentage difference between $C(\boldsymbol{\nu}^e)$ and the minimal cost, $C(\boldsymbol{\nu}^s)$, associated with the socially
optimal allocation of facility flow rates, $\boldsymbol{\nu}^s$. The following lemma provides a
characterization of $\boldsymbol{\nu}^e$ (of interest in its own right) which we will use in our
analysis.

Lemma 7.1 *An individually optimal vector of facility flow rates, $\boldsymbol{\nu}^e$, satisfies
the variational inequality,*

$$\sum_{j \in J} (\nu_j^e - \nu_j) G_j(\nu_j^e) \leq 0 \,, \tag{7.28}$$

for all $\boldsymbol{\nu} = (\boldsymbol{\nu}_j, j \in J) \in \mathcal{N}$.

Proof Suppose the system is operating with individually optimal route flow
rates, $\boldsymbol{\lambda}^e$, and facility flow rates, $\boldsymbol{\nu}^e$. Then each customer chooses route r with
probability

$$p_r^e := \lambda_r^e / \lambda \,, \ r \in R \,.$$

The Nash-equilibrium property asserts that no individual customer has an
incentive to deviate unilaterally from these probabilities. This implies that all
routes with $\lambda_r > 0$ are "shortest routes" in the sense that they achieve the
minimum,

$$\min_{r \in R} \sum_{j:j \in r} G_j(\nu_j^e) = \pi(\lambda) \,.$$

The total cost of the individually optimal solution is therefore given by

$$C(\boldsymbol{\nu}^e) = \lambda \pi(\lambda) \,.$$

The Nash-equilibrium property therefore implies that, assuming the cost per
unit of flow at each facility j remains equal to $G_j(\nu_j^e)$, any other choice of route
probabilities, $\{p_r, r \in R\}$ (and therefore any other allocation, $\boldsymbol{\nu}$, of flows to
facilities) will result in a larger cost, because it will be using routes other than
the shortest. That is, for all $\boldsymbol{\nu} \in \mathcal{N}$,

$$\sum_{j \in J} \nu_j G_j(\nu_j^e) \geq \sum_{j \in J} \nu_j^e G_j(\nu_j^e) = C(\boldsymbol{\nu}^e) = \lambda \pi(\lambda) \,.$$

Therefore, for all $\boldsymbol{\nu} \in \mathcal{N}$,

$$\sum_{j \in J} (\nu_j^e - \nu_j) G_j(\nu_j^e) \leq 0 \,.$$

∎

The essential insight in this proof is that an individually optimal flow allocation must solve the linearization of the original minimum-cost problem,
in which the cost per unit flow at facility j is constant and equal to $G_j(\nu_j^e)$.

But the solution to a min-cost network-flow problem with linear costs consists in sending all flow on routes with minimum total cost per unit flow, i.e., on "shortest" routes. And we know that all routes with positive flow in an individually optimal flow allocation use only shortest routes.

It follows from Lemma 7.1 that, for any $\nu \in \mathcal{N}$,

$$
\begin{aligned}
C(\nu^e) &= \sum_{j \in J} \nu_j^e G_j(\nu_j^e) \\
&\leq \sum_{j \in J} \nu_j G_j(\nu_j^e) \\
&= \sum_{j \in J} \nu_j G_j(\nu_j) + \sum_{j \in J} \nu_j (G_j(\nu_j^e) - G_j(\nu_j)) \\
&= C(\nu) + \sum_{j \in J} \nu_j (G_j(\nu_j^e) - G_j(\nu_j)) ,
\end{aligned}
$$

so that

$$
C(\nu^e) \leq C(\nu) + \sum_{j \in J} \nu_j (G_j(\nu_j^e) - G_j(\nu_j)) . \tag{7.29}
$$

The following theorem shows how this inequality can be used to bound the difference between $C(\nu^e)$ and $C(\nu^s)$.

Theorem 7.2 *Suppose there exists a constant $\sigma < 1$ such that*

$$
\sum_{j \in J} \nu_j (G_j(\nu_j^e) - G_j(\nu_j)) \leq \sigma C(\nu^e) , \tag{7.30}
$$

for all $\nu \in \mathcal{N}$. Then $C(\nu^e) \leq (1 - \sigma)^{-1} C(\nu^s)$.

Proof For any $\nu \in \mathcal{N}$, using (7.29) and (7.30) we have

$$
\begin{aligned}
C(\nu^e) &\leq C(\nu) + \sum_{j \in J} \nu_j (G_j(\nu_j^e) - G_j(\nu_j)) \\
&\leq C(\nu) + \sigma C(\nu^e) .
\end{aligned}
$$

Since this inequality holds for all $\nu \in \mathcal{N}$, it holds in particular for the socially optimal vector, ν^s. Thus

$$
C(\nu^e) \leq C(\nu^s) + \sigma C(\nu^e) ,
$$

from which the desired result follows. ∎

To apply Theorem 7.2, we need to find a constant σ which is an upper bound for the ratio of $\sum_{j \in J} \nu_j (G_j(\nu_j^e) - G_j(\nu_j))$ to $C(\nu^e)$. The following corollaries may be helpful.

Corollary 7.3 *For each $j \in J$, suppose there exists a constant $\sigma_j < 1$ such that*

$$
\frac{\nu_j (G_j(\nu_j^e) - G_j(\nu_j))}{\nu_j^e G_j(\nu_j^e)} \leq \sigma_j , \tag{7.31}
$$

for all $0 \leq \nu_j \leq \nu_j^e$. Then $\sigma := \max_{j \in J} \sigma_j$ satisfies the conditions of Theorem 7.2 and hence $C(\nu^e) \leq (1 - \sigma)^{-1} C(\nu^s)$.

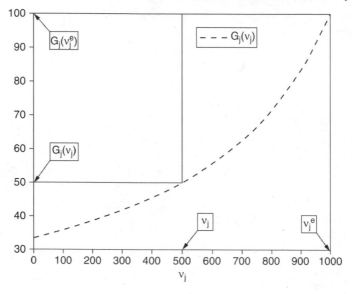

Figure 7.4 *Illustration of Theorem 7.2*

Proof Since $G_j(\cdot)$ is nondecreasing and (7.31) holds for all $0 \leq \nu_j \leq \nu_j^e$, (7.31) holds for all $\nu_j \geq 0$, $j \in J$. Therefore, it holds for all $\boldsymbol{\nu} = (\nu_j, j \in J) \in \mathcal{N}$, because $\mathcal{N} \subset \{\boldsymbol{\nu}|\nu_j \geq 0, j \in J\}$. It then follows that (7.30) holds for all $\boldsymbol{\nu} \in \mathcal{N}$, since

$$\frac{\sum_{j \in J} \nu_j(G_j(\nu_j^e) - G_j(\nu_j))}{C(\boldsymbol{\nu}^e)} = \frac{\sum_{j \in J} \nu_j(G_j(\nu_j^e) - G_j(\nu_j))}{\sum_{j \in J} \nu_j^e G_j(\nu_j^e)}$$

$$\leq \max_{j \in J} \frac{\nu_j(G_j(\nu_j^e) - G_j(\nu_j))}{\nu_j^e G_j(\nu_j^e)}.$$

∎

This corollary is useful in applications where it is easier to find separate upper bounds on each ratio, $\nu_j(G_j(\nu_j^e) - G_j(\nu_j))/\nu_j^e G_j(\nu_j^e)$ than to find directly an upper bound on the ratio of the sums. Of course, the resulting value of σ will not be tight in general.

Figure 7.4 illustrates the relation between $\nu_j^e G_j(\nu_j^e)$ (the area of the larger rectangle) and $\nu_j(G_j(\nu_j^e) - G_j(\nu_j))$ (the area of the smaller rectangle) for a particular value of ν_j, $0 < \nu_j < \nu_j^e$. In this illustration, the waiting-cost function $G_j(\nu_j)$ is strictly convex and increasing, with $G_j(0) > 0$. This is the case, for example, with a steady-state $M/M/1$ queue with a linear waiting-cost function, in which $G_j(\nu_j) = h_j/(\mu_j - \nu_j)$.

Corollary 7.4 *Suppose $H_j(\nu_j) = \nu_j G_j(\nu_j)$ is strictly convex in $\nu_j \geq 0$, for all facilities $j \in J$. Then the function, $\nu_j(G_j(\nu_j^e) - G_j(\nu_j))$ is strictly concave in $\nu_j \in [0, \nu_j^e]$ and therefore its maximum, denoted ν_j^*, is the unique solution* *to the optimality equation,*

$$G_j(\nu_j^e) = G_j(\nu_j)) + \nu_j G_j'(\nu_j) .\qquad(7.32)$$

Since $\sum_{j\in J} \nu_j(G_j(\nu_j^e) - G_j(\nu_j))$ is separable in $\boldsymbol{\nu} = (\nu_j, j \in J)$, it follows that the ratio,

$$\frac{\sum_{j\in J} \nu_j(G_j(\nu_j^e) - G_j(\nu_j))}{C(\boldsymbol{\nu}^e)} ,$$

attains its maximum at $\boldsymbol{\nu} = (\nu_j^*, j \in J)$. Let

$$\sigma := \frac{\sum_{j\in J} \nu_j^*(G_j(\nu_j^e) - G_j(\nu_j^*))}{C(\boldsymbol{\nu}^e)} = \frac{\sum_{j\in J}(\nu_j^*)^2 G_j'(\nu_j^*)}{C(\boldsymbol{\nu}^e)} .\qquad(7.33)$$

If $\sigma < 1$ then σ satisfies the conditions of Theorem 7.2 and hence $C(\boldsymbol{\nu}^e) \le (1-\sigma)^{-1}C(\boldsymbol{\nu}^s)$.

Note that (7.32) is equivalent to the necessary and sufficient condition for a socially optimal arrival rate in a single facility with waiting-cost function, $G_j(\cdot)$, and linear utility function with reward coefficient, $r = G_j(\nu_j^e)$.

The following corollary combines Corollaries 7.3 and 7.4.

Corollary 7.5 *Suppose $H_j(\nu_j) = \nu_j G_j(\nu_j)$ is strictly convex in $\nu_j \ge 0$, for all facilities $j \in J$. For each $j \in J$ let*

$$\sigma_j := \frac{\nu_j^*(G_j(\nu_j^e) - G_j(\nu_j^*))}{\nu_j^e G_j(\nu_j^e)} ,\qquad(7.34)$$

where ν_j^* is the unique solution to (7.32). If $\sigma_j < 1$ for all $j \in J$, then $\sigma := \max_{j\in J} \sigma_j$ satisfies the conditions of Theorem 7.2 and hence $C(\boldsymbol{\nu}^e) \le (1-\sigma)^{-1}C(\boldsymbol{\nu}^s)$.

If $G_j(\nu_j)$ is convex, $j \in J$, the following corollary provides an alternative approach for calculating σ.

Corollary 7.6 *Suppose $G_j(\nu_j)$ is convex in $\nu_j \ge 0$, for all facilities $j \in J$. Then it follows that*

$$\frac{\nu_j(G_j(\nu_j^e) - G_j(\nu_j))}{\nu_j^e G_j(\nu_j^e)} \le \frac{\nu_j^e G_j'(\nu_j^e)}{4G_j(\nu_j^e)} =: \hat{\sigma}_j , \ j \in J .$$

If $\hat{\sigma}_j < 1$ for all $j \in J$, then $\hat{\sigma} := \max_{j\in J}\{\hat{\sigma}_j\}$ satisfies the conditions of Theorem 7.2 and hence $C(\boldsymbol{\nu}^e) \le (1-\sigma)^{-1}C(\boldsymbol{\nu}^s)$.

Proof By the convexity of $G_j(\cdot)$ we have

$$G_j(\nu_j^e) - G_j(\nu_j) \le (\nu_j^e - \nu_j)G_j'(\nu_j^e) ,$$

for all $\nu_j \in [0, \nu_j^e]$. The desired result then follows from the fact that

$$\nu_j(\nu_j^e - \nu_j) \le (\nu_j^e)^2/4 , \ \nu_j \in [0, \nu_j^e] .$$

∎

This upper bound, though not as tight as the previous ones, has the advantage

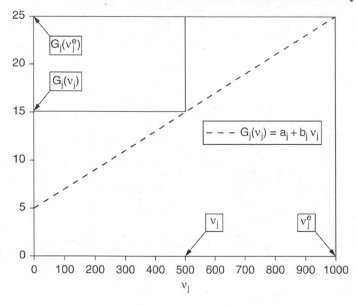

Figure 7.5 *Illustration of Derivation of Upper Bound for Affine Waiting-Cost Function*

that it does not require calculation of ν_j^*, $j \in J$. However, in order to have $\hat{\sigma}_j < 1$, the function $G_j(\nu_j)$ must not be "too convex." In Example 1 below we apply this bound to the case of affine waiting cost functions.

Example 1   Suppose the facility waiting cost functions are affine: $G_j(\nu_j) = a_j + b_j \nu_j$, where $a_j \geq 0$, $b_j > 0$, $j \in J$. Figure 7.5 illustrates the derivation of the upper bound in this case. The area of the smaller rectangle can be at most one quarter of the area of the larger rectangle. (This can be seen directly, or as an application of Corollary 7.6, using the fact that $G_j(\nu_j) \geq b_j \nu_j = \nu_j G_j'(\nu_j)$.) Thus we can set $\sigma = 1/4$ and apply Theorem 7.2 to obtain the inequality,

$$C(\boldsymbol{\nu}^e) \leq (1 - \sigma)^{-1} C(\boldsymbol{\nu}^s) = (4/3) C(\boldsymbol{\nu}^s) \, .$$

Example 2   Suppose each facility behaves like an $M/M/1$ queue in steady state with a linear waiting-cost function. Then for facility $j \in J$ we have

$$G_j(\nu_j) = \frac{h_j}{\mu_j - \nu_j} \, , \quad 0 \leq \nu_j < \mu_j \, ,$$

where μ_j is the service rate and h_j is the waiting cost coefficient at facility j. In this case, $H_j(\nu_j) = h_j \nu_j / (\mu_j - \nu_j)$, which is strictly convex in $0 \leq \nu_j < \mu_j$. From (7.32) we have

$$\nu_j^* = \mu_j - \sqrt{\mu_j (\mu_j - \nu_j^e)} \, .$$

(Note that the waiting-cost coefficient, h_j, does not appear in this expression.)

Let us apply Corollary 7.4. Substituting for ν_j^* in equation (7.33) and simplifying yields

$$\sigma = \frac{\sum_{j \in J} \left(\sqrt{\mu_j/(\mu_j - \nu_j^e)} - 1 \right)^2}{\sum_{j \in J} \left(h_j \nu_j^e/(\mu_j - \nu_j^e) \right)}.$$

If $\sigma < 1$, then Theorem 7.2 applies and hence $C(\boldsymbol{\nu}^e) \leq (1 - \sigma)^{-1} C(\boldsymbol{\nu}^s)$. Note also that in this case the upper bound, σ, is tight in the sense that the inequality (7.30) in Theorem 7.2 is satisfied with equality (over all $\nu_j \geq 0$, $j \in J$).

As an alternative, we can apply Corollary 7.5. Substituting for ν_j^* in equation (7.34) and simplifying yields

$$\sigma_j = \left(\sqrt{\frac{\mu_j}{\nu_j^e}} - \sqrt{\frac{\mu_j}{\nu_j^e} - 1} \right)^2.$$

Let $\rho_j^e := \nu_j^e/\mu_j$, $j \in J$. Then $\sigma_j = \phi(\rho_j^e)$, where

$$\phi(\rho) := \left(\sqrt{\rho^{-1}} - \sqrt{\rho^{-1} - 1} \right)^2 = \frac{2(1 - \sqrt{1 - \rho})}{\rho} - 1.$$

Note that $\phi(\rho)$ is strictly increasing in $\rho \in [0, 1)$ and approaches 1 as $\rho \to 1$. Let

$$\rho^e := \max_{j \in J} \rho_j^e.$$

Then

$$\sigma := \max_{j \in J} \sigma_j = \phi(\rho^e) \tag{7.35}$$

satisfies the conditions of Theorem 7.2. Note that, once again, the waiting-cost coefficients, h_j, $j \in J$, do not appear explicitly in this expression. However, the individually optimal flow rate, ν_j^e, at facility j will in general depend on the waiting-cost coefficient, h_j (as well as the waiting-cost coefficients, h_k, $k \neq j$, at some or all of the other facilities).

The implication of equation (7.35) is that we need only identify the facility j with the maximal traffic intensity, ρ_j^e, under the individually optimal allocation of flows in order to find a value of σ that satisfies the conditions of Theorem 7.2.

Figure 7.6 graphs $\phi(\rho)$ for $0 \leq \rho < 1$. In Table 7.7 we have exhibited the value of $\sigma = \phi(\rho^e)$ and the corresponding upper bound, $(1 - \sigma)^{-1}$, on $C(\boldsymbol{\nu}^e)/C(\boldsymbol{\nu}^s)$ for various values of ρ^e.

As expected, the upper bound on $C(\boldsymbol{\nu}^e)/C(\boldsymbol{\nu}^s)$ increases dramatically as ρ^e approaches 1. For small values of ρ^e, however, the upper bound gives considerably more modest estimates of the price of anarchy. For example, at $\rho^e = 0.30$, the cost of the toll-free i.o. allocation can be no more than 10% larger than the cost of the s.o. allocation; at $\rho^e = 0.50$, no more than 21% larger; at $\rho^e = 0.75$, no more than 50%. Even at $\rho^e = 0.90$, $C(\boldsymbol{\nu}^e)$ can be at most slightly more than twice as large as $C(\boldsymbol{\nu}^s)$.

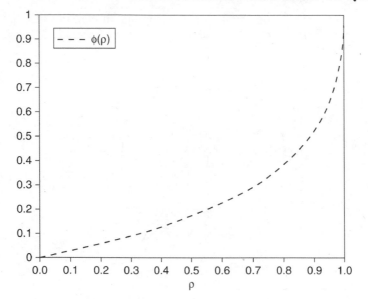

Figure 7.6 *Graph of* $\phi(\rho)$

ρ^e	σ	$(1-\sigma)^{-1}$
0.10	0.0263	1.0270
0.20	0.0557	1.0590
0.30	0.0889	1.0976
0.40	0.1270	1.1455
0.50	0.1716	1.2071
0.55	0.1970	1.2454
0.60	0.2251	1.2906
0.65	0.2566	1.3452
0.70	0.2922	1.4129
0.75	0.3333	1.5000
0.80	0.3820	1.6180
0.85	0.4417	1.7910
0.90	0.5195	2.0811
0.95	0.6345	2.7361
0.98	0.7522	4.0355
0.99	0.8182	5.5000

Figure 7.7 *Table: Values of* $\sigma = \phi(\rho^e)$ *and* $(1-\sigma)^{-1}$

Contrast these results with those for the case of an affine waiting-cost function, in which the upper bound (much cruder, because it is independent of the i.o. rates ν_j^e) is $4/3$. For example, when ρ^e is less than about 0.65, the upper bound for the $M/M/1$ example (which has a strictly concave waiting-cost function) is smaller than that for the affine case.

Of course, the upper bounds we have derived in this section can be quite conservative, since they are independent of the topology of the network. As an extreme illustration, consider a network of parallel facilities, each with the same waiting-cost function. In this case, it is obvious (by symmetry) that the individually optimal and socially optimal arrival rates coincide and hence $C(\boldsymbol{\nu}^e) = C(\boldsymbol{\nu}^s)$, regardless of the value of the traffic intensity. To illustrate how conservative the bounds can be in nonsymmetric systems, let us consider again the special case of a network of n parallel facilities, which was the subject of Chapter 6. We shall focus on $M/M/1$ facilities.

Example 3   Parallel $M/M/1$ Facilities Consider a system consisting of n independent parallel facilities, with facility j behaving as an $M/M/1$ queue in steady state with service rate μ_j, $j \in J$. There is a single class of customers arriving at fixed rate λ. The decision variables are the arrival rates, ν_j, $j \in J$, at the various facilities, where $\sum_{j\in J} \nu_j = \lambda$. The waiting cost per customer at facility j is linear, with waiting-cost coefficient h_j, so that

$$G_j(\nu_j) = \frac{h_j}{\mu_j - \nu_j} , \ j \in J .$$

In Section 6.1.2.3 of Chapter 6 we analyzed the the ratio, $C(\boldsymbol{\nu}^e)/C(\boldsymbol{\nu}^s)$, for the case $h_j = 1$, $j \in J$, emphasizing heavy traffic, that is, the behavior of the ratio as $\lambda \to \mu$, where $\mu = \sum_{j \in J} \mu_j$ (the total service rate). Using the explicit expressions for the individually and socially optimal arrival rates at the facilities, we showed that

$$\lim_{\lambda \uparrow \mu} C(\boldsymbol{\nu}^e)/C(\boldsymbol{\nu}^s) = n\mu/(\sum_{j\in J} \sqrt{\mu_j})^2 \leq n ,$$

where n is the number of parallel facilities.

Extending this analysis to the present model, it can easily be shown that

$$\frac{C(\boldsymbol{\nu}^e)}{C(\boldsymbol{\nu}^s)} = \frac{\left(\sum_{j\in J} h_j\right)\left(\sum_{j\in J}\mu_j\right) - (\mu - \lambda)\sum_{j\in J} h_j}{\left(\sum_{j\in J}\sqrt{h_j\mu_j}\right)^2 - (\mu - \lambda)\sum_{j\in J} h_j} ,$$

provided λ is large enough so that all n facilities have positive arrival rates. From this expression we see that the ratio, $C(\boldsymbol{\nu}^e)/C(\boldsymbol{\nu}^s)$ actually *decreases* as $\lambda \to \mu$, approaching the (finite) heavy-traffic limit,

$$\frac{\left(\sum_{j\in J} h_j\right)\left(\sum_{j\in J}\mu_j\right)}{\left(\sum_{j\in J}\sqrt{h_j\mu_j}\right)^2} .$$

By contrast, the upper bound $(1 - \sigma)^{-1}$ derived above increases to infinity as $\lambda \to \mu$.

Let us look in more detail at the derivation of σ in the case of parallel facilities. To keep the exposition simple, we shall again assume that λ is large enough that all facilities have positive arrival rates, under both individual and social optimization. First note that in this case the variational inequality (7.28) holds with equality, that is,

$$\sum_{j \in J} (\nu_j^e - \nu_j) G_j(\nu_j^e) = 0 ,$$

for all $\boldsymbol{\nu} \in \mathcal{N}$. This is true because

$$\sum_{j \in J} \nu_j = \lambda , \text{ for all } \boldsymbol{\nu} \in \mathcal{N} .$$

It follows that (7.29) also holds with equality for all $\boldsymbol{\nu} \in \mathcal{N}$:

$$C(\boldsymbol{\nu}^e) = C(\boldsymbol{\nu}) + \sum_{j \in J} \nu_j (G_j(\nu_j^e) - G_j(\nu_j)) , \text{ for all } \boldsymbol{\nu} \in \mathcal{N} .$$

For the individually optimal allocation, we have

$$G_j(\nu_j) = \pi , \ j \in J ,$$

so that $C(\boldsymbol{\nu}^e) = \lambda\pi$. For the upper bound, σ, derived in Corollary 7.4, we therefore have (using (7.33))

$$\begin{aligned} \sigma &= \frac{\sum_{j \in J} \nu_j^* (G_j(\nu_j^e) - G_j(\nu_j^*))}{C(\boldsymbol{\nu}^e)} \\ &= \frac{\sum_{j \in J} \nu_j^* (\pi - G_j(\nu_j^*))}{\lambda\pi} . \end{aligned}$$

Thus we see that the upper bound is derived by solving, for each facility j, a social optimization problem with linear utility in which the reward coefficient is π. But π is the imputed reward that induces individually optimizing customers to join each facility j at a rate ν_j^e such that $\sum_{j \in J} \nu_j^e = \lambda$. We know from the analysis in Chapter 6, however, that the Lagrangean relaxation of the social optimization problem requires an imputed reward (Lagrange multiplier) α that is strictly larger than π. (See Lemma 6.5 in Section 6.1.2.3.) Using π rather than $\alpha > \pi$ in the social optimization problem leads to facility arrival rates that are uniformly smaller than the individually optimal rates and therefore sum to a quantity strictly smaller than λ. The result is an upper bound on the difference between $C(\boldsymbol{\nu}^e)$ and $C(\boldsymbol{\nu}^s)$ that is based on a systematic underestimate of $C(\boldsymbol{\nu}^s)$.

This interpretation suggests an explanation for why the upper bound, $(1 - \sigma)^{-1}$, on the ratio, $C(\boldsymbol{\nu}^e)/C(\boldsymbol{\nu}^s)$, increases to infinity as $\lambda \to \mu$ in the case of parallel $M/M/1$ facilities, whereas the ratio itself actually decreases.

7.5 Facility Optimal Arrival Rates and Routes

The modeling and analysis of facility optimal arrival rates and routes in a single-class network of queues is, for the most part, a straightforward extension of the case of parallel facilities, which we considered in Chapter 6. We shall consider both cooperating and competing facilities and concentrate on models, techniques, and results that illustrate what can change when one generalizes from parallel facilities to an arbitrary network.

7.5.1 Cooperating Facilities

Suppose a single operator (or a team of cooperating operators) sets the tolls, δ_j, for all facilities $j \in J$. The goal is to maximize the total profit. We assume that the costs of operating the facilities are fixed. Then maximizing the total profit is equivalent to maximizing the total revenue,

$$\sum_{j \in J} \nu_j \delta_j \ ,$$

received from the tolls paid by the entering customers.

Assuming (as usual) that the arriving customers are individual optimizers, any particular set of values δ_j, $j \in J$, for the tolls will result in arrival rates λ_r, $r \in R$, that satisfy the equilibrium conditions, (7.2)-(7.5). To keep the exposition simple, we focus on the case where the total arrival rate, λ, is positive. Then the equilibrium conditions may be written in the following form:

$$U'(\lambda) \ = \ \pi \ ,$$

$$\sum_{r \in R} \lambda_r \ = \ \lambda \ ,$$

$$\sum_{j:j \in r} (\delta_j + G_j(\nu_j)) \ \geq \ \pi \ , r \in R \ ,$$

$$\lambda_r \left(\sum_{j:j \in r} (\delta_j + G_j(\nu_j)) - \pi \right) \ = \ 0 \ , r \in R \ ,$$

$$\sum_{r:j \in r} \lambda_r \ = \ \nu_j \ , j \in J \ ,$$

$$\lambda_r \ \geq \ 0 \ , r \in R \ .$$

Thus, we have the following formulation of the facility optimization problem:

$$\max_{\{\pi, \lambda; \lambda_r, r \in R; \nu_j, \delta_j, j \in J\}} \ \sum_{j \in J} \nu_j \delta_j$$

$$\text{s.t.} \quad \lambda_r \left(\sum_{j:j \in r} (\delta_j + G_j(\nu_j)) - \pi \right) = 0 \ , r \in R \ ,$$

$$\sum_{j:j\in r} (\delta_j + G_j(\nu_j)) \geq \pi \, , \; r \in R \, ,$$

$$\sum_{r:j\in r} \lambda_r = \nu_j \, , \; j \in J \, ,$$

$$\sum_{r\in R} \lambda_r = \lambda \, ,$$

$$U'(\lambda) = \pi \, ,$$

$$\lambda_r \geq 0 \, , \; r \in R \, .$$

Using the third set of constraints, we can write the objective function in equivalent form as

$$\sum_{r\in R} \lambda_r \sum_{j:j\in r} \delta_j \, .$$

Summing the first set of constraints (the complementary-slackness conditions) over $r \in R$, subtracting this sum (which equals zero) from the objective function, and simplifying leads to the following equivalent problem:

$$\max_{\{\pi, \lambda; \lambda_r, r\in R; \nu_j, \delta_j, j\in J\}} \sum_{r\in R} \lambda_r (\pi - \sum_{j:j\in r} G_j(\nu_j))$$

$$\textbf{(F)} \qquad \text{s.t.} \qquad \lambda_r \left(\sum_{j:j\in r} (\delta_j + G_j(\nu_j)) - \pi \right) = 0 \, , \; r \in R \, ,$$

$$\sum_{j:j\in r} (\delta_j + G_j(\nu_j)) \geq \pi \, , \; r \in R \, ,$$

$$\sum_{r:j\in r} \lambda_r = \nu_j \, , \; j \in J \, ,$$

$$\sum_{r\in R} \lambda_r = \lambda \, ,$$

$$U'(\lambda) = \pi \, ,$$

$$\lambda_r \geq 0 \, , \; r \in R \, .$$

Now consider this problem without the first two sets of constraints. Using the last equality constraint to eliminate π from the objective function, we have the following (relaxed) version of the facility optimization problem:

$$\max_{\{\lambda; \lambda_r, r\in R; \nu_j, j\in J\}} \lambda U'(\lambda) - \sum_{r\in R} \lambda_r \sum_{j:j\in r} G_j(\nu_j)$$

$$\textbf{(F')} \qquad \text{s.t.} \qquad \sum_{r\in R} \lambda_r = \lambda \, ,$$

$$\sum_{r:j\in r} \lambda_r = \nu_j \, ,$$

$$\lambda_r \geq 0 \, , \; r \in R \, .$$

This problem is in the same form as the social optimization problem $(\mathbf{S})$, with the utility function $U(\lambda)$ replaced by $\tilde{U}(\lambda) = \lambda U'(\lambda)$. If $\tilde{U}(\lambda)$ is concave and nondecreasing, then we can apply the techniques and results derived for the social optimization problem to this relaxed version of the facility optimization problem. If $\tilde{U}(\lambda)$ is not concave and nondecreasing, then the same techniques and many of the results still apply, with the proviso that the KKT conditions may no longer have a unique solution (cf. Section 2.1.3 of Chapter 2 and Section 6.1.3 of Chapter 6).

It remains to examine the relationship between the relaxed version $(\mathbf{F'})$ and the original version $(\mathbf{F})$ of the facility optimization problem. To this end let $\xi_r := \sum_{j:j\in r} \delta_j$, $r \in R$. That is, ξ_r is the total toll paid by a customer who chooses route r. Note that the first two sets of constraints in Problem $(\mathbf{F})$ (the only place where the variables, δ_j, $j \in J$, appear) can be written equivalently as:

$$\xi_r \geq \pi - \sum_{j:j\in r} G_j(\nu_j) , \ r \in R ; \tag{7.36}$$

$$= \pi - \sum_{j:j\in r} G_j(\nu_j) , \text{ if } \lambda_r > 0 , \ r \in R . \tag{7.37}$$

Thus, if we were free to choose the route tolls, ξ_r, independently, we could solve Problem $(\mathbf{F})$ by first solving the relaxed Problem $(\mathbf{F'})$ and then choosing the ξ_r to satisfy (7.36) and (7.37). (This is the approach we were able to use in the case of a single facility in Chapter 2 and in the case of parallel facilities in Chapter 6.) The optimal value of the objective function for both problems would therefore be the same.

But in fact we cannot choose the route tolls independently, since they must satisfy the constraints,

$$\xi_r = \sum_{j:j\in r} \delta_j , \ r \in R ,$$

where $\delta_j \geq 0$, $j \in J$. That is, there must exist a set of nonnegative facility tolls such that the total toll on each route equals the sum of the facility tolls at all the facilities on that route. Whether or not this is true depends on both the topology of the network and the values of the route tolls, which are determined (nonuniquely) by (7.36) and (7.37). Therefore, in general this requirement puts additional constraints on the feasible values of the route arrival rates, λ_r, $r \in R$.

To summarize:

1. If tolls are charged for routes, then the facility-optimal route arrival rates, λ_r, $r \in R$, may be found by solving Problem $(\mathbf{F'})$. Optimal route tolls may then be found by setting $\xi_r = U'(\lambda) - \sum_{j:j\in r} G_j(\nu_j)$, $r \in R$. All nonnegative route arrival rates are feasible for Problem $(\mathbf{F'})$, since the equality constraints only serve to define λ and ν_j, $j \in J$, in terms of the λ_r, $r \in R$.

2. If tolls are charged at the facilities, then the facility-optimal route arrival

rates, λ_r, $r \in R$, and tolls, δ_j, $j \in J$, may be found by solving Problem **(F)**. Route arrival rates are feasible for Problem **(F)** if and only if they belong to the set

$$\mathcal{L} := \{(\lambda_r, r \in R) : \exists \delta_j, j \in J, \text{ s.t. } (7.36) \text{ and } (7.37) \text{ are satisfied }, r \in R\},$$

with $\xi_r := \sum_{j:j \in r} \delta_j$, $r \in R$; $\nu_j := \sum_{r:j \in r} \lambda_r$, $j \in J$; $\lambda := \sum_{r \in R} \lambda_r$; and $\pi := U'(\lambda)$.

It follows that in general the optimal value of the objective function for Problem **(F)** may be smaller than the optimal value of the objective function for the relaxed Problem **(F′)**. Put a different way, the cooperating facilities will always be better off (or at least no worse off) charging a toll for each route rather than charging a toll for each facility.

7.5.1.1 Cooperating Facilities: Fixed Total Arrival Rate

In this subsection we briefly consider the cooperative facility optimal solution in the case of a fixed total arrival rate λ. The problem takes the following form:

$$\max_{\{\pi; \lambda_r, r \in R; \nu_j, \delta_j, j \in J\}} \sum_{j \in J} \nu_j \delta_j$$

$$\text{s.t.} \quad \pi \leq \sum_{j:j \in r} (\delta_j + G_j(\nu_j)) , \ r \in R ,$$

$$\pi = \sum_{j:j \in r} (\delta_j + G_j(\nu_j)) , \text{ if } \lambda_r > 0 , \ r \in R ,$$

$$\nu_j = \sum_{r:j \in r} \lambda_r , \ j \in J ,$$

$$\sum_{r \in R} \lambda_r = \lambda ,$$

$$\lambda_j \geq 0 , \ j \in J .$$

The same arguments as we used for parallel facilities in Section 6.1.3.2 of Chapter 6 lead to the same conclusion: the cooperating facilities may earn an arbitrarily large revenue by choosing an arbitrarily large value of the full price, π, and then choosing the facility tolls so that $\sum_{r \in R} \lambda_r = \lambda$. We leave the details to the reader. As in the special case of parallel facilities, the economic motivation for this result is clear. Since the facilities are cooperating, they are acting as a monopolistic service provider, facing a completely inelastic demand.

7.5.1.2 Cooperating Facilities: Heavy-Tailed Rewards

What happens in the setting of a single-class network if $\lambda U'(\lambda) \to \infty$ as $\lambda \to 0$? For the special case of parallel facilities we saw that the facility operator can earn an arbitrarily large profit by choosing an arbitrarily small arrival rate (equivalently, charging an arbitrarily large toll) and bonding all traffic to the facility j with the smallest zero-flow waiting cost $G_j(0)$.

Using similar arguments, we can extend this result to an arbitrary single-class network of queues. Once again we shall find it convenient to work with a formulation of the facility optimization problem in terms of the arrival rates, λ_r, $r \in R$, as the decision variables. We shall begin with the relaxed version, Problem $(\mathbf{F'})$, which we rewrite here for convenience:

$$\max_{\{\lambda; \lambda_r, r \in R; \nu_j, j \in J\}} \quad \lambda U'(\lambda) - \sum_{r \in R} \lambda_r \sum_{j:j \in r} G_j(\nu_j)$$

$$(\mathbf{F'}) \qquad \text{s.t.} \qquad \sum_{r \in R} \lambda_r = \lambda \,,$$

$$\sum_{r:j \in r} \lambda_r = \nu_j \,,$$

$$\lambda_r \geq 0 \,, \; r \in R \,.$$

Recall that in general an optimal solution to Problem $(\mathbf{F'})$ can be implemented by charging appropriate route tolls, ξ_r, $r \in R$, but not necessarily by charging facility tolls, δ_j, $j \in J$. For the result we seek, however, both implementations are feasible, as we shall demonstrate.

Consider a fixed, arbitrary λ. Since λ is fixed, the optimal λ_j, $j \in J$, corresponding to this value of λ will solve the following minimization problem:

$$\min_{\{\lambda_r, r \in R\}} \quad \sum_{r \in R} \lambda_r \sum_{j:j \in r} G_j(\nu_j)$$

$$\text{s.t.} \qquad \sum_{r \in R} \lambda_r = \lambda \,,$$

$$\sum_{r:j \in r} \lambda_r = \nu_j \,,$$

$$\lambda_r \geq 0 \,, \; r \in R \,.$$

The necessary and sufficient KKT conditions for this problem are:

$$\alpha \;\leq\; \sum_{j:j \in r} (G_j(\nu_j) + \lambda_r G'_j(\nu_j)) + \sum_{s \neq r} \lambda_s \sum_{j:j \in r, j \in s} G'_j(\nu_j) \,, \; r \in R \,,$$

$$\alpha \;=\; \sum_{j:j \in r} (G_j(\nu_j) + \lambda_r G'_j(\nu_j)) + \sum_{s \neq r} \lambda_s \sum_{j:j \in r, j \in s} G'_j(\nu_j) \,, \; \text{if } \lambda_r > 0 \,, \; r \in R \,,$$

$$\lambda \;=\; \sum_{r \in R} \lambda_r \,; \; \nu_j = \sum_{r:j \in r} \lambda_r \,, \; j \in J \,; \; \lambda_r \geq 0 \,, \; r \in R \,.$$

Let $\tilde{G}_r(0) := \sum_{j:j \in r} G_j(0)$, $r \in R$. Assume the routes are ordered so that

$$\tilde{G}_1(0) < \tilde{G}_r(0) \,, \; r \neq 1 \,.$$

(For simplicity we assume strict inequality.)

Theorem 7.7 *For sufficiently small $\lambda > 0$, the cooperative facility optimal allocation of arrival rates is*

$$\lambda_1 = \lambda \; ; \; \lambda_r = 0 \; , \; r \neq 1 \; .$$

Moreover, this allocation can be implemented by charging a toll, $\delta_j = U'(\lambda) - \tilde{G}_r(\lambda)$ at any one of the facilities, $j \in 1$, and a zero toll at each of the other facilities.

Proof It suffices to show that this allocation satisfies the above KKT conditions for sufficiently small λ. For this particular allocation, these conditions are satisfied if

$$\alpha \;\; = \;\; \tilde{G}_1(\lambda) + \lambda\tilde{G}_1'(\lambda)) \; ,$$
$$\alpha \;\; < \;\; \tilde{G}_r(0) \; , \; r \neq 1$$

Since $\tilde{G}_1(\lambda) + \lambda G'(\lambda_1)$ is a continuous function of λ, for sufficiently small λ we have

$$\alpha = \tilde{G}_1(\lambda) + \lambda\tilde{G}_1'(\lambda) < \tilde{G}_r(0) \; , \; r \neq 1 \; ,$$

so that the KKT conditions are satisfied. ∎

As a corollary of this theorem, we have

Corollary 7.8 *Suppose $\lambda U'(\lambda) \to \infty$ as $\lambda \to 0$. Then the cooperative facility optimal solution has an unbounded objective function as $\lambda \to 0$. The optimal solution consists in allocating the total arrival rate, λ, to route 1 (i.e., the route r with the minimal no-flow total waiting cost, $\tilde{G}_r(0)$) and then letting $\lambda \to 0$.*

7.5.2 Competing Facilities

Suppose now that each facility has a separate operator, who sets the toll at that facility and whose goal is to maximize the total profit at that facility. Again we assume that the costs of operating the facilities are fixed, so that maximizing the profit at facility j is equivalent to maximizing the revenue,

$$\nu_j \delta_j \; ,$$

received from the tolls paid by the customers who use that facility.

Assuming (as usual) that the arriving customers are individual optimizers, any particular set of values, δ_j, $j \in J$, for the tolls will result in route arrival rates, λ_r, $r \in R$, and facility flow rates, ν_j, $j \in J$, that satisfy the equilibrium conditions, (7.2)-(7.5). Each of the (competing) revenue-maximizing facility operators seeks to find a value for the toll at its facility that maximizes its revenue, with the equilibrium conditions as constraints. We seek a Nash equilibrium solution for the tolls, that is, a set of values for δ_j, $j \in J$, such that, at each facility j, δ_j maximizes the revenue, $\nu_j\delta_j$, at that facility, assuming that δ_k, $k \neq j$, do not change. In other words, no facility operator can benefit (i.e., achieve a larger revenue at its facility) by unilaterally changing the value of its toll.

Focusing on a particular facility j, the set of tolls, $\{\delta_k, k \in J\}$, satisfies the Nash-equilibrium characterization with respect to that facility, if δ_j maximizes $\nu_j \delta_j$, with the above i.o. equilibrium conditions as constraints (in which δ_k, $k \neq j$, are fixed). That is, δ_j achieves the maximum in the following problem:

$$\max_{\{\pi, \lambda, \delta_j; \lambda_r, r \in R; \nu_k, k \in J\}} \nu_j \delta_j$$

$$(\mathbf{F}_j) \qquad \text{s.t.} \qquad \lambda_r \left(\sum_{k:k \in r} (\delta_k + G_k(\nu_k)) - \pi \right) = 0 , \ r \in R ,$$

$$\sum_{k:k \in r} (\delta_k + G_k(\nu_k)) \geq \pi , \ r \in R ,$$

$$\sum_{r:k \in r} \lambda_r = \nu_k , \ k \in J ,$$

$$\sum_{r \in R} \lambda_r = \lambda ,$$

$$U'(\lambda) = \pi ,$$

$$\lambda_r \geq 0 , \ r \in R .$$

with δ_k, $k \neq j$, fixed.

Rather than continue to analyze this problem in its general form, we shall just consider a special case which illustrates some characteristics of a general network which are not present in the case of parallel facilities.

7.5.2.1 Competing Facilities: Queues in Series

An interesting special case of a single-class network is one in which there is only one route, which contains all facilities $j \in J$. Although our network model does not distinguish the order of the facilities on a route, we shall refer to this as the case of *queues in series*. Again, to keep the exposition simple we shall assume that the arrival rate λ is positive.

The optimization problem for facility j in the case of a variable total arrival rate λ with utility function $U(\lambda)$ may be written as:

$$\max_{\{\delta_j, \lambda\}} \lambda \delta_j$$

$$(\mathbf{F}_j) \qquad \text{s.t.} \qquad U'(\lambda) = \sum_{k \in J} (\delta_k + G_k(\lambda)) ,$$

$$\lambda \geq 0 ,$$

with δ_k, $k \neq j$, fixed. The set of tolls, $\{\delta_k, k \in J\}$, constitutes a Nash equilibrium for the competing facility operators if, together with the total arrival rate, they simultaneously satisfy the optimization problems $(\mathbf{F}_j)$ for all $j \in J$.

For each $j \in J$ and given δ_k, $k \neq j$, define

$$\tilde{G}_j(\lambda) := G_j(\lambda) + \sum_{k \neq j} (\delta_k + G_k(\lambda)) .$$

Then Problem ($\mathbf{F}_j$) can be written equivalently as

$$\max_{\{\lambda \geq 0\}} \lambda U'(\lambda) - \lambda \tilde{G}_j(\lambda) ,$$

with δ_k, $k \neq j$, fixed and $\delta = U'(\lambda) - \tilde{G}_j(\lambda)$. This problem is in the same form as the facility optimization problem for a single facility (cf. Section 2.1.3 of Chapter 2). The KKT condition for $\lambda > 0$ to be optimal for this problem is

$$U'(\lambda) + \lambda U''(\lambda) = \tilde{G}_j(\lambda) + \lambda \tilde{G}'_j(\lambda) .$$

In order for λ and δ_j, $j \in J$, to constitute a Nash equilibrium for the competitive facility operators, this condition must hold simultaneously for all facilities $j \in J$. In particular, the r.h.s. of this equation must equal some constant, α, which does not depend on j. This implies that

$$\alpha = \sum_{k \neq j} \delta_k + \sum_{k \in J}(G_k(\lambda) + \lambda G'_k(\lambda)) , \; j \in J ,$$

so that $\sum_{k \neq j} \delta_k$ must also be independent of j. This can be true only if the facility tolls, δ_k, are the same at all facilities $k \in J$.

Let $\xi := \sum_{k \in J} \delta_k$, so that each $\delta_k = \xi/n$, where n is the total number of facilities. Thus the competitive facility optimization problem for queues in series can be solved by first solving the single-facility problem,

$$\max_{\{\lambda \geq 0\}} \lambda U'(\lambda) - \lambda \tilde{G}(\lambda) ,$$

where $\tilde{G}(\lambda) := \sum_{k \in J} G_k(\lambda)$, and then setting $\xi = U'(\lambda) - \tilde{G}(\lambda)$ and $\delta_k = \xi/n$, $k \in J$. It follows that in this case the competitive facility-optimal solution is also optimal for the cooperative problem. (The reverse is not true, inasmuch as *any* set of facility tolls, δ_k, $k \in J$, such that $\sum_{k \in J} \delta_k = \xi$, is optimal for the cooperative problem.)

7.6 Endnotes

Section 6.1.2.1

Much of the early research on optimal arrival rates and routing in single-class networks was done by researchers in the theory of road traffic flow. Wardrop [196] pointed out the distinction between social and individual optimization in this context. Indeed, in road traffic flow theory an individually optimal allocation of flows (*traffic assignment*) is typically referred to as a *Wardrop equilibrium*. An influential early reference is the book by Beckmann, McGuire, and Winsten [17]. Algorithms for numerical calculation of socially and individually optimal flow allocations were developed by Dafermos and colleagues in a series of papers beginning with [50]. For a comprehensive treatment of the mathematical theory of road traffic networks, see Sheffi [178].

Researchers in telecommunications recognized that the general model for allocation of flows in a network could also be applied to communication networks. Bertsekas and Gallager [20] is a good comprehensive reference for early research in this area. For a more recent reference, emphasizing pricing, see Courcoubetis and Weber [44]. For a comprehensive survey of game models in telecommunications, see Altman et al. [6].

Section 7.4.1

The original paper by Braess (in German) is [28]. For discussions in English of Braess's paradox and its implications, see, e.g., Murchland [149], Frank [67], Dafermos and Nagourney [49], and Cohen and Kelly [40]. Our example network in Section 7.4.1 is based on a network presented in Correa et al. [41].

The examples in Section 7.4.3, showing that the marginal cost of a socially optimal allocation can be less than the average cost of an individually optimal allocation, are new, to the best of my knowledge.

Section 7.4.5

The first papers to find bounds on the difference between the costs of an individually optimal and a socially optimal allocation which are independent of the network topology (and, apparently, the first to coin the term "price of anarchy") were by Roughgarden and Tardos (see, e.g., [163], [166], [164], [165]). The simple derivation of the price-of-anarchy inequality (7.29), using the variational inequality (7.28), is taken from Correa et al. [41]. An excellent reference on use of variational inequalities in the analysis of traffic networks is Nagourney [150].

Stidham [190] extends the analysis of parallel queueing facilities to a network of queues and compares the heavy-traffic behavior of the ratio of individually optimal to socially optimal costs to bounds derived from the price-of-anarchy approach of Roughgarden and Tardos.

Section 7.5.2.1

Veltman and Hassin [195] consider a variant of the model discussed in this section. They use the term, *complementary products*, to characterize their model, reflecting the property that an increase in price at one facility tends to reduce, rather than increase, the demand at the other facility.

CHAPTER 8

Multiclass Networks of Queues

In this chapter we continue our discussion of how some of the models and techniques presented in the earlier chapters can be extended to networks of queues. We now consider a network of queues with more than one class of customers.

Our model is a multiclass generalization of the models in Chapter 7 in which the arrival rate and the routing of jobs of a single class were the decision variables. Here one must choose the arrival rate and job routings for each of several classes of customers. The model may also be considered as a generalization – from a single facility to a network – of the multiclass model introduced in Chapter 4.

8.1 General Model

As in the case of the single-class model of Chapter 7, we consider a network consisting of a set J of facilities and a set R of routes. Each route $r \in R$ consists of a subset of facilities (or resources), and we use the notation $j \in r$ to indicate that facility j is on route r.

There are now m distinct classes of customers, labelled $i \in M$, where $M := \{1, 2, \ldots, m\}$. Let λ_i denote the arrival rate (flow) of class i (a decision variable), $i \in M$. The set of feasible values for λ_i is denoted by A_i. Our default assumption will be that $A_i = [0, \infty)$, $i \in M$. Class i earns a utility, $U_i(\lambda_i)$, per unit time when the arrival rate for class i is λ_i. As usual, we assume that $U_i(\lambda_i)$ is nondecreasing, differentiable, and concave in $\lambda_i \in A_i$.

For each class i there may be several available routes through the network. The set of available routes for class i is denoted R_i ($R_i \subseteq R$). Each class-i job that enters the system must be assigned to one of the routes, $r \in R_i$. Let λ_{ir} denote the class-i flow assigned to route r, $r \in R_i$. The flows, λ_{ir}, $r \in R_i$, $i \in M$, are decision variables, subject to the constraint that, for each class i, the total class-i flow must equal λ_i:

$$\sum_{r \in R_i} \lambda_{ir} = \lambda_i , \ i \in M .$$

Let ν_j denote the total flow at facility j: the sum of the flows on all the routes that use that facility. That is,

$$\nu_j = \sum_{i \in M} \sum_{r \in R_i : j \in r} \lambda_{ir} , \ j \in J .$$

We assume that at each facility j the waiting cost for each class is a function of the total flow ν_j at that facility. Let $G_{ij}(\nu_j)$ denote the average waiting cost of a class-i job at facility j. We assume that $G_{ij}(\nu_j)$ takes values in $[0, \infty]$ and is strictly increasing and differentiable in ν_j, $0 \le \nu_j < \infty$.

Remark 1   A more general model would allow the waiting-cost functions, $G_{ij}(\cdot)$, to depend on the vector, $(\lambda_{ir}, i \in M, r \in R_i : j \in r)$, rather than just on ν_j, the sum of its components. Such a model would allow for a priority discipline and/or class-dependent service rates at a facility. The techniques for analyzing a multiclass, single-facility system with these features were developed in detail in Chapter 4. These techniques may be applied in the network setting as well, but to keep the exposition simple we will not do so in this chapter, but instead leave it to the reader to make these generalizations.

Given the flows, ν_j, at the various facilities, the total waiting cost incurred by a class-i job that follows a particular route r is the sum of the resulting waiting costs at the facilities on that route:

$$\sum_{j:j\in r} G_{ij}(\nu_j) \, , \ r \in R_i \, .$$

There may also be a toll δ_j which is charged to each customer who uses facility j, $j \in J$, and/or a toll ξ_i which is charged to each class-i customer who enters the system, $i \in M$. In this case the total cost (full price) for a class-i job that enters the system and uses route $r \in R_i$ is given by

$$\xi_i + \sum_{j:j\in r} (\delta_j + G_{ij}(\nu_j)) \, .$$

Remark 2   A more general model would allow the toll at each facility to depend on the class, but we shall not need this generality, since we are confining our attention to models in which the waiting cost for each class at each facility depends only on the total flow at that facility.

Example 1   Linear Waiting Costs Suppose the average waiting time of a job at facility j is independent of class and a function, $W_j(\nu_j)$, of the total flow at j. We assume that $W_j(\nu_j)$ takes values in $[0, \infty]$ and is strictly increasing and differentiable in ν_j, $0 \le \nu_j < \infty$. Assume that a job of class i incurs a waiting cost $h_i \cdot t$ if the waiting time experienced by the job equals t, $t \ge 0$. Then

$$G_{ij}(\nu_j) = h_i \cdot W_j(\nu_j) \, , \ i \in M \, , \ j \in J \, ,$$

and the total waiting cost incurred by a class-i job that follows route $r \in R_i$ is given by

$$h_i \cdot \sum_{j:j\in r} W_j(\nu_j) \, ,$$

$i \in M$.

As an example of a queueing model that instantiates these assumptions, consider a network of *FIFO M/M/1* facilities operating in steady state. Let μ_j denote the service rate at facility j, $j \in J$, and suppose a *FIFO* queue discipline is used at each facility. Then

$$W_j(\nu_j) = \frac{1}{\mu_j - \nu_j} \,,\ 0 \le \nu_j < \mu_j \,, \tag{8.1}$$

with $W_j(\nu_j) = \infty$ for $\nu_j \ge \mu_j$, $j \in J$.

As we have noted previously, the formula (8.1) for $W_j(\nu_j)$ also holds for a network of *M/GI/1* facilities each of which operates under the processor-sharing (*PS*) or last-in, first-out, preemptive-resume (*LIFO-PR*) discipline.

As in our previous models for selection of arrival rates, the solution to the decision problem depends on who is making the decision and what criteria are being used. The decision may be made by the individual customers, each concerned only with its own net utility (*individual optimality*), by an agent for each class concerned with the net benefit received by that class (*class optimality*), by an agent interested in maximizing the aggregate net utility to all customers (*social optimality*), or by a system operator interested in maximizing profit (*facility optimality*).

Before considering several special cases of the general model, we first establish the existence (and in some cases uniqueness) of solutions for each of these optimality criteria in the setting of the general model, under weak conditions on the utility and waiting-cost functions. We also show that many of the results from the previous chapter concerning the Price of Anarchy extend with appropriate modifications to the general multiclass model.

8.1.1 Individually Optimal Arrival Rates

The equilibrium conditions for an individually optimal allocation are

$$U_i'(\lambda_i) \ \le\ \xi_i + \sum_{j:j\in r} (G_{ij}(\nu_j) + \delta_j) \,,\ i \in M \,,\ r \in R_i \,, \tag{8.2}$$

$$U_i'(\lambda_i) \ =\ \xi_i + \sum_{j:j\in r} (G_{ij}(\nu_j) + \delta_j) \,,\ \text{if } \lambda_{ir} > 0 \,,\ i \in M \,,\ r \in R_i \,, \tag{8.3}$$

$$\nu_j \ =\ \sum_{i\in M}\sum_{r\in R_i:j\in r} \lambda_{ir} \,,\ j \in J \,, \tag{8.4}$$

$$\lambda_i \ =\ \sum_{r\in R_i} \lambda_{ir} \,,\ i \in M \,, \tag{8.5}$$

$$\lambda_{ir} \ \ge\ 0 \,,\ i \in M \,,\ r \in R_i \,. \tag{8.6}$$

8.1.1.1 Existence and Uniqueness of Individually Optimal Solution

Under a mild technical assumption, we can show that there exists a unique solution to the equilibrium conditions (8.2)–(8.6).

Assumption 1 There exists a compact convex set $\tilde{A} \subseteq A = \prod_{i \in M} A_i$ such that

1. $G_{ij}(\nu_j) < \infty$ for all $i \in M$, $j \in J$, and $\boldsymbol{\lambda} = (\lambda_{i,r}, i \in M, r \in R_i) \in \tilde{A}$,

2. for all $i \in M$, $r \in R_i$, $U_i'(\lambda_i) < \xi_i + \sum_{j:j \in r}(\delta_j + G_{ij}(\nu_j))$ for all $\boldsymbol{\lambda} \in A - \tilde{A}$,

where $\nu_j = \sum_{i \in M} \sum_{r \in R_i : j \in r} \lambda_{ir}$, $j \in J$.

Theorem 8.1 *Under Assumption 1 there exists a unique solution to the equilibrium conditions which characterize the vector $\boldsymbol{\lambda}^e$ of individually optimal arrival rates.*

The proof of this theorem is deferred until the next section, since it depends on constructing a class optimization problem that is equivalent to the individual optimization problem.

For an example in which this assumption is satisfied, see the discussion in the next section.

8.1.2 Class-Optimal Arrival Rates

We now consider the problem from the point of view of the manager of each class. The class manager for class i is concerned with maximizing the net benefit, $B_i(\boldsymbol{\lambda})$, received per unit time by jobs of class i, where

$$B_i(\boldsymbol{\lambda}) = U_i(\lambda_i) - \sum_{r \in R_i} \lambda_{ir} \sum_{j:j \in r} (\xi_i + \delta_j + G_{ij}(\nu_j)) , \ i \in M , \quad (8.7)$$

with $\lambda_i = \sum_{r \in R_i} \lambda_{ir}$, $i \in M$, and $\nu_j = \sum_{i \in M} \sum_{r \in R_i : j \in r} \lambda_{ir}$, $j \in J$. Equivalently, we can write $B_i(\boldsymbol{\lambda})$ as

$$B_i(\boldsymbol{\lambda}) = U_i(\lambda_i) - (\lambda_i \xi_i + \sum_{j \in J} H_{ij}(\boldsymbol{\lambda})) ,$$

where

$$H_{ij}(\boldsymbol{\lambda}) := \sum_{r \in R_i : j \in r} \lambda_{ir}(\delta_j + G_{ij}(\nu_j)) . \quad (8.8)$$

That is, $H_{ij}(\boldsymbol{\lambda})$ is the cost per unit time (including tolls) incurred by class-i jobs at facility j.

Thus, the goal of the class manager for class i is to solve the following maximization problem:

$$\max_{\{\lambda_i; \lambda_{ir}, r \in R_i; \nu_j, j \in J\}} \quad U_i(\lambda_i) - (\lambda_i \xi_i + \sum_{j \in J} H_{ij}(\boldsymbol{\lambda}))$$

$$(\mathbf{C}_i) \qquad \text{s.t.} \qquad \lambda_i = \sum_{r \in R_i} \lambda_{ir} ,$$

$$\nu_j = \sum_{k \in M} \sum_{s \in R_k : j \in s} \lambda_{ks} , \ j \in J ,$$

$$\lambda_{ir} \geq 0 , \ r \in R_i ,$$

where λ_{ks}, $s \in R_k$, $k \neq i$, are fixed. A vector (λ) is class optimal if it is a Nash equilibrium for the classes, that is, if λ simultaneously solves Problem (C_i) for each class $i \in M$.

The *KKT* conditions for Problem (C_i) with $H_{ij}(\lambda)$ defined by (8.8) are:

$$U_i(\lambda_i) \leq \xi_i + \sum_{j:j\in r}[\delta_j + G_{ij}(\nu_j) + \sum_{s\in R_i:j\in s}\lambda_{is}G'_{ij}(\nu_j)] , \; r \in R_i ,$$

$$U_i(\lambda_i) = \xi_i + \sum_{j:j\in r}[\delta_j + G_{ij}(\nu_j) + \sum_{s\in R_i:j\in s}\lambda_{is}G'_{ij}(\nu_j)] ,$$

$$\text{if } \lambda_{ir} > 0 , \; r \in R_i$$

$$\lambda_i = \sum_{r\in R_i}\lambda_{ir} ,$$

$$\nu_j = \sum_{k\in M}\sum_{s\in R_k:j\in s}\lambda_{ks} , \; j \in J ,$$

$$\lambda_{ir} \geq 0 , \; r \in R_i .$$

In order for the solution to be class optimal, therefore, these conditions must hold simultaneously for all classes $i \in M$. Note that the class-optimal solution may be induced by charging individually optimizing class-i customers an additional toll at facility j $(j \in J)$ equal to

$$\sum_{s\in R_i:j\in s}\lambda_{is}G'_{ij}(\nu_j) ,$$

which is the external effect, on class-i customers only, of a marginal increase in the class-i arrival rate at facility j.

8.1.2.1 Existence and Uniqueness of Class-Optimal Solution

To determine conditions under which there exists a unique class-optimal allocation, we shall use an extension of the approach used in Section 4.2.2.1 of Chapter 4. We need the following assumptions:

Assumption 2 There exists a compact polyhedral convex set, $\tilde{A} \subseteq A := \prod_{i\in M} A_i$, such that

1. $G_{ij}(\nu_j) < \infty$ for all $i \in M$, $j \in J$, and $\lambda = (\lambda_{ir}, i \in M, r \in R_i) \in \tilde{A}$, with $\nu_j = \sum_{i\in M}\sum_{r\in R_i:j\in r}\lambda_{ir}, j \in J$;

2. for each $\lambda \in A$, there exists a $\lambda' = (\lambda'_{ir}, i \in M, r \in R_i) \in \tilde{A}$ at which the only active constraints among those defining $\tilde{A}$ are (possibly) some of the nonnegativity constraints, such that $B_i(\lambda') \geq B_i(\lambda)$ for all $i \in M$.

Assumption 3 For all $i \in M$, $\sum_{j\in J} H_{ij}(\lambda)$ is a jointly convex function of $(\lambda_{ir}, r \in R_i)$, for each set of fixed values of λ_{ks}, $s \in R_k$, $k \neq i$, such that $\lambda \in \tilde{A}$.

Using the same approach as in Section 4.2.2.1 of Chapter 4, we can prove the following theorem, as a consequence of Theorems 2 and 6 of Rosen [161].

Theorem 8.2 *Under Assumptions 2 and 3, the solution to the class optimization problem exists and defines a unique Nash equilibrium.*

To see how this theorem applies in a particular example of a multiclass network, consider the case in which each facility j operates as an $M/M/1$ queue in steady state with a *FIFO* discipline and a linear waiting-cost function. In this case, $G_{ij}(\nu_j) = h_i \cdot W_j(\nu_j)$, where

$$W_j(\nu_j) = \frac{1}{\mu_j - \nu_j} \ , \ 0 \le \nu_j < \mu_j \ , \tag{8.9}$$

with $W_j(\nu_j) = \infty$ for $\nu_j \ge \mu_j$. Let ϵ_j be given, with $0 < \epsilon_j < \mu_j, j \in J$. Define

$$\hat{A} := \{\boldsymbol{\lambda} : \nu_j = \sum_{k \in M} \sum_{s \in R_k : j \in s} \lambda_{ks} \le \mu_j - \epsilon_j \ , \ j \in J\} \ .$$

Since $\hat{A}$ is defined by linear inequalities, it is closed and convex. It is bounded, since it is contained in the bounded set

$$\bar{A} := \{\boldsymbol{\lambda} : \sum_{r \in R_k} \lambda_{ks} \le \max_{j \in J} \mu_j \ , \ k \in M\} \ .$$

Then Assumption 2 (1) holds. Since $W_j(\nu_j) \uparrow \infty$ as $\nu_j \uparrow \mu_j$, we can choose an $\epsilon_j > 0$ sufficiently small that Assumption 2 (2) holds. Finally, Assumption 3 holds since $W_j(\cdot)$ is convex and increasing.

8.1.2.2 Proof of Theorem 8.1

We shall use an extension of the argument in Section 4.2.2.1 of Chapter 4 to construct a class optimization problem which is equivalent to the individual optimization problem. The existence and uniqueness of the individually optimal solution then follows from the existence and uniqueness of the solution to this class optimization problem.

Consider a particular class i and the maximization problem,

$$\max_{\{\lambda_i, \lambda_{ir}, r \in R_i\}} \quad U_i(\lambda_i) - (\lambda_i \xi_i + \sum_{j \in J} \int_0^{\nu_j} (\delta_j + G_{ij}(\nu))d\nu)$$

$$\text{s.t.} \quad \lambda_i = \sum_{r \in R_i} \lambda_{ir} \ ,$$

$$\nu_j = \sum_{k \in M} \sum_{s \in R_k : j \in s} \lambda_{ks} \ , \ j \in J \ ,$$

$$\lambda_{ir} \ge 0 \ , \ r \in R_i \ ,$$

where λ_{ks}, $s \in R_k$, $k \ne i$, are fixed. This problem is a modified version of Problem $(\mathbf{C}_i)$ in a class optimization problem. To be specific, the objective function for the problem can be written as

$$U_i(\lambda_i) - (\lambda_i \xi_i + \sum_{j \in J} H_{ij}(\boldsymbol{\lambda})) \, ,$$

where $H_{ij}(\boldsymbol{\lambda})$ (the cost per unit time incurred by class-i jobs at facility j) now takes the form,

$$H_{ij}(\boldsymbol{\lambda}) = \int_0^{\nu_j} (\delta_j + G_{ij}(\nu)) d\nu \, .$$

The KKT conditions for this problem are:

$$U_i(\lambda_i) \leq \xi_i + \sum_{j:j \in r} (\delta_j + G_{ij}(\nu_j)) \, , \ r \in R_i \, ,$$

$$U_i(\lambda_i) = \xi_i + \sum_{j:j \in r} (\delta_j + G_{ij}(\nu_j)) \, , \text{ if } \lambda_{ir} > 0 \, , \ r \in R_i \, ,$$

$$\lambda_i = \sum_{r \in R_i} \lambda_{ir} \, ,$$

$$\nu_j = \sum_{k \in M} \sum_{s \in R_k : j \in s} \lambda_{ks} \, , \ j \in J \, ,$$

$$\lambda_{ir} \geq 0 \, , \ r \in R_i \, .$$

In order for the solution to be class optimal (that is, a Nash equilibrium for the classes) these conditions must hold simultaneously for all classes $i \in M$. But this is true if and only if the equilibrium conditions for an individually optimal solution are satisfied.

Suppose Assumption 1 holds. Then Assumption 2 holds for the equivalent class optimization problem. Since $G_{ij}(\cdot)$ is increasing and ν_j is a linear function of $\boldsymbol{\lambda}$, $H_{ij}(\boldsymbol{\lambda})$ is jointly convex in $(\lambda_{ir}, r \in R_i)$, for any fixed values of $(\lambda_{ks}, k \in M, s \in R_k)$. Therefore Assumption 3 holds for the equivalent class optimization problem and it follows from Theorem 8.2 that the equivalent class optimization problem has a unique Nash equilibrium. Therefore the individually optimal problem also has a unique solution. This completes the proof of Theorem 8.1.

8.1.3 Socially Optimal Arrival Rates

A *socially optimal* allocation of arrival rates, $\boldsymbol{\lambda}^s = (\lambda_{ir}^s, i \in M, r \in R_i)$, is defined as one that maximizes the aggregate net benefit to all classes. It may thus be found by solving the following optimization problem:

$$\max_{\{\lambda_i, \lambda_{ir}, i \in M, r \in R_i; \nu_j, j \in J\}} \sum_{i \in M} U_i(\lambda_i) - \sum_{i \in M} \sum_{r \in R_i} \lambda_{ir} \sum_{j:j \in r} G_{ij}(\nu_j)$$

$$(\mathbf{P}) \qquad \text{s.t.} \qquad \nu_j = \sum_{i:i \in M} \sum_{r \in R_i : j \in r} \lambda_{ir} \, , \ j \in J \, ,$$

$$\lambda_i = \sum_{r \in R_i} \lambda_{ir} \, , \ i \in M \, ,$$

$$\lambda_{ir} \geq 0 \, , \, i \in M \, , \, r \in R_i \, .$$

A socially optimal allocation of arrival rates $\boldsymbol{\lambda}^s = (\lambda_i^s, i \in M)$ must satisfy the following *KKT* conditions, which are necessary for an optimal solution to Problem **(P)**:

$$U_i'(\lambda_i) \; \leq \; \sum_{j:j \in r} (G_{ij}(\nu_j) + \delta_j) \, , \, i \in M \, , \, r \in R_i \, , \tag{8.10}$$

$$U_i'(\lambda_i) \; = \; \sum_{j:j \in r} (G_{ij}(\nu_j) + \delta_j) \, , \, \text{if } \lambda_{ir} > 0 \, , \, i \in M \, , \, r \in R_i \, , \tag{8.11}$$

$$\nu_j \; = \; \sum_{i \in M} \sum_{r \in R_i : j \in r} \lambda_{ir} \, , \, j \in J \, , \tag{8.12}$$

$$\lambda_i \; = \; \sum_{r \in R_i} \lambda_{ir} \, , \, i \in M \, , \tag{8.13}$$

$$\delta_j \; = \; \sum_{k \in M} \sum_{s \in R_k : j \in s} \lambda_{ks} G_{kj}'(\nu_j) \, , \, j \in S \, , \tag{8.14}$$

$$\lambda_{ir} \; \geq \; 0 \, , \, i \in M \, , \, r \in R_i \, . \tag{8.15}$$

Note that δ_j as given by (8.14) is the external effect of a marginal increase in flow at facility j on the waiting costs of the existing flows of all classes at facility j, $j \in J$. Thus, as expected, the *KKT* necessary conditions for a socially optimal allocation coincide with the equilibrium conditions (8.2)–(8.6) for an individually optimal solution with $\xi_i = 0$, $i \in M$, provided that the toll, δ_j, charged at each facility j equals the external effect – evaluated at the socially optimal allocation, $\boldsymbol{\lambda}^s$.

We have already seen in Chapter 4 that, even in the special case of the single-facility model, the total waiting-cost function,

$$\sum_{i \in M} \sum_{r \in R_i} \lambda_{ir} \sum_{j:j \in r} G_{ij}(\nu_j) \, ,$$

is not in general jointly convex in $\boldsymbol{\lambda} = (\lambda_i, i \in M)$, in spite of the convexity properties of its components. Hence the objective function for Problem **(P)** may not be jointly concave in $\boldsymbol{\lambda}$ and the first-order *KKT* conditions may not be sufficient for a socially optimal allocation.

In the single-facility setting we observed that it was the heterogeneity of the waiting costs among the different classes that created the lack of joint convexity of the total waiting-cost function. When the classes are homogeneous in their waiting costs, however, the total waiting-cost function for the single-facility problem is always jointly convex. As we shall see, this property extends to the multiclass network model. (See Section 8.4 below.)

8.1.4 Facility-Optimal Arrival Rates

Now we consider the system from the point of view of revenue-maximizing facility operators who choose the tolls at the facilities. Assuming that the arriving customers are individual optimizers, a particular choice of tolls, δ_j, for the facilities $j \in J$ will result in a vector, $\boldsymbol{\lambda} = (\lambda_{ir}, i \in M, r \in R_i)$, of arrival rates that satisfy the equilibrium conditions, (8.2)–(8.6), for an individually optimal allocation. We consider only the case of cooperating facilities.

8.1.4.1 Cooperating Facilities

Suppose a single operator (or a collective of cooperating facility operators) chooses the tolls at all the facilities. For this case the analysis of facility-optimal arrival rates for a multiclass, single-facility queueing system (see Section 4.2.4 of Chapter 4) extends in a straightforward way to a multiclass network of queues with fixed routes.

The revenue-maximizing operator thus seeks values of δ_j, $j \in J$, and ξ_i, $i \in M$, that solve the following optimization problem:

$$\max_{\{\xi_i,\lambda_i,\lambda_{ir},i\in M, r\in R_i; \delta_j, \nu_j, j\in J\}} \sum_{i\in M} \lambda_i \xi_i + \sum_{j\in J} \nu_j \delta_j$$

$$\text{s.t.} \quad \lambda_{ir}\left(U_i'(\lambda_i) - \xi_i - \sum_{j:j\in r}(\delta_j + G_{ij}(\nu_j))\right) = 0 \,, \ i \in M \,, \ r \in R_i \,,$$

$$U_i'(\lambda_i) \leq \xi_i + \sum_{j:j\in r}(\delta_j + G_{ij}(\nu_j)) \,, \ i \in M \,, \ r \in R_i \,,$$

$$\lambda_i = \sum_{r\in R_i} \lambda_{ir} \,, \ i \in M \,,$$

$$\nu_j = \sum_{i\in M} \sum_{r\in R_i : j\in r} \lambda_{ir} \,, \ j \in J \,,$$

$$\lambda_{ir} \geq 0 \,, \ i \in M \,, \ r \in R_i \,.$$

Summing the l.h.s. of the first equality constraint over $r \in R_i$ and $i \in M$ and adding the sum (which equals zero) to the objective function and rearranging terms, we obtain the equivalent formulation,

$$\max_{\{\lambda_i,\lambda_{ir},i\in M, r\in R_i; \nu_j, j\in J\}} \sum_{i\in M} \lambda_i U_i'(\lambda_i) - \sum_{i\in M}\sum_{r\in R_i} \lambda_{ir} \sum_{j:j\in r} G_{ij}(\nu_j)$$

(F)
$$\text{s.t.} \quad \lambda_i = \sum_{r\in R_i} \lambda_{ir} \,, \ i \in M \,,$$

$$\nu_j = \sum_{i\in M} \sum_{r\in R_i : j\in r} \lambda_{ir} \,, \ j \in J \,,$$

$$\lambda_{ir} \geq 0 \,, \ i \in M \,, \ r \in R_i \,,$$

with

$$\xi_i + \sum_{j:j\in r} \delta_j \ \geq \ U_i'(\lambda_i) - \sum_{j:j\in r} G_{ij}(\nu_j) \,, \ i \in M \,, \ r \in R_i \qquad (8.16)$$

$$= \ U_i'(\lambda_i) - \sum_{j:j\in r} G_{ij}(\nu_j) \,,$$

$$\text{if } \lambda_{ir} > 0 \, , \; i \in M \, . \, , \; r \in R_i \qquad (8.17)$$

We shall presently exhibit a set of tolls, ξ_i, $i \in M$, δ_j, $j \in J$, which satisfy these conditions.

As in the single-facility model (see Chapter 4), Problem **(F)** is in the same form as the social optimization problem, but with $U_i(\lambda_i)$ replaced by $\lambda_i U'(\lambda_i)$, $i \in M$.

Let $\boldsymbol{\lambda}^f = (\lambda_{ir}^f, i \in M, r \in R_i)$ denote a solution to Problem **(F)**. The vector $\boldsymbol{\lambda}^f$ satisfies the following necessary *KKT* conditions for optimality:

$$U_i'(\lambda_i) + \lambda_i U_i''(\lambda_i) \; \leq \; \sum_{j:j \in r} \left(G_{ij}(\nu_j) + \sum_{k \in M} \sum_{s \in R_k : j \in s} \lambda_{ks} G_{kj}'(\nu_j) \right) ,$$
$$i \in M \, , \; r \in R_i \, , \qquad (8.18)$$

$$U_i'(\lambda_i) + \lambda_i U_i''(\lambda_i) \; = \; \sum_{j:j \in r} \left(G_{ij}(\nu_j) + \sum_{k \in M} \sum_{s \in R_k : j \in s} \lambda_{ks} G_{kj}'(\nu_j) \right) ,$$
$$\text{if } \lambda_{ir} > 0 \, , \; i \in M \, , \; r \in R_i \, , \qquad (8.19)$$

$$\nu_j \; = \; \sum_{i \in M} \sum_{r \in R_i : j \in r} \lambda_{ir} \, , \; j \in J \, , \qquad (8.20)$$

$$\lambda_i \; = \; \sum_{r \in R_i} \lambda_{ir} \, , \; i \in M \, , \qquad (8.21)$$

$$\lambda_{ir} \; \geq \; 0 \, , \; i \in M \, , \; r \in R_i \, . \qquad (8.22)$$

It is easily verified that the following tolls satisfy conditions (8.16) and (8.17):

$$\xi_i^f \; := \; -\lambda_i U''(\lambda_i) \, , \; i \in M \, ;$$
$$\delta_j \; := \; \sum_{k \in M} \sum_{s \in R_k : j \in s} \lambda_{ks} G_{kj}'(\nu_j) \, , \; j \in J \, .$$

Note that the facility tolls coincide with the socially optimal facility tolls and the class tolls are the same as for the single-facility, multiclass model under facility optimization (see Chapter 4).

The objective function for Problem **(F)** may not be concave in $\boldsymbol{\lambda}$, even under the (operative) assumptions that $U_i(\lambda_i)$ is concave, because $\lambda_i U_i'(\lambda_i)$ may not be concave. Thus in this case, in contrast to the socially optimal problem, concavity of the objective function may fail even if the class-i waiting-cost per unit time, $\sum_{r \in R_i} \lambda_{ir} \sum_{j:j \in r} G_{ij}(\nu_j)$, is convex in $\boldsymbol{\lambda}$ for all $i \in M$. In this case, the *KKT* conditions may have multiple solutions.

8.1.5 Comparison of S.O. and Toll-Free I.O. Solutions

In our analysis of single-class networks of queues, we compared the socially optimal and the toll-free individually optimal allocations (see Section 7.4 of Chapter 7). Prominent among the results were Braess's Paradox and the Price

of Anarchy. The examples we presented to illustrate Braess's Paradox are, of
course, valid in the context of a multiclass network of queues, since a single-
class network is a special case of a multiclass network.

We now show how the results concerning the Price of Anarchy extend to
multiclass networks.

8.1.5.1 Price of Anarchy

Suppose the total arrival rate, λ_i, is fixed for each class $i \in M$. As in the
case of a single-class network, we begin with the formulation of the social
optimization problem with fixed class arrival rates, λ_i, $i \in M$, as a constrained
minimization problem:

$$\min_{\{\lambda_{ir}, i \in M, r \in R_i; \nu_j, j \in J\}} \quad \sum_{i \in M} \sum_{r \in R_i} \lambda_{ir} \sum_{j:j \in r} G_{ij}(\nu_j)$$

$$\text{s.t.} \quad \sum_{r \in R} \lambda_{ir} = \lambda_i \, , \; i \in M \, ,$$

$$\nu_j = \sum_{i \in M} \sum_{r \in R_i: j \in r} \lambda_{ir} \, ,$$

$$\lambda_{ir} \geq 0 \, , \; i \in M \, , \; r \in R_i \, .$$

Now define

$$\nu_{ij} := \sum_{r \in R_i: j \in r} \lambda_{ir} \, , \; i \in M \, , \; j \in J \, . \tag{8.23}$$

Thus, ν_{ij} is the total class-i flow at facility j. (Note that $\nu_{ij} = 0$ if there is
no route $r \in R_i$ that uses facility j.) Using this definition, we can rewrite the
above maximization problem in equivalent form as

$$\min_{\{\lambda_{ir}, i \in M, r \in R_i; \nu_{ij}, i \in M, j \in J\}} \quad \sum_{j \in J} \sum_{i \in M} \nu_{ij} G_{ij}(\nu_j)$$

$$\text{s.t.} \quad \sum_{r \in R} \lambda_{ir} = \lambda_i \, , \; i \in M \, , \tag{8.24}$$

$$\nu_{ij} = \sum_{r \in R_i: j \in r} \lambda_{ir} \, , \tag{8.25}$$

$$\lambda_{ir} \geq 0 \, , \; i \in M \, , \; r \in R_i \, , \tag{8.26}$$

with $\nu_j = \sum_{i \in M} \nu_{ij}$, $j \in J$. Let $\boldsymbol{\nu} := (\nu_{ij}, i \in M, j \in J)$ and define the feasible
set $\mathcal{N}$ as the set of all $\boldsymbol{\nu}$ for which there exists a $\boldsymbol{\lambda} = (\lambda_{ir}, i \in M, r \in R_i)$
such that $\boldsymbol{\nu}$ and $\boldsymbol{\lambda}$ satisfy the constraints, (8.24), (8.25), and (8.26). (Note
the difference between the definitions of $\boldsymbol{\nu}$ and $\mathcal{N}$ here and in Section 7.4.5
in Chapter 7.) Then the social optimization problem may be rewritten with
decision variables, $\boldsymbol{\nu} = (\nu_{ij}, i \in M, j \in J)$, as

$$\min_{\{\boldsymbol{\nu} \in \mathcal{N}\}} \quad C(\boldsymbol{\nu}) := \sum_{j \in J} \sum_{i \in M} \nu_{ij} G_{ij}(\nu_j) \, ,$$

with $\nu_j = \sum_{i \in M} \nu_{ij}$, $j \in J$. Let $\boldsymbol{\nu}^s = (\nu_{ij}^s, i \in M, j \in J)$ denote the solution to this problem, that is, the socially optimal vector of class-facility flow rates.

Let $\boldsymbol{\nu}^e = (\nu_{ij}^e, i \in M, j \in J)$ denote the vector of class-facility flow rates corresponding to an individually optimal allocation, $\boldsymbol{\lambda}^e = (\lambda_{ir}^e, i \in M, r \in R)$, of flow rates of the various classes on the various routes.

Lemma 8.3 *An individually optimal vector of facility flow rates, $\boldsymbol{\nu}^e$, satisfies the variational inequality,*

$$\sum_{j \in J} \sum_{i \in M} (\nu_{ij}^e - \nu_{ij}) G_{ij}(\nu_j^e) \leq 0 \,,$$

for all $\boldsymbol{\nu} = (\nu_{ij}, i \in M, j \in J) \in \mathcal{N}$.

Proof Suppose the system is operating with individually optimal route flow rates, $\boldsymbol{\lambda}^e$, and corresponding class-facility flow rates, $\boldsymbol{\nu}^e$. Then each customer of class i chooses route r with probability

$$p_{ir}^e := \lambda_{ir}^e / \lambda_i \,, \quad i \in M \,, \; r \in R_i \,.$$

The Nash-equilibrium property asserts that no individual customer has an incentive to deviate unilaterally from these probabilities. This implies that, for each class $i \in M$, all routes $r \in R_i$ with $\lambda_{ir}^e > 0$ are "shortest routes" in the sense that they achieve the minimum cost per unit of flow,

$$\min_{r \in R_i} \sum_{j: j \in r} G_{ij}(\nu_j^e) \,,$$

when the cost per unit of class-i flow at facility j is fixed at $G_{ij}(\nu_j^e)$. For each class $i \in M$, therefore, assuming the cost per unit of class-i flow at each facility j remains equal to $G_{ij}(\nu_j^e)$, any other choice of route probabilities, $\{p_{ir}, r \in R_i\}$, (and hence of flows, $\{\lambda_{ir}, r \in R_i\}$) will result in a larger cost for class i, because it will be using routes other than the shortest. Thus, for all class-i flow allocations, $\{\lambda_{ir}, r \in R_i\}$,

$$\sum_{r \in R_i} \lambda_{ir} \sum_{j: j \in r} G_{ij}(\nu_j^e) \geq \sum_{r \in R_i} \lambda_{ir}^e \sum_{j: j \in r} G_{ij}(\nu_j^e) \,.$$

Summing over all $i \in M$, we have

$$\sum_{i \in M} \sum_{r \in R_i} \lambda_{ir} \sum_{j: j \in r} G_{ij}(\nu_j^e) \geq \sum_{i \in M} \sum_{r \in R_i} \lambda_{ir}^e \sum_{j: j \in r} G_{ij}(\nu_j^e) \,,$$

or, equivalently, using (8.23),

$$\sum_{j \in J} \sum_{i \in M} (\nu_{ij}^e - \nu_{ij}) G_{ij}(\nu_j^e) \leq 0 \,.$$

$\blacksquare$

It follows from Lemma 8.3 that, for any $\boldsymbol{\nu} \in \mathcal{N}$,

$$C(\boldsymbol{\nu}^e) \;=\; \sum_{j \in J} \sum_{i \in M} \nu_{ij}^e G_{ij}(\nu_j^e)$$

$$\leq \sum_{j \in J} \sum_{i \in M} \nu_{ij} G_{ij}(\nu_j^e)$$

$$= \sum_{j \in J} \sum_{i \in M} \nu_{ij} G_{ij}(\nu_j) + \sum_{j \in J} \sum_{i \in M} \nu_{ij}(G_{ij}(\nu_j^e) - G_{ij}(\nu_j))$$

$$= C(\nu) + \sum_{j \in J} \sum_{i \in M} \nu_{ij}(G_{ij}(\nu_j^e) - G_{ij}(\nu_j)),$$

and hence

$$C(\nu^e) - C(\nu) \leq \sum_{j \in J} \nu_j(\bar{G}_j(\nu_j^e) - \underline{G}_j(\nu_j)), \tag{8.27}$$

where

$$\bar{G}_j(\nu_j) := \max_{i \in M} G_{ij}(\nu_j), \ \underline{G}_j(\nu_j) := \min_{i \in M} G_{ij}(\nu_j), \ \nu_j \geq 0, \ j \in J.$$

Using the example of a linear waiting-cost function, we now show how this inequality can be used to bound the percentage difference between $C(\nu^e)$ and $C(\nu^s)$.

Example 2   Linear Waiting Costs (revisited). Suppose the waiting cost for each class at each facility is linear in the average waiting time at that facility:

$$G_{ij}(\nu_j) = h_i \cdot W_j(\nu_j), \ i \in M, \ j \in J.$$

Let $\bar{h} := \max_{i \in M} h_i$, $\underline{h} := \min_{i \in M} h_i$. Then

$$\bar{G}_j(\nu_j) = \bar{h} \cdot W_j(\nu_j), \ \underline{G}_j(\nu_j) = \underline{h} \cdot W_j(\nu_j), \ j \in J.$$

Moreover, for all $\nu \in \mathcal{N}$,

$$\underline{h} \sum_{j \in J} \nu_j W_j(\nu_j) \leq C(\nu) \leq \bar{h} \sum_{j \in J} \nu_j W_j(\nu_j).$$

Let $\tau := \bar{h}/\underline{h}$.

Theorem 8.4 *Suppose there exists a constant $\sigma < \tau^{-1} \leq 1$ such that*

$$\max_{j \in J} \sup_{0 \leq \nu_j \leq \nu_j^e} \frac{\nu_j(W_j(\nu_j^e) - W_j(\nu_j))}{\nu_j^e W_j(\nu_j^e)} \leq \sigma. \tag{8.28}$$

Then

$$\frac{C(\nu^e)}{C(\nu^s)} \leq \frac{\tau}{1 - \tau\sigma}.$$

Proof   For any $\nu \in \mathcal{N}$, using (8.27) and (8.28), we have

$$C(\nu^e) - C(\nu) \leq \sum_{j \in J} \nu_j(\bar{G}_j(\nu_j^e) - \underline{G}_j(\nu_j))$$

$$\leq \bar{h} \sum_{j \in J} \nu_j(W_j(\nu_j^e) - W_j(\nu_j))\mathbf{1}\{\nu_j \leq \nu_j^e\}$$

$$+ (\bar{h} - \underline{h}) \sum_{j \in J} \nu_j W_j(\nu_j)$$

$$\leq \quad \bar{h}\sigma \sum_{j\in J} \nu_j^e W_j(\nu_j^e) + (\bar{h} - \underline{h}) \sum_{j\in J} \nu_j W_j(\nu_j)$$

$$\leq \quad (\bar{h}/\underline{h})\sigma C(\boldsymbol{\nu}^e) + ((\bar{h} - \underline{h})/\underline{h})C(\boldsymbol{\nu}) \ ,$$

from which it follows that

$$(1 - \tau\sigma)C(\boldsymbol{\nu}^e) \leq \tau C(\boldsymbol{\nu}) \ .$$

Since this inequality holds for all $\boldsymbol{\nu} \in \mathcal{N}$, it holds in particular for the socially optimal vector, $\boldsymbol{\nu}^s$. Thus

$$\frac{C(\boldsymbol{\nu}^e)}{C(\boldsymbol{\nu}^s)} \leq \frac{\tau}{1 - \tau\sigma} \ .$$

∎

The techniques discussed in Section 7.4.5 of Chapter 7 for deriving the constant σ can be used here in the context of the model with linear waiting costs. Of course, the upper bound in Theorem 8.4 only applies if $\sigma < \tau^{-1}$. If the class waiting-cost coefficients, h_i, $i \in M$, are very heterogeneous – that is, if τ is significantly larger than one – then σ must be significantly less than one. Obviously, the bound is most useful when the classes are not very homogeneous.

8.2 Fixed Routes: Optimal Solutions

Now suppose that for each class i there is a single fixed route used by all jobs of that class. For economy of notation we shall use the index i to denote both the class and its associated fixed route. In particular, we shall write $j \in i$ when facility j lies on the route used by all class-i jobs. Now the only decision to be made for each class i is to choose its arrival rate, λ_i.

The total flow at facility j is now given by

$$\nu_j = \sum_{i:j\in i} \lambda_i \ , \ j \in J \ . \tag{8.29}$$

Given the total flows, ν_j, at the various facilities, the total waiting cost incurred by each class-i job is now given by

$$\sum_{j:j\in i} G_{ij}(\nu_j) \ .$$

The *full price*, π_i, for class i is therefore

$$\pi_i = \xi_i + \sum_{j:j\in i} (\delta_j + G_{ij}(\nu_j)) \ .$$

Since this model is a special case of the general model introduced in the previous section, the results there concerning existence and uniqueness of individually optimal and class optimal solutions still apply.

Remark 3 An alternative approach to this model starts from the multiclass model of Chapter 4, replacing the single service facility is replaced by a network of facilities. Much of the analysis and many of the results from Chapter 4 apply to this more general setting. In fact, by appropriately defining a total waiting cost function, $\hat{G}_i(\boldsymbol{\lambda})$, and a total toll, $\hat{\delta}_i$, for class i, we can regard the network model as a special case of the general multiclass, single-facility model of Section 4.1 of Chapter 4.

Specifically, let the total waiting-cost function and toll for class i be defined by ($i \in M$)

$$\hat{G}_i(\boldsymbol{\lambda}) \quad := \quad \sum_{j:j\in i} G_{ij}(\nu_j) \,, \tag{8.30}$$

$$\hat{\delta}_i \quad := \quad \xi_i + \sum_{j:j\in i} \delta_j \,, \tag{8.31}$$

where $\boldsymbol{\lambda} = (\lambda_i, i \in M)$ and $\nu_j = \sum_{i:j\in i} \lambda_i, \, j \in J$. Then the full price for each entering job of class i is

$$\pi_i = \hat{\delta}_i + \hat{G}_i(\boldsymbol{\lambda}) \,,$$

which takes the same form as the full price for a general multiclass, single-facility model as defined in Section 4.1 of Chapter 4. In effect, we are treating the network as a kind of "super" facility. We can do this because the single-facility model of Section 4.1 of Chapter 4 allows the waiting cost for each class to depend on the entire arrival-rate vector, $\boldsymbol{\lambda}$.

8.2.1 Individually Optimal Arrival Rates

Let $\boldsymbol{\lambda}^e = (\lambda_i^e, i \in M)$ denote the individually optimal vector of arrival rates. The equilibrium conditions satisfied by $\boldsymbol{\lambda}^e$ are:

$$U_i'(\lambda_i) \quad \leq \quad \xi_i + \sum_{j:j\in i} (\delta_j + G_{ij}(\nu_j)) \,, \tag{8.32}$$

$$U_i'(\lambda_i) \quad = \quad \xi_i + \sum_{j:j\in i} (\delta_j + G_{ij}(\nu_j)) \,, \text{ if } \lambda_i > 0 \,, \tag{8.33}$$

$$\nu_j \quad = \quad \sum_{i:j\in i} \lambda_i \,, \tag{8.34}$$

$$\lambda_i \quad \geq \quad 0 \,, \, i \in M \,. \tag{8.35}$$

As remarked above, these conditions have a unique solution, provided the assumptions of Theorem 8.1 hold, since the model with fixed routes is a special case of the general model.

8.2.2 Class-Optimal Arrival Rates

The class manager for class i is concerned with maximizing the net benefit, $B_i(\boldsymbol{\lambda})$, received per unit time by jobs of class i, where

$$B_i(\boldsymbol{\lambda}) = U_i(\lambda_i) - \lambda_i \left(\xi_i + \sum_{j:j\in i} (\delta_j + G_{ij}(\nu_j)) \right) , \qquad (8.36)$$

$i \in M$, with $\nu_j = \sum_{k:j\in k} \lambda_k$, $j \in J$.

The following necessary *KKT* conditions must hold simultaneously for all $i \in M$ in order for an allocation $\boldsymbol{\lambda} = (\lambda_i, i \in M)$ to be class optimal.

$$U_i'(\lambda_i) \leq \xi_i + \sum_{j:j\in i} (\delta_j + G_{ij}(\nu_j)) + \lambda_i \sum_{j:j\in i} G_{ij}'(\nu_j) , \qquad (8.37)$$

$$U_i'(\lambda_i) = \xi_i + \sum_{j:j\in i} (\delta_j + G_{ij}(\nu_j)) + \lambda_i \sum_{j:j\in i} G_{ij}'(\nu_j) , \text{ if } \lambda_i > 0 , \ (8.38)$$

$$\nu_j = \sum_{k:j\in k} \lambda_k , \ j \in J , \qquad (8.39)$$

$$\lambda_i \geq 0 , \ i \in M . \qquad (8.40)$$

Under the assumptions of Theorem 8.2, these conditions have a unique solution.

8.2.3 Socially Optimal Arrival Rates

Consider a flow allocation, $\boldsymbol{\lambda} = (\lambda_i, i \in M)$. The net benefit per unit time for class i is

$$B_i(\boldsymbol{\lambda}) = U_i(\lambda_i) - \lambda_i \sum_{j:j\in i} G_{ij}(\nu_j) . \qquad (8.41)$$

A *socially optimal* allocation of arrival rates $\boldsymbol{\lambda}^s = (\lambda_i^s, i \in M)$ to the classes is defined as one that maximizes the aggregate net benefit to all classes. It may thus be found by solving the following optimization problem:

$$\max \quad \sum_{i\in M} U_i(\lambda_i) - \sum_{i\in M} \lambda_i \sum_{j:j\in i} G_{ij}(\nu_j)$$

$$\textbf{(P)} \qquad \text{s.t.} \quad \nu_j = \sum_{i:j\in i} \lambda_i , \ j \in J ,$$

$$\lambda_i \geq 0 , \ i \in M .$$

A socially optimal allocation of arrival rates $\boldsymbol{\lambda}^s = (\lambda_i^s, i \in M)$ must satisfy the following *KKT* conditions, which are necessary for an optimal solution to Problem **(P)**:

$$U_i'(\lambda_i) \leq \sum_{j:j\in i} (G_{ij}(\nu_j) + \delta_j) , \qquad (8.42)$$

$$U_i'(\lambda_i) \;=\; \sum_{j:j\in i} \left(G_{ij}(\nu_j) + \delta_j\right) \,, \text{ if } \lambda_i > 0 \,, \tag{8.43}$$

$$\nu_j \;=\; \sum_{i:j\in i} \lambda_i \,, \tag{8.44}$$

$$\delta_j \;=\; \sum_{k:j\in k} \lambda_k G_{kj}'(\nu_j) \,, \tag{8.45}$$

$$\lambda_i \;\geq\; 0 \,, \; i \in M \,. \tag{8.46}$$

Although the model with fixed routes has a simpler structure than the general model, the *KKT* conditions may still not be sufficient for a global maximum of the objective function in Problem **(P)**. Indeed, we have already shown that the total waiting-cost function,

$$\sum_{i\in M} \lambda_i \sum_{j:j\in i} G_{ij}(\nu_j) \,,$$

is not in general jointly convex in $\boldsymbol{\lambda} = (\lambda_i, i \in M)$, in spite of the convexity properties of its components, even in the special case of a single-facility model with waiting costs dependent only on total flow at the facility (see Section 4.4.3 of Chapter 4). Hence the objective function for Problem **(P)** may not be jointly concave in $\boldsymbol{\lambda}$ and the first-order *KKT* conditions may not be sufficient for a socially optimal allocation.

8.2.4 Facility-Optimal Arrival Rates

Now we consider the system from the point of view of revenue-maximizing facility operators who choose the tolls, δ_j, at the facilities $j \in J$, and the tolls, ξ_i, for each entering customer of class i, $i \in M$.

Again we consider only the case of cooperating facilities.

8.2.4.1 Cooperating Facilities

As in the general model, the facility optimization problem for cooperating facilities may be solved by first solving an arrival-rate optimization problem. This problem now takes the form:

$$\max_{\{\lambda_i, i\in M; \nu_j, j\in J\}} \quad \sum_{i\in M} \lambda_i U_i'(\lambda_i) - \sum_{i\in M}\sum_{j:j\in i} G_{ij}(\nu_j)$$

(F')

$$\text{s.t.} \quad \nu_j = \sum_{i:j\in i} \lambda_i \,, \; j \in J \,,$$

$$\lambda_i \geq 0 \,, \; i \in M \,.$$

Let $\boldsymbol{\lambda}^f = (\lambda_i^f, i \in M)$ denote an optimal solution to Problem **(F')**. The vector $\boldsymbol{\lambda}^f$ satisfies the following necessary *KKT* conditions for optimality:

$$U_i'(\lambda_i) + \lambda_i U_i''(\lambda_i) \;\leq\; \sum_{j:j\in i} \left(G_{ij}(\nu_j) + \sum_{k:j\in k} \lambda_k G_{kj}'(\nu_j) \right) \,,$$

$$i \in M \ , \ r \in R_i \ , \qquad\qquad (8.47)$$

$$U_i'(\lambda_i) + \lambda_i U_i''(\lambda_i) \ = \ \sum_{j:j\in r} \left(G_{ij}(\nu_j) + \sum_{k:j\in k} \lambda_k G_{kj}'(\nu_j) \right) ,$$

$$\text{if } \lambda_i > 0 \ , \ i \in M \ , \qquad (8.48)$$

$$\nu_j \ = \ \sum_{i:i\in j} \lambda_i \ , \ j \in J \ , \qquad\qquad (8.49)$$

$$\lambda_i \ \geq \ 0 \ , \ i \in M \ . \qquad\qquad\qquad (8.50)$$

The facility optimal tolls are given by

$$\xi_i^f \ = \ -\lambda_i^f U''(\lambda_i^f) \ , \ i \in M \ ,$$

$$\delta_j^f \ = \ \sum_{k:j\in k} \lambda_k^f G_{kj}'(\nu_j^f) \ , \ j \in J \ .$$

Again the facility tolls coincide with the socially optimal facility tolls and the class tolls are the same as for the single-facility multiclass model (see Chapter 6). And again the KKT conditions may have multiple solutions, both because the total waiting-cost function, $\lambda_i \sum_{j:j\in i} G_{ij}(\nu_j)$, may not be jointly convex in $\boldsymbol{\lambda}$, and because $\lambda_i U'(\lambda_i)$ may not be concave in λ_i, $i \in M$.

8.3 Fixed Routes: Dynamic Adaptive Algorithms

We now turn our attention to the application of dynamic adaptive algorithms to a multiclass network with fixed routes. The extension of these algorithms from the multiclass, single-facility model of Chapter 4 is straightforward. We consider continuous-time algorithms only.

8.3.1 Continuous-Time Dynamic Algorithms

Suppose the arrival rates and tolls evolve in continuous time. For each $i \in M$, let $\lambda_i(t)$ denote the arrival rate for class i at time t, $t \geq 0$. The vector of arrival rates at time t is $\boldsymbol{\lambda}(t) = (\lambda_i(t), i \in M)$. Let $\delta_j(t)$ denote the toll charged at facility j at time t, $j \in J$. Let $\xi_i(t)$ denote the toll charged at time t to each entering class-i customer, $i \in M$. Let

$$\nu_j(t) = \sum_{j:j\in i} \lambda_i(t) \ , \ t \geq 0 \ ,$$

the total flow at facility j at time t. Let ω_i be a positive constant. Consider the system of differential equations ($i \in M$),

$$\frac{d}{dt}\lambda_i(t) = \omega_i \left(U_i'(\lambda_i(t)) - [\xi_i(t) + \sum_{j:j\in i} (G_{ij}(\nu_j(t)) + \delta_j(t))] \right) . \qquad (8.51)$$

As usual, if the current value of $\lambda_i(t)$ is zero and the r.h.s. of the differential equation is negative, it is understood that the algorithm keeps the flow at zero rather than reducing it further.

Define

$$\hat{\delta}_i(t) := \xi_i(t) + \sum_{j:j\in i} \delta_j(t) \ , \ t \geq 0 \ , \ i \in M \ ,$$

so that $\hat{\delta}_i(t)$ is the total toll charged to an entering class-i customer at time t. Using the definition (8.30), we have

$$\hat{G}_i(\boldsymbol{\lambda}(t)) = \sum_{j:j\in i} G_{ij}(\boldsymbol{\lambda}(t)) \ ; \ i \in M \ , \ t \geq 0 \ .$$

That is, $\hat{G}_i(\boldsymbol{\lambda}(t))$ is the total waiting cost incurred by a class-i customer who enters at time t. As we shall see presently, we can directly apply the results for the continuous-time dynamic algorithm developed in Section 4.3.1 of Chapter 4 to this algorithm for a multiclass network with fixed routes, by simply replacing $\delta_i(t)$ and $G_i(\boldsymbol{\lambda}(t))$ by $\hat{\delta}_i(t)$ and $\hat{G}_i(\boldsymbol{\lambda}(t))$, respectively.

As in the single-facility case, the specific behavior of the algorithm depends on how we choose the tolls – in this case, $\delta_j(t)$, $j \in J$, and $\xi_i(t)$, $i \in M$. If $\delta_j(t) = \delta_j$, $j \in J$, and $\xi_i(t) = \xi$, $i \in M$ (fixed tolls), then the algorithm converges to the individually optimal arrival rate vector, $\boldsymbol{\lambda}^e$, associated with the fixed tolls, δ_j, $j \in J$, and ξ, $i \in M$. On the other hand, if we choose the tolls equal to the appropriate (time-varying) values for class optimality, social optimality, or facility optimality, then the algorithm converges (not surprisingly) to a class-optimal, socially optimal, or facility optimal vector of arrival rates, respectively. These assertions are proved in the following four sections.

8.3.1.1 Algorithm with Fixed Toll: Individual Optimality

Consider the dynamical system governed by the differential equation (8.51), and suppose that the toll charged each job processed at facility j is a fixed quantity, δ_j, $j \in J$, and the toll charged each entering job of class i is a fixed quantity, ξ_i, $i \in M$. The system of differential equations is now

$$\frac{d}{dt}\lambda_i(t) = \omega_i \left(U_i'(\lambda_i(t)) - [\xi_i + \sum_{j:j\in i}(G_{ij}(\nu_j(t)) + \delta_j)] \right) \ , \ i \in M \ , \quad (8.52)$$

where $\nu_j(t) = \sum_{k:j\in k} \lambda_k(t)$, $j \in J$.

A stationary point $\boldsymbol{\lambda}$ of this system, at which the r.h.s. of each of the differential equations (8.51) equals zero (or is less than or equal to zero, if $\lambda_i = 0$), satisfies (8.32)–(8.35), and is therefore a Nash equilibrium for individually optimizing customers when entering customers of class i are charged the toll $\hat{\delta}_i = \xi_i + \sum_{j:j\in i}\delta_j$, $i \in M$.

The following theorem is an application of Theorem 4.4 of Chapter 4.

Theorem 8.5 *Under Assumption 1, the dynamic algorithm (8.52) converges*

*globally to the unique solution to the equilibrium conditions (8.32)–(8.35),
which characterize the vector $\boldsymbol{\lambda}^e$ of individually optimal arrival rates.*

8.3.1.2 Algorithm with Time-Varying Toll: Class Optimality

Let δ_j, $j \in J$, and ξ_i, $i \in M$, be given, fixed tolls. Consider the dynamical system governed by the differential equation (8.51), and suppose that each entering class-i job is charged the fixed toll, ξ_i, and each class-i job processed at facility j is charged a time-varying toll

$$\delta_j(t) = G'_{ij}(\nu_j(t)) + \delta_j ,$$

$i \in M$, $j \in J$. The system of differential equations is now ($i \in M$)

$$\frac{d}{dt}\lambda_i(t) = \omega_i \left(U'_i(\lambda_i(t)) - \xi_i - \sum_{j:j\in i} \left[G'_{ij}(\nu_j(t)) + \delta_j \right] \right) . \qquad (8.53)$$

Recall that by Theorem 4.2 there is a unique class-optimal solution under Assumptions 2 and 3, which satisfies the equilibrium conditions, (8.37)–(8.40).

Using the definitions of $\hat{\delta}_i(t)$ and $\hat{G}_i(\boldsymbol{\lambda}(t))$, we can apply Theorem 4.5 of Chapter 4 to yield the following theorem:

Theorem 8.6 *Under Assumptions 2 and 3 the dynamic algorithm (8.51) converges globally to the unique solution to the equilibrium conditions (8.37)–(8.40), which characterize the vector $\boldsymbol{\lambda}^c$ of class optimal arrival rates.*

8.3.1.3 Algorithm with Time-Varying Toll: Social Optimality

Consider the dynamical system governed by the differential equation (8.51), and suppose now that each job processed at facility j is charged a time-varying toll,

$$\delta_j(t) = \sum_{k:j\in k} \lambda_k(t) G'_{kj}(\nu_j(t)) , \qquad (8.54)$$

$j \in J$, and the toll, $\xi_i(t)$, charged each entering customer of class i is zero, $i \in M$. The system of differential equations is now ($i \in M$)

$$\frac{d}{dt}\lambda_i(t) = \omega_i \left(U'_i(\lambda_i(t)) - \sum_{j:j\in i} [G_{ij}(\nu_j(t)) + \sum_{k:j\in k} \lambda_k(t) G'_{kj}(\nu_j(t))] \right) , \quad (8.55)$$

where $\nu_j(t) = \sum_{k:j\in k} \lambda_k(t)$.

In this version of the dynamic algorithm, the total class-i toll, $\hat{\xi}_i(t) = \sum_{j:j\in i} \delta_j(t)$, is the overall external effect of class-i flow: the marginal increase in the total waiting cost of all existing flows in all classes at all facilities $j \in i$ as a result of a marginal increase in λ_i.

The stability analysis of the algorithm in this case parallels that for social optimality in the multiclass, single-facility model (cf. Section 4.3.1.3 of Chapter 4). Using the definition, $\hat{G}_i(\boldsymbol{\lambda}) = \sum_{j:j\in i} G_{ij}(\nu_j)$, for the total class-$i$

waiting cost, we can write the objective function for social optimality in the multiclass network model with fixed routes as

$$\mathcal{U}(\boldsymbol{\lambda}) = \sum_{i \in M} U_i(\lambda_i) - \sum_{i \in M} \lambda_i \hat{G}_i(\boldsymbol{\lambda}) \ .$$

Consequently, it follows from the analysis in Section 4.3.1.3 of Chapter 4 that the algorithm converges (with strictly increasing aggregate utility) to a solution to the KKT conditions. We have observed that there may in general be several such solutions, each of which is a Nash equilibrium for individually optimizing customers when charged the appropriate tolls. Which equilibrium is approached depends on the starting point.

8.3.1.4 Algorithm with Time-Varying Toll: Facility Optimality

Consider the dynamical system governed by the differential equation (8.51), and suppose now that each job processed at facility j is charged a time-varying toll,

$$\delta_j(t) = \sum_{k:j \in k} \lambda_k(t) G'_{kj}(\nu_j(t)) \ , \tag{8.56}$$

$j \in J$, where $\nu_j(t) = \sum_{k:j \in k} \lambda_k(t)$. In addition, each class-i job that enters the system is charged a time-varying toll,

$$\xi_i(t) = -\lambda_i(t) U''(\lambda_i(t)) \ , \ i \in M \ . \tag{8.57}$$

The system of differential equations is now ($i \in M$)

$$\frac{d}{dt}\lambda_i(t) \quad = \quad \omega_i(U'_i(\lambda_i(t)) - \lambda_i(t)U''_i(\lambda_i(t))$$

$$- \sum_{j:j \in i}[G_{ij}(\nu_j(t)) + \sum_{k:j \in k}\lambda_k(t)G'_{kj}(\nu_j(t))]) \ . \tag{8.58}$$

At a positive stationary point of this system, we have

$$U'_i(\lambda_i) + \lambda_i U''_i(\lambda_i) - \sum_{j:j \in i}[G_{ij}(\nu_j) + \sum_{k:j \in k}\lambda_k G'_{kj}(\nu_j)] = 0 \ , \tag{8.59}$$

which are the necessary KKT conditions for $\boldsymbol{\lambda} = (\lambda_i, i \in M)$ to be be facility optimal. By an argument similar that used in Chapter 4 for the single-facility multiclass problem (using the equivalence between the facility optimization problem and a social optimization problem with modified utility functions, $\tilde{U}_i(\lambda_i) := \lambda_i U'_i(\lambda_i)$), one can show that the dynamic algorithm defined by the differential equations (8.58) converges (with strictly increasing aggregate utility) to a solution to the KKT conditions for a facility optimal allocation. However, since the modified utility function, $\tilde{U}(\boldsymbol{\lambda})$, need not be concave (as we observed) the solution to (3.39) to which the algorithm converges need not be *globally* optimal for the facility optimization problem, even if the waiting cost function is concave in $\boldsymbol{\lambda}$ (which, as we have already observed, may need not be the case).

8.4 Fixed Routes: Homogeneous Waiting Costs

In this section we consider the model with a fixed route for each class in the special case of homogeneous waiting costs. We concentrate on individual and social optimization.

Suppose the average waiting cost incurred by a job at facility j is $G_j(\nu_j)$, regardless of class, for all $j \in J$. For given values of the facility tolls, δ_j, $j \in J$, and class tolls, ξ_i, $i \in M$, the equilibrium conditions for an individually optimal allocation are

$$U_i'(\lambda_i) \leq \xi_i + \sum_{j:j\in i}(\delta_j + G_j(\nu_j)) \, , \, i \in M \, ,$$

$$U_i'(\lambda_i) = \xi_i + \sum_{j:j\in i}(\delta_j + G_j(\nu_j)) \, , \, \text{if } \lambda_i > 0 \, , \, i \in M \, ,$$

$$\nu_j = \sum_{i:j\in i}\lambda_i \, , \, j \in J \, ,$$

$$\lambda_i \geq 0 \, , \, i \in M \, .$$

The objective function for social optimization problem becomes

$$\mathcal{U}(\boldsymbol{\lambda}) = \sum_{i\in M}U_i(\lambda_i) - \sum_{i\in M}\lambda_i\sum_{j:j\in i}G_j(\nu_j)$$

$$= \sum_{i\in M}U_i(\lambda_i) - \sum_{j\in J}\left(\sum_{i:j\in i}\lambda_i\right)G_j(\nu_j)$$

$$= \sum_{i\in M}U_i(\lambda_i) - \sum_{j\in J}\nu_j G_j(\nu_j) \, ,$$

where

$$\nu_j = \sum_{i:j\in i}\lambda_i \, , \, j \in J \, .$$

Now we make the following assumption.

Assumption 4   For all $j \in J$, the function $H_j(\nu_j) := \nu_j G_j(\nu_j)$ is convex in $\nu_j \geq 0$.

Since ν_j is a linear function of $\boldsymbol{\lambda}$, it follows from this assumption that the total waiting-cost function,

$$\hat{G}_i(\boldsymbol{\lambda}) = \sum_{j\in J}\nu_j G_j(\nu_j) \, ,$$

is jointly convex in $\boldsymbol{\lambda}$ and hence $\mathcal{U}(\boldsymbol{\lambda})$ is jointly concave in $\boldsymbol{\lambda}$.

Note that Assumption 4 is satisfied in many queueing models. For example, when waiting costs at each facility are linear (with h_j = waiting cost per job per unit time at facility j) and facility j is modeled as a *FIFO M/M/1* queue with service rate μ_j, operating in steady state, we have

$$G_j(\nu_j) = \frac{1}{\mu_j - \nu_j} \ , \ 0 \le \nu_j < \mu_j \ ,$$

with $G_j(\nu_j) = \infty$ for $\nu_j \ge \mu_j$. In this case $G_j(\cdot)$ is convex and nondecreasing and therefore $H_j(\cdot)$ is convex.

Thus, when the waiting costs at each facility are homogeneous across the classes, the objective function for social optimization is jointly concave and therefore has a unique global maximum, which is the unique solution to the KKT conditions

$$U_i'(\lambda_i) \ \le \ \sum_{j:j\in i} (G_j(\nu_j) + \delta_j)$$

$$U_i'(\lambda_i) \ = \ \sum_{j:j\in i} (G_j(\nu_j) + \delta_j) \ , \text{ if } \lambda_i > 0 \ ,$$

where $(j \in J)$

$$\nu_j \ = \ \sum_{i:j\in i} \lambda_i \ ,$$

$$\delta_j \ = \ \sum_{k:j\in k} \lambda_k G_j'(\nu_j) \ .$$

8.4.1 Continuous-Time Dynamic Algorithm

The differential equations for social optimization now take the form $(i \in M)$:

$$\frac{d}{dt}\lambda_i(t) = \omega_i \left(U_i'(\lambda_i(t)) - \sum_{j:j\in i} [G_j(\nu_j(t)) + \sum_{k:j\in k} \lambda_k(t)G_j'(\nu_j(t))] \right) \ , \quad (8.60)$$

where $\nu_j(t) = \sum_{k:j\in k} \lambda_k(t)$.

Since the objective function for social optimization is now concave in $\boldsymbol{\lambda}$, the KKT conditions have a unique solution, which is the unique global maximum. It follows that the algorithm converges (with strictly increasing aggregate utility) to this unique solution.

8.5 Variable Routes: Homogeneous Waiting Costs

In this section we consider the model with a choice of routes for each class in the special case of homogeneous waiting costs. Again we concentrate on individual and social optimization.

Suppose the average waiting cost incurred by a job at facility j is $G_j(\nu_j)$, regardless of class, for all $j \in J$. For given values of the facility tolls, δ_j, $j \in J$, and class tolls, ξ_i, $i \in M$, the equilibrium conditions for an individually optimal allocation are:

$$U_i'(\lambda_i) \leq \xi_i + \sum_{j:j\in r}[\delta_j + G_j(\nu_j)] \ , \ i \in M \ , \ r \in R_i \ , \ , \tag{8.61}$$

$$U_i'(\lambda_i) = \xi_i + \sum_{j:j\in r}[\delta_j + G_j(\nu_j)] \ , \ \text{if } \lambda_{ir} > 0 \ , \ i \in M \ , \ r \in R_i \ , \ , \tag{8.62}$$

$$\nu_j = \sum_{i\in M}\sum_{r\in R_i:j\in r}\lambda_{ir} \ , \ j \in J \ , \tag{8.63}$$

$$\lambda_i = \sum_{r\in R_i}\lambda_{ir} \ , \ i \in M \ , \tag{8.64}$$

$$\lambda_{ir} \geq 0 \ , \ i \in M \ , \ r \in R_i \ . \tag{8.65}$$

The objective function for social optimization problem becomes

$$\begin{aligned}
\mathcal{U}(\boldsymbol{\lambda}) &= \sum_{i\in M}U_i(\lambda_i) - \sum_{i\in M}\sum_{r\in R_i}\lambda_{ir}\sum_{j:j\in r}G_j(\nu_j) \\
&= \sum_{i\in M}U_i(\lambda_i) - \sum_{j\in J}(\sum_{i\in M}\sum_{r\in R_i:j\in r}\lambda_{ir})G_j(\nu_j) \\
&= \sum_{i\in M}U_i(\lambda_i) - \sum_{j\in J}\nu_j G_j(\nu_j) \ .
\end{aligned}$$

Thus the social optimization problem can be written as:

$$\max_{\{\lambda_i,\lambda_{ir},r\in R_i;\nu_j,j\in J\}} \quad \mathcal{U}(\boldsymbol{\lambda}) = \sum_{i\in M}U_i(\lambda_i) - \sum_{j\in J}\nu_j G_j(\nu_j)$$

$$\text{s.t.} \quad \nu_j = \sum_{i\in M}\sum_{r\in R_i:j\in r}\lambda_{ir} \ , \ j \in J \ ,$$

$$\lambda_i = \sum_{r\in R_i}\lambda_{ir} \ , \ i \in M \ ,$$

$$\lambda_{ir} \geq 0 \ , \ i \in M \ , \ r \in R_i \ .$$

As in the case of a single-class network, let us make the following assumption.

Assumption 5 For all $j \in J$, the function $H_j(\nu_j) := \nu_j G_j(\nu_j)$ is convex in $\nu_j \geq 0$.

Since ν_j is a linear function of $\boldsymbol{\lambda} = (\lambda_{ir}, i \in M, r \in R_i)$, it follows from this assumption that the total waiting cost per unit, $\sum_{j\in J}\nu_j G_j(\nu_j)$, is jointly convex in $\boldsymbol{\lambda}$ and hence $\mathcal{U}(\boldsymbol{\lambda})$ is jointly concave in $\boldsymbol{\lambda}$.

Recall that Assumption 5 is satisfied in many queueing models. For example, when waiting costs at each facility are linear (with $h = $ waiting cost per job per unit time in the system) and facility j is modeled as a *FIFO M/M/1* queue with service rate μ_j, operating in steady state, we have

$$G_j(\nu_j) = \frac{h}{\mu_j - \nu_j} \ , \ 0 \leq \nu_j < \mu_j \ ,$$

with $G_j(\nu_j) = \infty$ for $\nu_j \geq \mu_j$. In this case $G_j(\cdot)$ is convex and nondecreasing and therefore $H_j(\cdot)$ is convex.

Thus, when the waiting costs at each facility are homogeneous across the classes, the objective function for social optimization is jointly concave and therefore has a unique global maximum, which is the unique solution to the *KKT* conditions

$$U_i'(\lambda_i) \leq \sum_{j:j\in r} (G_j(\nu_j) + \delta_j) \,, \, i \in M \,,\, r \in R_i \,, \tag{8.66}$$

$$U_i'(\lambda_i) = \sum_{j:j\in r} (G_j(\nu_j) + \delta_j) \,, \text{ if } \lambda_{ir} > 0 \,,\, i \in M \,,\, r \in R_i \,, \tag{8.67}$$

$$\nu_j = \sum_{i\in M}\sum_{r\in R_i:j\in r} \lambda_{ir} \,,\, j \in J \,, \tag{8.68}$$

$$\lambda_i = \sum_{r\in R_i} \lambda_{ir} \,,\, i \in M \,, \tag{8.69}$$

$$\delta_j = \nu_j G_j'(\nu_j) \,,\, j \in S \,, \tag{8.70}$$

$$\lambda_{ir} \geq 0 \,,\, i \in M \,,\, r \in R_i \,. \tag{8.71}$$

8.5.1 Price of Anarchy

We now show how the results concerning the Price of Anarchy for multiclass networks (cf. Section 8.1.5.1) simplify in the special case of homogeneous waiting costs. In fact, as we shall see, the results from Chapter 7 for a single-class network extend directly to a multiclass network with homogeneous waiting costs.

As in the case of a single-class network, we begin with the formulation of the social optimization problem with fixed class arrival rates, λ_i, $i \in M$, as a constrained minimization problem:

$$\min_{\{\lambda_r, r\in R_i, i\in M; \nu_j, j\in J\}} \sum_{j\in J} \nu_j G_j(\nu_j)$$

$$\text{s.t.} \quad \sum_{r\in R_i} \lambda_{ir} = \lambda_i \,,\, i \in M \,, \tag{8.72}$$

$$\nu_j = \sum_{i\in M}\sum_{r\in R_i:j\in r} \lambda_{ir} \,,\, j \in J \,, \tag{8.73}$$

$$\lambda_{ir} \geq 0 \,,\, i \in M \,,\, r \in R_i \,. \tag{8.74}$$

Let $\nu := (\nu_j, j \in J)$ and define the feasible set $\mathcal{N}$ as the set of all ν for which there exists a $\lambda = (\lambda_{ir}, i \in M, r \in R_i)$ such that ν and λ satisfy the constraints, (8.72), (8.73), and (8.74). Then the social optimization problem may be rewritten with decision variables, $\nu = (\nu_j, j \in J)$, as follows:

$$\min_{\{\nu\in\mathcal{N}\}} C(\nu) := \sum_{j\in J} \nu_j G_j(\nu_j) \,.$$

We now have the social optimization problem in the same format as in the case of a single-class network (cf. Section 7.4.5 of Chapter 7). Let $\boldsymbol{\nu}^s = (\nu_j^s, j \in J)$ denote the socially optimal vector of facility flow rates.

Now let $\boldsymbol{\nu}^e = (\nu_j^e, j \in J)$ denote the vector of facility flow rates corresponding to an individually optimal allocation, $\boldsymbol{\lambda}^e = (\lambda_{ir}^e, i \in M, r \in R_i)$, of flow rates of the various classes on the various routes. The same argument as used in the proof of Lemma 8.3 for the general multiclass model leads in this special case to the following lemma:

Lemma 8.7 *An individually optimal vector of facility flow rates, $\boldsymbol{\nu}^e$, satisfies the variational inequality,*

$$\sum_{j \in J} (\nu_j^e - \nu_j) G_j(\nu_j^e) \le 0 ,$$

for all $\boldsymbol{\nu} = (\boldsymbol{\nu}_j, j \in J) \in \mathcal{N}$.

Note that this variational inequality is simpler than the one for the general multiclass model. Indeed, it is formally identical to the variational inequality derived in Chapter 7 for a single-class network, although (as noted above) the definition of $\mathcal{N}$ is different here.

From this point we can use exactly the same arguments as used in Chapter 7 to establish the following theorem:

Theorem 8.8 *Suppose there exists a constant $\sigma < 1$ such that*

$$\max_{j \in J} \sup_{0 \le \nu_j \le \nu_j^e} \frac{\nu_j (G_j(\nu_j^e) - G_j(\nu_j))}{\nu_j^e G_j(\nu_j^e)} \le \sigma . \tag{8.75}$$

Then $C(\boldsymbol{\nu}^e) \le (1 - \sigma)^{-1} C(\boldsymbol{\nu}^s)$.

The techniques discussed in Section 7.4.5 of Chapter 7 for deriving the constant σ can be used in this setting as well.

8.6 Endnotes

Section 8.4.1

In a series of papers, [105], [109], [106], [107], [108], Kelly et al. propose a variant of this dynamic algorithm as a model for an adaptive pricing algorithm for control of traffic in a communication network. In their model, the facilities are links in the communication network between nodes, which serve as switches routers. The cost function, $G_j(\nu_j)$, may represent an explicit cost of waiting (as in our model), or it may perform the role of a penalty function designed to "shape" the traffic so that links do not become overloaded.

APPENDIX A

Scheduling a Single-Server Queue

A.1 Strong Conservation Laws

In this section we provide a brief overview of the theory of strong conservation laws, which provides a basis for the characterization of the achievable region for scheduling problems. (See [21] for more details.)

The abstract setting for the theory of strong conservation laws is a system consisting of m *activities* and a set Φ of *admissible scheduling rules*. A scheduling rule specifies which activity to engage in at each point in time, and this choice may depend on the current state and history (but not the future behavior) of the system. In applications to queueing systems, an activity will typically correspond to serving a particular class of customers. The definition of an admissible scheduling rule may involve some restrictions on the choice of activity at certain points in time. For example, in a queueing application we may or may not be allowed to interrupt a service until it is completed (preemption or nonpreemption). At each point in time, an activity may be either *on* or *off*, and we assume that this information is a part of the state description.

Among admissible scheduling rules, we are particularly interested in *strict priority rules*, which give strict preference to certain activities over others. More specifically, corresponding to each permutation, ψ, of $M = \{1, 2, \ldots, m\}$, there is a strict priority rule, $\phi(\psi)$, which at each point in time engages in the activity i with the smallest index $\psi(i)$ among all activities that are currently on. In other words, the class i with $\psi(i) = 1$ has priority over all other classes, followed by the class j with $\psi(j) = 2$, and so forth.

Associated with each admissible scheduling rule, $\phi \in \Phi$, there is a *performance measure*, x_i^ϕ, for each activity i. In our applications x_i^ϕ will typically be proportional to the average waiting time in the system or in the queue for a class-i customer. The performance vectors, $\boldsymbol{x}^\phi = (x_1^\phi, \ldots, x_m^\phi)$, associated with the admissible scheduling rules, $\phi \in \Phi$, are said to satisfy *strong conservation laws* if there is a constant, α_S, associated with each subset $S \subset M = \{1, 2, \ldots, m\}$ of activities such that

$$\sum_{j \in M} x_j^\phi = \alpha_M \text{ for all } \phi \in \Phi, \tag{A.1}$$

and for all $S \subset M$,

$$\sum_{j \in S} x_j^\phi \geq \alpha_S \text{ for all } \phi \in \Phi, \tag{A.2}$$

343

$$\sum_{j \in S} x_j^{\phi(\psi)} \;=\; \alpha_S \text{ for all } \psi \text{ such that } \{\psi_1, \psi_2, \ldots, \psi_{|S|}\} = S \,. \quad \text{(A.3)}$$

In words, the performance vectors from admissible scheduling strategies are said to satisfy strong conservation laws if the total performance over all classes is invariant under all admissible scheduling strategies and the total performance over the classes in any subset $S \subset M$ is minimized by any strict-priority rule giving priority to those jobs over jobs of classes in S^c. (We call such a rule an S-rule.)

The *achievable region* is denoted by $\mathcal{R}$ and is defined as follows:

$$\mathcal{R} = \{\boldsymbol{x} = (x_1, \ldots, x_m) : \sum_{j \in M} x_j = \alpha_M \,;\, \sum_{j \in S} x_j \geq \alpha_S \,,\, \text{for all } S \subset M\} \,.$$

$$\text{(A.4)}$$

Note that $\mathcal{R}$ is a convex polytope in m-dimensional Euclidean space, defined by a finite number of linear inequalities.

Thus we see that when conservation laws apply, the performance vector $\boldsymbol{x}^\phi$ from an admissible scheduling rule ϕ belongs to the achievable region $\mathcal{R}$. Moreover, it can be shown that the performance vectors of the strict-priority rules $\phi(\psi)$, corresponding to the permutations ψ of $\{1, 2, \ldots, m\}$, constitute the extreme points of $\mathcal{R}$.

The following theorem applies when strong conservation laws are satisfied.

Theorem A.1 *Given a set of performance vectors $\{x^\phi : \phi \in \Phi\}$ that satisfies strong conservation laws, the optimal solution to a problem of the form*

$$\min\left\{ \sum_{j=1}^{N} c_j x_j^\phi \,,\, \phi \in \Phi \right\} \quad \text{(A.5)}$$

is attained by the strict-priority rule that assigns priority to the job classes in the order of decreasing cost coefficients. That is, the strict-priority rule $\phi(\psi)$, where

$$c_{\psi_1} \geq c_{\psi_2} \geq \cdots \geq c_{\psi_N} \,, \quad \text{(A.6)}$$

is optimal over all $\phi \in \Phi$.

For a proof, see, e.g., [21].

A.2 Work-Conserving Scheduling Systems

Now we begin to apply this theory to queueing systems. We start with a $G/G/1$ queue, that is, a single-server system with general (not necessarily i.i.d.) interarrival and service times.

Definition A *work-conserving scheduling system* (*WCSS*) consists of
(i) a bivariate random sequence, $\boldsymbol{\Theta} = \{(\boldsymbol{A}^{(n)}, \boldsymbol{S}^{(n)}), n = 1, 2, \ldots\}$, with a given probability measure P, where $\boldsymbol{A}^{(n)}$ and $\boldsymbol{S}^{(n)}$ are the arrival instant and work requirement, respectively, of job n, $n \geq 1$ ($\boldsymbol{A}^{(0)} = 0$);
(ii) a single server who works (at unit rate) on one customer at a time;

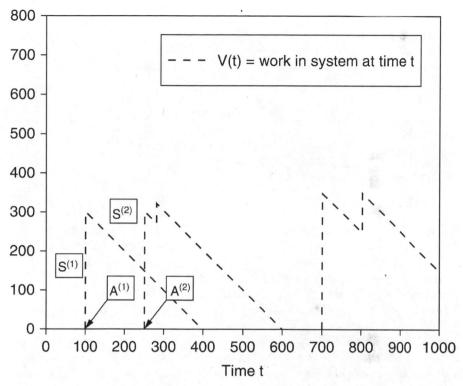

Figure A.1 *Graph of $V(t)$: Work in System*

(iii) a set of *nonanticipative* and *nonidling* scheduling rules, $\Phi = \{\phi\}$.

A scheduling rule is *nonanticipative* if the decision about which job to process at time t depends only on $\{(A^{(n)}, S^{(n)}), n = 1, 2, \ldots, A(t)\}$, where $A(t) = \max\{n : A^{(n)} \leq t\}$, and possibly on decisions taken before time t. It is *nonidling* if the server is always busy when there is at least one job in the system.

The scheduling rules in $\phi \in \Phi$ may be preemptive or nonpreemptive, the essential property being that the probability measure P governing Θ must be the same for all scheduling rules $\phi \in \Phi$. That is, the arrival instants and the work required by each job must be independent of the scheduling rule. (Note that, if a particular scheduling rule, $\phi \in \Phi$, is preemptive, then it must be preemptive resume in order for this requirement to be satisfied.)

The *work* in the system at time $t \geq 0$, denoted by $V(t)$, is defined as the sum of the remaining work requirements (service times) of all the customers in the system at time t. Figure A.1 illustrates the evolution of $V(t)$ over time for a particular realization of $\Theta = \{(A^{(n)}, S^{(n)}), n = 1, 2, \ldots\}$. At each arrival point, $A^{(n)}$, the graph of $V(t)$ takes an upward jump of magnitude $S^{(n)}$. Between these jumps, the graph of $V(t)$ decreases at unit rate, as the server performs work at unit rate, as long as $V(t) > 0$. The probability measure governing the evolution of $V(t)$ is therefore completely determined by the probability measure P governing Θ, which in turn does not depend on the scheduling rule. Hence $\{V(t), t \geq 0\}$ is invariant with respect to scheduling rules $\phi \in \Phi$, both in terms of its sample-path evolution and probabilistically. It follows that the limiting average,

$$\lim_{t \to \infty} t^{-1} \int_0^t V(s)ds ,$$

(when it exists) is also invariant. When $\{V(t), t \geq 0\}$ is ergodic, that is, when

$$\lim_{t \to \infty} t^{-1} \int_0^t V(s)ds = \mathrm{E}[V] , \text{ w.p.1} ,$$

for a suitably chosen random variable V, then it follows that the expectation $\mathrm{E}[V]$ is also invariant. For example, when the *WCSS* under study is a *GI/GI/1* queue (see the next section), the invariant expectation in question is the steady-state expected work in the system.

A customer (along with its associated work) passes through various states while in the system. When a customer enters the system, it enters the *primary queue* and remains there until it enters *service* for the first time. If preemption is allowed, then the customer may also spend time in the *secondary queue*, which consists of customers who are not currently in service but who have received some service and have been preempted at least once. (When preemption is not allowed, the secondary queue is always empty: $Y(t) = 0$, for all $t \geq 0$. In this case we shall refer to the primary queue simply as the *queue*.)

The work in the system can be classified similarly. We can write

$$V(t) = U(t) + Y(t) + Z(t) , t \geq 0 ,$$

where $U(t)$ is the *work in the primary queue* (the sum of the work requirements of all waiting customers who have not yet received any service), $Y(t)$ is the *work in the secondary queue* (the sum of the remaining work requirements of all waiting customers who have received some service), and $Z(t)$ is the *work in service* (the remaining work of the customer in service) at time $t \geq 0$. Furthermore, when $\{U(t), t \geq 0\}$, $\{Y(t), t \geq 0\}$, and $\{Z(t), t \geq 0\}$ are ergodic, so that

$$\lim_{t \to \infty} t^{-1} \int_0^t U(s)ds = \mathrm{E}[U] , \text{ w.p.1} ,$$

$$\lim_{t \to \infty} t^{-1} \int_0^t Y(s)ds = \mathrm{E}[Y] , \text{ w.p.1} ,$$

$$\lim_{t \to \infty} t^{-1} \int_0^t Z(s)ds = \mathrm{E}[Z] , \text{ w.p.1} ,$$

then we have

$$\mathrm{E}[V] = \mathrm{E}[U] + \mathrm{E}[Y] + \mathrm{E}[Z] . \tag{A.7}$$

A.2.1 Example: GI/GI/1 *Queue*

Consider a *WCSS* in which the interarrival times $\{A^{(n)} - A^{(n-1)}, n \geq 1\}$ and work requirements $\{S^{(n)}, n \geq 1\}$ are mutually independent sequences of i.i.d. random variables (a *GI/GI/1* queue). Let A and S have the common distributions of $A^{(n)} - A^{(n-1)}$, $n \geq 1$, and $S^{(n)}$, $n \geq 1$, respectively. Let $\lambda := (\mathrm{E}[A])^{-1}$ (the arrival rate), $\mu := (\mathrm{E}[S])^{-1}$, and $\rho := \lambda/\mu$.

It follows from a standard sample-path argument and the strong law of large numbers (cf., e.g., Chapter 6 of El-Taha and Stidham [60]) that $\{Z(t), t \geq 0\}$ is ergodic, and (w.p.1)

$$\lim_{t \to \infty} t^{-1} \int_0^t Z(s) ds = \mathrm{E}[Z] = \lambda \mathrm{E}[S^2]/2 \ , \ \text{w.p.1} \ , \tag{A.8}$$

which is invariant with respect to the scheduling rule $\phi \in \Phi$.

Definition A scheduling rule ϕ is *regenerative* if the decision about which job to process next does not use any information from previous busy cycles and coincides with the decision that would be made in the first busy cycle, given the same information about previous arrival instants, work requirements, and decisions during the current busy cycle.

Examples of regenerative scheduling rules are the strict priority queue disciplines, and the first-in, first-out (*FIFO*), last-in, first-out (*LIFO*), and service-in-random-order (*SIRO*) queue disciplines.

Suppose that $\rho = \lambda \mathrm{E}[S] < 1$. Consider an arbitrary regenerative scheduling rule, $\phi \in \Phi$. The stochastic processes $\{U(t)\}$, $\{Y(t)\}$, and $\{V(t)\}$ are regenerative with respect to the beginnings of successive busy periods and the length of a busy cycle and the number of customers served in a busy cycle both have finite mean. Let U, Y, and V have the steady-state distributions of $U(t)$, $Y(t)$, and $V(t)$, respectively. We have the following theorem.

Theorem A.2 *Consider a* GI/GI/1 WCSS *in which the scheduling rules,* $\phi \in \Phi$, *are regenerative. Suppose* $\rho = \lambda \mathrm{E}[S] < 1$. *Then* $\{U(t), t \geq 0\}$, $\{Y(t), t \geq 0\}$, *and* $\{V(t), t \geq 0\}$ *are all ergodic, with*

$$\mathrm{E}[V] = \mathrm{E}[U + Y] + \lambda \mathrm{E}[S^2]/2 \ . \tag{A.9}$$

Both $\mathrm{E}[V]$ *and* $\mathrm{E}[U + Y]$ *are invariant with respect to the scheduling rule* $\phi \in \Phi$.

Proof   Invariance of $\mathrm{E}[V]$ follows from the fact that we have a *WCSS*. Equation (A.9) and the invariance of $\mathrm{E}[U + Y]$ then follow from (A.7) and (A.8).   ∎

Since the *FIFO* queue discipline is an example of a regenerative scheduling rule, we can assume a *FIFO* discipline for the purpose of evaluating the invariant expectations in the formula (A.9). Suppose then that scheduling rule is *FIFO* and let $D^{(n)}$ denote the delay (waiting time in the queue) of customer n, $n \geq 1$. Suppose $\rho < 1$. Then the stochastic process, $\{D^{(n)}\}$, is regenerative.

Let D denote the steady-state version of $D^{(n)}$. Since the scheduling rule is FIFO, $D^{(n)}$ and $S^{(n)}$ are independent. It follows by a sample-path argument (cf., e.g., Chapter 6 of El-Taha and Stidham [60]) that

$$E[V] = \lambda \left(E[S]E[D] + E[S^2]/2\right) = \rho E[D] + \lambda E[S^2]/2 . \qquad (A.10)$$

Note that, in formula (A.10), $E[D]$ is the expected delay associated specifically with the FIFO discipline. Although $E[V]$, $E[S]$, and $E[S^2]$ are invariant among all regenerative rules $\phi \in \Phi$, in general $E[D]$ is not invariant.

To use this formula to calculate the invariant quantity, $E[V]$, we must be able first to calculate the expected steady-state delay $E[D]$ under a FIFO queue discipline. We can do this when the arrival process is Poisson, as shown below.

A.2.1.1 M/GI/1 *Queue*

Recall that in the FIFO case, the work in the system, $V(t)$, is also the virtual delay (waiting time in the queue) at time t. Under the additional assumption that the arrivals are from a Poisson process with parameter λ, it follows from PASTA (cf. Chapter 6 of [60]) that the virtual and actual delays have the same distribution and hence $E[V] = E[D]$, which implies (using (A.10)) that

$$E[D] = \rho E[D] + \lambda E[S^2]/2 .$$

Therefore, for the FIFO queue discipline,

$$E[V] = E[D] = \frac{\lambda E[S^2]}{2(1 - \rho)} . \qquad (A.11)$$

(This is the *Pollaczek-Khintchine* formula.) Thus we have the following corollary of Theorem A.2.

Corollary A.3 *Consider an* M/GI/1 *WCSS in which the scheduling rules $\phi \in \Phi$ are regenerative. Suppose $\rho = \lambda E[S] < 1$. Then*

$$E[V] \quad = \quad \left(\frac{1}{1 - \rho}\right) \frac{\lambda E[S^2]}{2} , \qquad (A.12)$$

$$E[U + Y] \quad = \quad \left(\frac{\rho}{1 - \rho}\right) \frac{\lambda E[S^2]}{2} , \qquad (A.13)$$

for all $\phi \in \Phi$.

Note that, although the FIFO discipline is nonpreemptive, this result applies to *all* scheduling rules $\phi \in \Phi$, including rules that are nonpreemptive, preemptive-resume, or some combination of the two.

Returning to the general case of a GI/GI/1 *WCSS*, we now introduce the concept of a *service-time-independent* scheduling rule.

Definition A scheduling rule ϕ is *service-time independent* (*STI*) if the decision about which job to process next does not use any information about the work requirements of the jobs in the system.

Examples of service-time-independent scheduling rules are the first-in, first-out (*FIFO*), last-in, first-out (*LIFO*), and service-in-random-order (*SIRO*) queue disciplines.

Consider an arbitrary regenerative and *STI* rule, $\phi \in \Phi$. Let $\boldsymbol{D}^{(n)}$ denote the delay (waiting time in the primary queue), $n \geq 1$. Suppose that $\rho = \lambda \mathrm{E}[\boldsymbol{S}] < 1$. Then the stochastic processes $\{\boldsymbol{D}^{(n)}\}$, $\{\boldsymbol{U}(t)\}$, $\{\boldsymbol{Y}(t)\}$, and $\{\boldsymbol{V}(t)\}$ are regenerative with respect to the beginnings of successive busy periods and the length of a busy cycle and the number of customers served in a busy cycle both have finite mean. Let $\boldsymbol{D}, \boldsymbol{U}, \boldsymbol{Y}$, and $\boldsymbol{V}$ have the steady-state distributions of $\boldsymbol{D}^{(n)}$, $\boldsymbol{U}(t)$, $\boldsymbol{Y}(t)$, and $\boldsymbol{V}(t)$, respectively. The following theorem generalizes the results previously given for the *FIFO* queue discipline (cf. proof of equation (A.10)).

Theorem A.4 *Consider a* GI/GI/1 *WCSS in which the scheduling rules* $\phi \in \Phi$ *are regenerative and* STI. *Suppose* $\rho = \lambda \mathrm{E}[\boldsymbol{S}] < 1$. *Then*

$$\mathrm{E}[\boldsymbol{U}] = \lambda \mathrm{E}[\boldsymbol{S}] \mathrm{E}[\boldsymbol{D}] = \rho \mathrm{E}[\boldsymbol{D}] , \qquad (A.14)$$

and therefore

$$\mathrm{E}[\boldsymbol{V}] = \rho \mathrm{E}[\boldsymbol{D}] + \mathrm{E}[\boldsymbol{Y}] + \lambda \mathrm{E}[\boldsymbol{S}^2]/2 , \qquad (A.15)$$

for all $\phi \in \Phi$. *The quantities* $\mathrm{E}[\boldsymbol{V}]$ *and* $\rho \mathrm{E}[\boldsymbol{D}] + \mathrm{E}[\boldsymbol{Y}]$ *are invariant over all* $\phi \in \Phi$.

Proof Since the scheduling rules $\phi \in \Phi$ are *STI*, $\boldsymbol{D}^{(n)}$ and $\boldsymbol{S}^{(n)}$ are independent for all $n \geq 0$, for all $\phi \in \Phi$, it follows by a sample-path argument (cf., e.g., Chapter 6 of El-Taha and Stidham [60]) that (A.14) and (A.15) hold for all $\phi \in \Phi$. The invariance of $\mathrm{E}[\boldsymbol{V}]$ has already been established in the more general setting of a *WCSS*. The invariance of $\rho \mathrm{E}[\boldsymbol{D}] + \mathrm{E}[\boldsymbol{Y}]$ follows from (A.15) and the invariance of $\mathrm{E}[\boldsymbol{V}]$ and $\lambda \mathrm{E}[\boldsymbol{S}^2]/2$. ∎

Note that $\mathrm{E}[\boldsymbol{D}]$ and $\mathrm{E}[\boldsymbol{Y}]$ are *not* in general invariant over $\phi \in \Phi$, although the quantity $\rho \mathrm{E}[\boldsymbol{D}] + \mathrm{E}[\boldsymbol{Y}]$ is.

A.2.2 Example: Multi-Class GI/GI/1 Queue

Consider the example of a *GI/GI/1 WCSS* as described at Section A.2.1, in which the admissible scheduling rules, $\phi \in \Phi$, are regenerative. Now suppose there are m classes of customers, numbered $i = 1, \ldots, m$. Independent of the state and history of the system at the n-th arrival point $\boldsymbol{A}^{(n)}$, the n-th arriving customer belongs to class i with probability λ_i/λ, where $\lambda_i > 0$, $i \in M = \{1, \ldots, m\}$, and $\sum_{i \in M} \lambda_i = \lambda$. Let $\{\boldsymbol{S}_i^{(n)}, n \geq 1\}$ denote the sequence of work requirements for customers of class i, $i \in M$. We assume that, for each $i \in M$, $\{\boldsymbol{S}_i^{(n)}, n \geq 1\}$ is a sequence of i.i.d. random variables and that these sequences are independent among different classes. Let $\boldsymbol{S}_i$ have the common distribution function, $F_i(t)$, of $\boldsymbol{S}_i^{(n)}$, $n \geq 1$. Let $\mu_i := (\mathrm{E}[\boldsymbol{S}_i])^{-1}$, and $\rho_i := \lambda_i/\mu_i$, $i \in M$.

The following exercises show that these assumptions are compatible with our general assumptions for a *GI/GI/1 WCSS*, stated at the beginning of Section A.2.1.

Exercise 1 Show that the work requirements, $S^{(n)}, n \geq 1$, of the customers arriving at the time points, $A^{(n)}, n \geq 1$, are i.i.d., and distributed as a generic random variable S with distribution function $F(t)$ given by

$$F(t) = \sum_{i \in M} \left(\frac{\lambda_i}{\lambda}\right) F_i(t) \,,$$

and expectation

$$\mathrm{E}[S] = \sum_{i \in M} \left(\frac{\lambda_i}{\lambda}\right) \mathrm{E}[S_i] = \frac{1}{\lambda} \sum_{i \in M} \rho_i \,. \tag{A.16}$$

Exercise 2 Let $A_i^{(n)}$ denote the time point at which the n-th customer of class i arrives, $n \geq 1$, $i \in M$ ($A_i^{(0)} = 0$). Show that the class-i interarrival times, $\{A_i^{(n)} - A_i^{(n-1)}, n \geq 1\}$, are i.i.d., and distributed as a generic random variable A_i with distribution function, $G_i(t)$, given by

$$G_i(t) = \sum_{n=1}^{\infty} (1 - \lambda_i/\lambda)^{n-1}(\lambda_i/\lambda)\mathrm{P}\{A^{(n)} \leq t\} \,,$$

and expectation

$$\begin{aligned}
\mathrm{E}[A_i] &= \sum_{n=1}^{\infty} (1 - \lambda_i/\lambda)^{n-1}(\lambda_i/\lambda)\mathrm{E}[A^{(n)}] \\
&= \sum_{n=1}^{\infty} (1 - \lambda_i/\lambda)^{n-1}(\lambda_i/\lambda)(n/\lambda) \\
&= 1/\lambda_i \,.
\end{aligned}$$

Note that it follows from (A.16) that $\rho = \lambda\mathrm{E}[S] = \sum_{i \in M} \rho_i$.

Let $V_i(t)$, $U_i(t)$, and $Y_i(t)$ denote, respectively, the class-i work in the system at time t, the class-i work in the primary queue, and the class-i work in the secondary queue at time t, $t \geq 0$. Let $D_i^{(n)}$ denote the delay (waiting time in the primary queue) of the n-th class-i customer, $n \geq 1$. The same expressions, without the arguments t or n, will denote the corresponding steady-state versions, which are well defined for each scheduling rule $\phi \in \Phi$, provided $\rho = \sum_{i \in M} \rho_i < 1$. (These properties follow from the theory of regenerative processes.) The following theorem is a class-by-class analogue of Theorem A.4. The proof is similar.

Theorem A.5 *Consider a multiclass* GI/GI/1 WCSS *in which the scheduling rules* $\phi \in \Phi_m$ *are regenerative and service-time independent within each class* $i \in M$. *Suppose* $\rho = \sum_{i \in M} \rho_i < 1$. *Then*

$$\mathrm{E}[U_i] = \rho_i\mathrm{E}[D_i] \,, \tag{A.17}$$

$$\mathrm{E}[\boldsymbol{V}_i] \;=\; \rho_i\mathrm{E}[\boldsymbol{D}_i] + \mathrm{E}[\boldsymbol{Y}_i] + \lambda_i\mathrm{E}[\boldsymbol{S}_i^2]/2 \,, \tag{A.18}$$

$i \in M$, for all $\phi \in \Phi_m$.

Note that $\lambda_i\mathrm{E}[\boldsymbol{S}_i^2]/2$ (the expected steady-state class-i work in service) is invariant, whereas $\mathrm{E}[\boldsymbol{V}_i]$, $\mathrm{E}[\boldsymbol{U}_i]$, $\mathrm{E}[\boldsymbol{Y}_i]$, and $\mathrm{E}[\boldsymbol{D}_i]$ typically depend on the scheduling rule, for each $i \in M$. Note also that

$$\sum_{i \in M} \lambda_i\mathrm{E}[\boldsymbol{S}_i^2]/2 = \lambda \sum_{i \in M} \left(\frac{\lambda_i}{\lambda}\right) \mathrm{E}[\boldsymbol{S}_i^2]/2 = \lambda\mathrm{E}[\boldsymbol{S}^2]/2 \,.$$

Since $\boldsymbol{V}(t) = \sum_{i \in M} \boldsymbol{V}_i(t)$, $\boldsymbol{U}(t) = \sum_{i \in M} \boldsymbol{U}_i(t)$, and $\boldsymbol{Y}(t) = \sum_{i \in M} \boldsymbol{Y}_i(t)$, by summing (A.18) over $i \in M$ one recovers equation (A.15).

A.3 *GI/GI/1 WCSS* with Nonpreemptive Scheduling Rules

Consider again a *GI/GI/1 WCSS* as defined in Section A.2.1. In this section we focus on nonpreemptive regenerative scheduling rules. As we noted, when the scheduling rule is nonpreemptive, $\boldsymbol{Y}(t) = 0$ for all $t \geq 0$. As a result, many of the expressions derived in Section A.2 simplify and stronger invariance results apply.

We begin with the special case of service-time-independent rules.

A.3.1 *Service-Time-Independent Scheduling Rules*

The following corollary of Theorem A.4 is immediate.

Corollary A.6 *Consider a* GI/GI/1 WCSS *in which the scheduling rules* $\phi \in \Phi$ *are nonpreemptive, regenerative, and STI. Suppose* $\rho = \lambda\mathrm{E}[\boldsymbol{S}] < 1$. *Then*

$$\mathrm{E}[\boldsymbol{U}] = \lambda\mathrm{E}[\boldsymbol{S}]\mathrm{E}[\boldsymbol{D}] = \rho\mathrm{E}[\boldsymbol{D}] \,, \tag{A.19}$$

and

$$\mathrm{E}[\boldsymbol{V}] = \rho\mathrm{E}[\boldsymbol{D}] + \lambda\mathrm{E}[\boldsymbol{S}^2]/2 \,, \tag{A.20}$$

for all $\phi \in \Phi$. *The expectations* $\mathrm{E}[\boldsymbol{U}]$, $\mathrm{E}[\boldsymbol{V}]$ *and* $\mathrm{E}[\boldsymbol{D}]$ *are invariant over all* $\phi \in \Phi$.

Let $\boldsymbol{W}^{(n)} = \boldsymbol{D}^{(n)} + \boldsymbol{S}^{(n)}$ denote the waiting time in the system of customer n, $n \geq 1$. Let $\boldsymbol{W}$ have the steady-state distribution of $\boldsymbol{W}^{(n)}$. Then it follows from Corollary A.6 that $\boldsymbol{W} = \boldsymbol{D} + \boldsymbol{S}$ is also invariant over all $\phi \in \Phi$.

A.3.1.1 M/GI/1 *Queue*

Now suppose the arrival process is Poisson. Since the *FIFO* queue discipline is an example of a regenerative *STI* scheduling rule, we can assume a *FIFO* discipline for the purpose of evaluating *all* the invariant expectations in the formula (A.20). For an *M/GI/1 WCSS*, we have already seen (cf. Corollary A.3) that $\mathrm{E}[\boldsymbol{V}]$ is invariant among all regenerative $\phi \in \Phi$ and given by (A.12). (Recall that, without the restriction to nonpreemptive *STI* rules, $\mathrm{E}[\boldsymbol{U}]$ and $\mathrm{E}[\boldsymbol{D}]$ are not in general invariant.) Now, with the additional restriction to nonpreemptive *STI* scheduling rules, we have the following additional result.

Corollary A.7 *Consider an* M/GI/1 *WCSS in which the scheduling rules* $\phi \in \Phi$ *are nonpreemptive, regenerative, and STI. Suppose* $\rho = \lambda \mathrm{E}[S] < 1$. *Then*

$$\mathrm{E}[V] = \mathrm{E}[D] \;\; = \;\; \left(\frac{1}{1-\rho}\right) \frac{\lambda \mathrm{E}[S^2]}{2} \,, \tag{A.21}$$

$$\mathrm{E}[U] \;\; = \;\; \left(\frac{\rho}{1-\rho}\right) \frac{\lambda \mathrm{E}[S^2]}{2} \,, \tag{A.22}$$

for all $\phi \in \Phi$.

A.3.2 Multi-Class GI/GI/1 WCSS

Consider the multi-class *GI/GI/1 WCSS* introduced in Section A.2.2. Within the admissible set Φ, we now restrict attention to nonpreemptive regenerative scheduling rules which are service-time independent (e.g., *FIFO*) within each class i. The class of all such scheduling rules will be denoted Φ_m.

The following corollary of Theorem A.5 is immediate.

Corollary A.8 *Consider a multiclass* GI/GI/1 *WCSS in which the scheduling rules* $\phi \in \Phi_m$ *are regenerative and nonpreemptive, and service-time independent within each class* $i \in M$. *Suppose* $\rho = \sum_{i \in M} \rho_i < 1$. *Then*

$$\mathrm{E}[U_i] \;\; = \;\; \rho_i \mathrm{E}[D_i] \,, \tag{A.23}$$
$$\mathrm{E}[V_i] \;\; = \;\; \rho_i \mathrm{E}[D_i] + \lambda_i \mathrm{E}[S_i^2]/2 \,, \tag{A.24}$$

$i \in M$, *for all* $\phi \in \Phi_m$.

A.3.2.1 Conservation Laws for Multi-Class Queues under Nonpreemptive Rules

Consider an arbitrary scheduling rule $\phi \in \Phi_m$. We can write equation (A.24) in equivalent form as

$$\mathrm{E}[V_i] = \rho_i \mathrm{E}[D_i] + \rho_i \gamma_i \,, \tag{A.25}$$

where $\gamma_i := \mathrm{E}[S_i^2]/2\mathrm{E}[S_i]$. Summing both sides of this equation and equation (A.23) over $i \in M$ and combining the results, we conclude that the vector, $(\mathrm{E}[D_1], \dots, \mathrm{E}[D_m])$, of expected steady-state delays satisfies the *conservation law*,

$$\sum_{i \in M} \rho_i \mathrm{E}[D_i] = \mathrm{E}[U] = \mathrm{E}[V] - \sum_{i \in M} \rho_i \gamma_i \,, \tag{A.26}$$

in which the right-hand-side is invariant over $\phi \in \Phi_m$.

Since $\mathrm{E}[W_i] = \mathrm{E}[D_i] + \mathrm{E}[S_i]$, this leads to the following conservation law satisfied by the vector, $(\mathrm{E}[W_1], \dots, \mathrm{E}[W_m])$, of expected steady-state waiting times in the system:

$$\sum_{i \in M} \rho_i \mathrm{E}[W_i] = \mathrm{E}[U] + \sum_{i \in M} \rho_i \mathrm{E}[S_i] = \mathrm{E}[V] + \sum_{i \in M} \rho_i \left(\frac{1}{\mu_i} - \gamma_i\right) . \tag{A.27}$$

As we have seen, in the $M/GI/1$ case we can derive an explicit expression for $\mathrm{E}[\boldsymbol{V}]$, using the fact that it is invariant and therefore equal to the expected work in the system under the *FIFO* queue discipline, which in turn is given by the Pollaczek-Khintchine formula. Thus, for an $M/GI/1$ queue we have

$$\mathrm{E}[\boldsymbol{V}] = \frac{\lambda \mathrm{E}[\boldsymbol{S}^2]}{2(1-\rho)} = \frac{\sum_{i \in M} \rho_i \gamma_i}{1 - \rho} \, ,$$

and hence the conservation equation (A.27) becomes

$$\sum_{i \in M} \rho_i \mathrm{E}[\boldsymbol{W}_i] = \alpha_M := \left(\frac{\rho}{1 - \rho}\right) \sum_{i \in M} \rho_i \gamma_i + \sum_{i \in M} \frac{\rho_i}{\mu_i} \, . \qquad (A.28)$$

Finally in the case where the work requirements in each class are exponentially distributed, we have $\gamma_i = (\mu_i)^{-1}$, so that

$$\sum_{i \in M} \rho_i \mathrm{E}[\boldsymbol{W}_i] = \left(\frac{1}{1 - \rho}\right) \sum_{i \in M} \frac{\rho_i}{\mu_i} \, . \qquad (A.29)$$

Using similar arguments we can also derive a set of inequalities, one for each subset $S \subset M$, which must be satisfied by the vector, $(\mathrm{E}[\boldsymbol{W}_1], \ldots, \mathrm{E}[\boldsymbol{W}_m])$, for any $\phi \in \Phi_m$. The reasoning is as follows.

Consider a subset S of job classes ($S \subset M$). Define ($t \geq 0$)

$$\boldsymbol{V}_S(t) \ := \ \sum_{i \in S} \boldsymbol{V}_i(t) \, , \text{ and}$$

$$\boldsymbol{U}_S(t) \ := \ \sum_{i \in S} \boldsymbol{U}_i(t) \, .$$

That is, $\boldsymbol{V}_S(t)$ is the total S-work in the system and $\boldsymbol{U}_S(t)$ is the total S-work in the queue at time t, $t \geq 0$. Now consider a scheduling rule $\phi \in \Phi_m$ which gives (nonpreemptive) priority to S-jobs over non-S-jobs, and call such a scheduling rule an *S-rule*. Within the set of all S-rules, $\boldsymbol{V}_S(t)$ is invariant, and therefore so is its steady-state expectation,

$$\mathrm{E}[\boldsymbol{V}_S] = \sum_{i \in S} \mathrm{E}[\boldsymbol{V}_i] \, . \qquad (A.30)$$

In addition, among all scheduling rules $\phi \in \Phi_m$, $\mathrm{E}[\boldsymbol{V}_S]$ is minimized by any S-rule. Thus under any scheduling rule $\phi \in \Phi_m$, we have (using (A.30) and (A.25))

$$\mathrm{E}[\boldsymbol{V}_S] = \sum_{i \in S} \rho_i \mathrm{E}[\boldsymbol{D}_i] + \sum_{i \in S} \rho_i \gamma_i \geq \mathrm{E}_S[\boldsymbol{V}_S] \qquad (A.31)$$

where $\mathrm{E}_S[\boldsymbol{V}_S]$ is the invariant expected steady-state S-work in the system under any S-rule. On the other hand, from (A.17) we have

$$\mathrm{E}[\boldsymbol{V}_S] = \mathrm{E}[\boldsymbol{U}_S] + \sum_{i \in S} \rho_i \gamma_i \, ,$$

where $\mathrm{E}[\boldsymbol{U}_S] = \sum_{i \in S} \mathrm{E}[\boldsymbol{U}_i]$. Since ρ_i and γ_i, $i \in M$, are invariant with respect to the scheduling rule, it follows that the expected steady-state S-work in the queue, $E[\boldsymbol{U}_S]$, is also minimized among all scheduling rules $\phi \in \Phi_m$ by S-rules, under which it is invariant. Thus we also have

$$E[\boldsymbol{U}_S] = \sum_{i \in S} \rho_i E[\boldsymbol{D}_i] \geq E_S[\boldsymbol{U}_S] = E_S[\boldsymbol{V}_S] - \sum_{i \in S} \rho_i \gamma_i , \tag{A.32}$$

where $E_S[\boldsymbol{U}_S]$ is the invariant expected steady-state S-work in the queue under any S-rule.

Since $E_S[\boldsymbol{U}_S]$ is the invariant expected steady-state S-work in the queue under any S-rule, in particular it is the S-work in the queue under an S-rule that serves S-jobs in *FIFO* order. Thus we have

$$E_S[\boldsymbol{U}_S] = \rho_S d_S, \tag{A.33}$$

where $\rho_S = \sum_{i \in S} \rho_i$ and d_S is the expected steady-state delay of an S-job under this particular S-rule. It remains to derive an explicit expression for d_S.

Under an S-rule that serves S-jobs in *FIFO* order, the expected virtual delay for an S-job is the sum of the expected S-work in the queue (namely, $\rho_S d_S$) and the expected total work in service, which equals $\lambda E[\boldsymbol{S}^2]/2$. Therefore, the expected virtual delay for a S-job under this S-rule equals

$$\rho_S d_S + \lambda E[\boldsymbol{S}^2]/2 .$$

We now specialize to an $M/GI/1$ system, so that *PASTA* holds. Then the virtual and actual delays coincide and we have

$$d_S = \rho_S d_S + \lambda E[\boldsymbol{S}^2]/2 ,$$

which, when solved for d_S, yields

$$d_S = \frac{\lambda E[\boldsymbol{S}^2]}{2(1 - \rho_S)} = \frac{\sum_{i \in M} \rho_i \gamma_i}{1 - \rho_S} . \tag{A.34}$$

Combining (A.32), (A.33) and (A.34) we have the following inequality, satisfied by the vector of expected steady-state delays, $(E[\boldsymbol{D}_1], \ldots, E[\boldsymbol{D}_m])$, under any scheduling rule, $\phi \in \Phi_m$:

$$\sum_{i \in S} \rho_i E[\boldsymbol{D}_i] \geq \left(\frac{\rho_S}{1 - \rho_S} \right) \sum_{i \in M} \rho_i \gamma_i . \tag{A.35}$$

Moreover, (A.35) holds with equality for S-rules.

Using the relation $E[\boldsymbol{W}_i] = E[\boldsymbol{D}_i] + 1/\mu_i$ leads to the following inequality satisfied by the vector of expected steady-state waiting times in the system, $(E[\boldsymbol{W}_1], \ldots, E[\boldsymbol{W}_m])$, in an M/GI/1 system, for all $S \subset M$ and scheduling rules $\phi \in \Phi_m$:

$$\sum_{i \in S} \rho_i E[\boldsymbol{W}_i] \geq \alpha_S := \left(\frac{\sum_{i \in S} \rho_i}{1 - \sum_{i \in S} \rho_i} \right) \sum_{i \in M} \rho_i \gamma_i + \sum_{i \in S} \frac{\rho_i}{\mu_i} . \tag{A.36}$$

The inequality holds with equality for all S-rules. Finally in the *M/M/1* case, we have $\gamma_i = (\mu_i)^{-1}$, so that the inequality corresponding to the subset S becomes

$$\sum_{i \in S} \rho_i \mathrm{E}[\boldsymbol{W}_i] \geq \alpha_S := \left(\frac{\sum_{i \in S} \rho_i}{1 - \sum_{i \in S} \rho_i} \right) \sum_{i \in M} \frac{\rho_i}{\mu_i} + \sum_{i \in S} \frac{\rho_i}{\mu_i} . \qquad (A.37)$$

The following theorem summarizes these results.

Theorem A.9 *Consider an* M/GI/1 *WCSS with* m *classes. Let* Φ_m *denote the set of all scheduling rules,* $\phi \in \Phi$, *which are regenerative and nonpreemptive, and* STI *within each class* $i \in M$. *Then, under any* $\phi \in \Phi_m$, *the vector of expected steady-state waiting times in the system,* $(\mathrm{E}[\boldsymbol{W}_1], \ldots, \mathrm{E}[\boldsymbol{W}_m])$, *satisfies the linear system,*

$$\sum_{i \in M} \rho_i \mathrm{E}[\boldsymbol{W}_i] = \alpha_M$$

$$\sum_{i \in S} \rho_i \mathrm{E}[\boldsymbol{W}_i] \geq \alpha_S , \ S \subset M ,$$

where

$$\alpha_S := \left(\frac{\sum_{i \in S} \rho_i}{1 - \sum_{i \in S} \rho_i} \right) \sum_{i \in M} \rho_i \gamma_i + \sum_{i \in S} \frac{\rho_i}{\mu_i} , \ S \subseteq M .$$

Moreover, the constraint corresponding to the subset S *is satisfied with equality for any* S-*rule, that is, for any rule* $\phi \in \Phi_m$ *that gives priority to jobs in classes* $i \in S$ *over jobs in classes* $i \notin S$.

In other words, strong conservation laws hold for $\phi \in \Phi_m$ *(cf. Section A.1).*

A.4 *GI/GI/1* Queue: Preemptive-Resume Scheduling Rules

Consider again a *GI/GI/1 WCSS* as defined in Section A.2.1. In this section we focus on preemptive-resume regenerative scheduling rules. The results derived in Section A.2 apply in this case, but without further assumptions they do not lead to explicit expressions or conservation laws for the quantities of interest. The difficulty arises in evaluating $\mathrm{E}[\boldsymbol{Y}]$, the expected steady-state work in the secondary queue.

When the work requirements in each class are exponentially distributed, however, we can circumvent this difficulty by exploiting the memoryless property of the exponential distribution.

A.4.0.2 *Conservation Laws for Multi-Class Queues*

Consider the multi-class *GI/GI/1 WCSS* introduced in Section A.2.2. Within the admissible set Φ, we now restrict attention to preemptive-resume regenerative scheduling rules which are service-time independent (e.g., *FIFO*) within each class i. The class of all such scheduling rules will be denoted Φ_m.

Suppose that the class-i work requirements are exponentially distributed. Now, the expected remaining work of a class-i is $1/\mu_i$ regardless of how much service that job has received. Therefore, the steady-state expected class-i work in the system is given by

$$\mathrm{E}[\boldsymbol{V}_i] = \frac{\mathrm{E}[\boldsymbol{L}_i]}{\mu_i} = \frac{\lambda_i \mathrm{E}[\boldsymbol{W}_i]}{\mu_i} = \rho_i \mathrm{E}[\boldsymbol{W}_i] \,, \tag{A.38}$$

where $\boldsymbol{L}_i = \lambda \boldsymbol{W}_i$ is expected steady-state number of class-i jobs in the system.

Summing (A.38) over $i \in M$, we obtain the following *conservation law*,

$$\sum_{i \in M} \rho_i \mathrm{E}[\boldsymbol{W}_i] = \mathrm{E}[\boldsymbol{V}] \,, \tag{A.39}$$

in which the right-hand-side is invariant over $\phi \in \Phi_m$.

When the arrival streams are Poisson, we have an explicit expression for $\mathrm{E}[\boldsymbol{V}]$, given by (A.11), which reduces to

$$\mathrm{E}[\boldsymbol{V}] = \frac{\sum_{i \in M} \rho_i / \mu_i}{1 - \rho} \,,$$

when the work requirements are exponentially distributed in each class. In this case, under any $\phi \in \Phi_m$, the vector of expected steady-state waiting times in the system, $(\mathrm{E}[\boldsymbol{W}_1], \ldots, \mathrm{E}[\boldsymbol{W}_m])$, satisfies the conservation law,

$$\sum_{i \in M} \rho_i \mathrm{E}[\boldsymbol{W}_i] = \frac{\sum_{i \in M} \rho_i / \mu_i}{1 - \rho} \,. \tag{A.40}$$

Using similar arguments we can also derive a set of inequalities, one for each subset $S \subset M$, which must be satisfied by the vector, $(\mathrm{E}[\boldsymbol{W}_1], \ldots, \mathrm{E}[\boldsymbol{W}_m])$, of expected steady-state waiting times in the system, for any $\phi \in \Phi_m$. The reasoning is as follows.

Consider a subset S of job classes ($S \subset M$). As in the analysis of nonpreemptive rules, define ($t \geq 0$)

$$\boldsymbol{V}_S(t) := \sum_{i \in S} \boldsymbol{V}_i(t) \,, \ t \geq 0 \,.$$

That is, $\boldsymbol{V}_S(t)$ is the total S-work in the system at time t, $t \geq 0$. Now consider a preemptive-resume scheduling rule $\phi \in \Phi_m$ which gives priority to S-jobs over non-S-jobs, and call such a scheduling rule an S-*rule*. Within the set of all S-rules, $\boldsymbol{V}_S(t)$ is invariant, and therefore so is its steady-state expectation,

$$\mathrm{E}[\boldsymbol{V}_S] = \sum_{i \in S} \mathrm{E}[\boldsymbol{V}_i] \,. \tag{A.41}$$

In addition, among all scheduling rules $\phi \in \Phi_m$, $\mathrm{E}[\boldsymbol{V}_S]$ is minimized by any S-rule. Thus under any scheduling rule $\phi \in \Phi_m$, we have (using (A.41) and (A.38))

$$\mathrm{E}[\boldsymbol{V}_S] = \sum_{i \in S} \rho_i \mathrm{E}[\boldsymbol{W}_i] \geq \mathrm{E}_S[\boldsymbol{V}_S] \tag{A.42}$$

where $\mathrm{E}_S[\boldsymbol{V}_S]$ is the invariant expected steady-state S-work in the system under any S-rule.

For any S-rule, the total S-work, $\boldsymbol{V}_S(t)$, is independent of all non-S jobs. Therefore, $\mathrm{E}_S[\boldsymbol{V}_S]$ is also the (invariant) expected steady-state work in a system where the demand consists only of S jobs. Then, by considering such a system under a *FIFO* discipline, when the arrivals are Poisson we obtain

$$\mathrm{E}_S[\boldsymbol{V}_S] = \frac{\sum_{i \in S} \rho_i \gamma_i}{1 - \sum_{i \in S} \rho_i} . \qquad (A.43)$$

In the present case, in which work requirements in each class are exponentially distributed, we therefore have (from (A.42) and (A.43)) the following conservation law for each class $S \subset M$ and scheduling rule $\phi \in \Phi_m$:

$$\sum_{i \in S} \rho_i \mathrm{E}[\boldsymbol{W}_i] \geq \frac{\sum_{i \in S} \rho_i/\mu_i}{1 - \sum_{i \in S} \rho_i} . \qquad (A.44)$$

Moreover, for each $S \subset M$ this inequality is satisfied with equality by any S-rule.

The following theorem summarizes these results.

Theorem A.10 *Consider an* M/GI/1 *WCSS with* m *classes. Suppose the work requirements in each class are exponentially distributed. Let* Φ_m *denote the set of all scheduling rules* $\phi \in \Phi$ *which are preemptive-resume and regenerative, and STI within each class* $i \in M$. *Then, under any* $\phi \in \Phi_m$, *the vector of expected steady-state waiting times in the system,* $(\mathrm{E}[\boldsymbol{W}_1], \dots, \mathrm{E}[\boldsymbol{W}_m])$, *satisfies the linear system,*

$$\sum_{i \in M} \rho_i \mathrm{E}[\boldsymbol{W}_i] = \alpha_M$$

$$\sum_{i \in S} \rho_i \mathrm{E}[\boldsymbol{W}_i] \geq \alpha_S , \ S \subset M ,$$

where

$$\alpha_S := \left(\frac{1}{1 - \sum_{i \in S} \rho_i} \right) \sum_{i \in S} \frac{\rho_i}{\mu_i} , \ S \subseteq M .$$

Moreover, the constraint corresponding to the subset S *is satisfied with equality for any* S-*rule, that is, for any rule* $\phi \in \Phi_m$ *that gives priority to jobs in classes* $i \in S$ *over jobs in classes* $i \notin S$.

In other words, strong conservation laws hold for $\phi \in \Phi_m$ *(cf. Section A.1).*

A.5 Endnotes

For a comprehensive survey of conservation laws and the achievable-region approach to scheduling queues, see Bertsimas [21]. Some of the material in this appendix is based on Green and Stidham [78].

References

[1] P. Afeche and H. Mendelson. Pricing and priority auctions in queueing systems with a generalized delay cost structure. *Management Science*, 50:869–882, 2004.

[2] G. Allon and A. Federgruen. Competition in service industries with segmented markets. Technical Report, Graduate School of Business, Columbia University (forthcoming in *Management Science*), 2004.

[3] G. Allon and A. Federgruen. Competition in service industries. *Operations Research*, 55:37–55, 2007.

[4] G. Allon and A. Federgruen. Service competition with general queueing facilities. *Operations Research*, 56:827–849, 2008.

[5] E. Altman, T. Başar, and R. Srikant. Nash equilibria for combined flow control and routing in networks: asymptotic behavior for a large number of users. *IEEE Trans. Automatic Control*, 47:917–930, 2002.

[6] E. Altman, T. Boulogne, R. El-Azouzi, T. Jiménez, and L. Wynter. A survey on networking games in telecommunications. *Computers and Operations Research*, 33:286–311, 2006.

[7] N. Argon and S. Ziya. Priority assignment under imperfect information on customer type identities. Technical Report, Department of Statistics and Operations Research, University of North Carolina at Chapel Hill (forthcoming in *Manufacturing and Service Operations Management*), 2008.

[8] M. Armony and M. Haviv. Price and delay competition between two service providers. *European J. Operational Research*, 147:32–50, 2003.

[9] A. Azaron and S. Fatemi Ghomi. Optimal control of service rates and arrivals in Jackson networks. *European J. Operational Research*, 147:17–31, 2003.

[10] D. Bails and L. Peppers. *Business Fluctuations*. Prentice Hall, Englewood Cliffs, New Jersey, 1993. 2nd ed.

[11] K. Balachandran and S. Radhakrishnan. Extensions to class dominance characteristics. *Management Science*, 40:1353–1360, 1994.

[12] K. Balachandran and S. Radhakrishnan. Cost of congestion, operational efficiency and management accounting. *European J. Operational Research*, 89:237–245, 1996.

[13] K. Balachandran and M. Schaefer. Public and private optimization at a service facility with approximate information on congestion. *European J. Operational Research*, 4:195–202, 1978.

[14] K. Balachandran and M. Schaefer. Class dominance characteristics at a service facility. *Econometrica*, 47:515–519, 1979.

[15] K. Balachandran and M. Schaefer. Regulation by price of arrivals to a congested facility. *Cahiers du C.E.R.O.*, 21:149–154, 1979.

[16] M. Bazaraa, H. Sherali, and C. Shetty. *Nonlinear Programming: Theory and Algorithms*. John Wiley and Sons, New York, 3rd edition, 2006.

[17] M. Beckmann, C. McGuire, and C. Winsten. *Studies in the Economics of Transportation.* Yale University Press, New Haven, CT, 1956.

[18] C. Bell and S. Stidham. Individual versus social optimization in the allocation of customers to alternative servers. *Management Science*, 29:831–839, 1983.

[19] D. Bertsekas. *Network Optimization: Continuous and Discrete Models.* Athena Scientific, Nashua, New Hampshire, 1998.

[20] D. Bertsekas and R. Gallager. *Data Networks.* Prentice Hall, Englewood Cliffs, New Jersey, 1992. 2nd ed.

[21] D. Bertsimas. The achievable region method in the optimal control of queueing systems. *Queueing Systems: Theory and Applications*, 21:337–389, 1996.

[22] K. Bharath-Kumar and J. Jaffe. A new approach to performance-oriented flow control. *IEEE Trans. Communications*, COM-29(4):427–435, April 1981.

[23] J. Blackburn. *Time Based Competition.* Richard Irwin, Homewood, IL, 1991.

[24] A. Bovopoulos and A. Lazar. Decentralized algorithms for optimal flow control. In *Proceedings of the* 25th *Allerton Conference on Communications, Control and Computing*, pages 979–988, Univ. of Illinois at Urbana-Champaign, October 1987.

[25] A. Bovopoulos and A. Lazar. Synchronous and asynchronous iterative algorithms for load balancing. In *Proc.* 22nd *Annual Conference on Information Sciences and Systems*, Princeton, NJ, March 1988.

[26] A. Bovopoulos and A. Lazar. Asynchronous algorithms for optimal flow control of BCMP networks. Technical Report 89-10, Washington Univ., St. Louis, MO, February 1989.

[27] R. Bradford. Pricing, routing, and incentive compatibility in multiserver queues. *European J. Operational Research*, 89:226–236, 1996.

[28] D. Braess. Uber ein paradoxon der verkehrsplanung. *Unternehmensforschung*, 12:258–268, 1968.

[29] G. Cachon and M. Lariviere. Capacity choice and allocation: strategic behavior and supply chain performance. *Management Science*, 45:1091–1108, 1999.

[30] D. Cansever. Decentralized algorithms for flow control in networks. In *Proc.* 25th *Conference on Decision and Control*, Athens, Greece, December 1986.

[31] M. Carter and R. Maddock. *Rational Expectations.* MacMillan, London, 1984.

[32] X. Chao, H. Chen, and W. Li. Optimal control for a tandem network of queues with blocking. *Acta Mathematicae Applicatae Sinica*, 13:425–437, 1997.

[33] C. Chase, J. Serrano, and P. Ramadge. Periodicity and chaos from switched flow systems: contrasting examples of discretely controlled continuous systems. *IEEE Trans. Automatic Control*, 38, 1993.

[34] C. Chau and K. Sim. The price of anarchy for nonatomic congestion games with symmetric cost maps and elastic demands. *Operational Research Letters*, 31:327–334, 2003.

[35] D. Chazan and W. Miranker. Chaotic relaxation. *Linear Algebra and Its Applications*, 2:199–222, 1969.

[36] H. Chen and M. Frank. Monopoly pricing when customers queue. *IIE Transactions*, 36:569–581, 2004.

[37] H. Chen and Y. Wan. Price competition of make-to-order firms. *IIE Transactions*, 35:817–832, 2003.

[38] H. Chen and D. Yao. Optimal intensity control of a queueing system with state-dependent capacity limit. *IEEE Trans. Automatic Control*, 25:459–464, 1990.

[39] C. Chiarella. The cobweb model: Its instability and the onset of chaos. *Economic Modelling*, 5:377–384, 1988.

[40] J. Cohen and F. Kelly. A paradox of congestion in a queueing network. *J. Applied Probability*, 27:730–734, 1990.

[41] J. Correa, A. Schulz, and N. Stier-Moses. Computational complexity, fairness, and the price of anarchy of the maximum latency problem. In *Integer Programming and Combinatorial Optimization*, volume 3064, pages 59–73. Springer Berlin/Heidelberg, 2004. Lecture Notes in Computer Science.

[42] J. Correa, A. Schulz, and N. Stier-Moses. Selfish routing in capacitated networks. *Mathematics of Operations Research*, 29:961–976, 2004.

[43] J. Correa, A. Schulz, and N. Stier-Moses. On the inefficiency of equilibria in congestion games. In M. Junger and V. Kaibel, editors, *IPCO 2005, LNCS 3509*, pages 167–181, Berlin, 2005. Springer-Verlag.

[44] C. Courcoubetis and R. Weber. *Pricing Communication Networks*. Wiley, New York, 2003.

[45] D.R. Cox and W.L. Smith. *Queues*. John Wiley, New York, 1961.

[46] T. Crabill, D. Gross, and M. Magazine. A classified bibliography of research on optimal design and control of queues. *Operations Research*, 25:219–232, 1977.

[47] M. Cramer. Optimal customer selection in exponential queues. Technical Report ORC 71-24, Operations Research Center, University of California, Berkeley, 1971.

[48] S. Dafermos. Traffic equilibrium and variational inequalities. *Transportation Science*, 14:42–54, 1980.

[49] S. Dafermos and A. Nagourney. On some traffic equilibrium theory paradoxes. *Transportation Research*, 18B:101–110, 1984.

[50] S. Dafermos and F. Sparrow. The traffic assignment problem for a general network. *J. Res. U.S. Nat. Bureau Standards*, 73B:91–118, 1969.

[51] C. Davidson. Equilibrium in oligopolistic service industries: an economic application of queueing theory. *J. Business*, 61:347–367, 1988.

[52] A. De Vany. Uncertainty, waiting time, and capacity utilization: a stochastic theory of product quality. *J. Political Economy*, 84:523–541, 1976.

[53] R. Devaney. *An Introduction to Chaotic Dynamical Systems*. Addison-Wesley, Menlo Park, Calif., 1989.

[54] S. Dewan and H. Mendelson. User delay costs and internal pricing for a service facility. *Management Sci.*, 36:1502–1517, 1990.

[55] R. Dolan. Incentive mechanisms for priority queueing problems. *Bell J. Economics*, 9:421–436, 1978.

[56] D. Douligeris and R. Mazumdar. A game theoretic approach to flow control in an integrated environment with two classes of users. In *Proc. Computer Networking Symposium*, pages 214–221, Washington, DC, April 1988.

[57] D. Douligeris and R. Mazumdar. User optimal flow control in an integrated environment. In *Proc. Indo-US Workshop on Systems and Signals*, Bangalore, India, January 1988.

[58] D. Douligeris and R. Mazumdar. Efficient flow control in a multiclass telecommunications environment. *IEEE Proceedings-I*, 138(6):494–502, December 1991.

[59] N. Edelson and D. Hildebrand. Congestion tolls for poisson queueing processes. *Econometrica*, 43:81–92, 1975.

[60] M. El-Taha and S. Stidham. *Sample-Path Analysis of Queueing Systems.* Kluwer Academic Publishing, Boston, 1998.

[61] P. Embrechts, C. Klüppelberg, and T. Mikosch. *Modelling Extremal Events for Insurance and Finance.* Springer, New York, 1997.

[62] M. Feigenbaum. Universal behavior in nonlinear systems. *Physica*, 7D:16–39, 1983.

[63] S. Floyd. *TCP* and explicit congestion notification. *ACM Comp. Comm. Rev.*, 24:10–23, 1994.

[64] S. Floyd and V. Jacobson. Random early detection gateways for congestion avoidance. *IEEE/ACM Trans. Networking*, 1:397–413, 1997.

[65] G. Foschini. On heavy traffic diffusion analysis and dynamic routing in packet-switched networks. *Computer Performance*, pages 499–513, 1977. Chandy, K.M. and Reiser, M. (eds.), North-Holland.

[66] G. Foschini and J. Salz. A basic dynamic routing problem and diffusion. *IEEE Trans. Communication*, 26:320–327, 1978.

[67] M. Frank. The Braess paradox. *Mathematical Programming*, 20:283–302, 1981.

[68] K. Fridgeirsdottir and R. Akella. Product portfolio management in delay sensitive markets. Technical report, Department of Management Science and Engineering, Stanford University, 2001.

[69] K. Fridgeirsdottir and S. Chiu. A note on convexity of the expected delay cost in single-server queues. *Operations Research*, 53:568–570, 2005.

[70] E. Friedman and A. Landsberg. Short-run dynamics of multi-class service facilities. *Operations Research Letters*, 14:221–229, 1993.

[71] E. Friedman and A. Landsberg. Long-run dynamics of queues: stability and chaos. *Operations Research Letters*, 18:185–191, 1996.

[72] I. Frutos and J. Gallego. Multiproduct monopoly: a queueing approach. *Applied Economics*, 31:565–576, 1999.

[73] J. George and J.M. Harrison. Dynamic control of a queue with adjustable service rate. *Operations Research*, 49:720–731, 2001.

[74] S. Gilbert and Z. Weng. Incentive effects favor non-consolidating queues in a service system: the principal-agent perspective. *Management Science*, 44:1662–1669, 1998.

[75] A. Glazer and R. Hassin. Stable priority purchasing in queues. *Operations Research Letters*, 4:285–288, 1985.

[76] J.-M. Grandmont and P. Malgrange. Nonlinear economic dynamics: Introduction. In J.-M. Grandmont, editor, *Nonlinear Economic Dynamics*, Orlando, 1986. Academic Press.

[77] W. Grassmann. The economic service rate. *J. Operational Research Society*, 30:149–155, 1979.

[78] T. Green and S. Stidham. Sample-path conservation laws, with applications to scheduling queues and fluid systems. *Queueing Systems: Theory and Applications*, 36:175–200, 2000.

[79] D. Gross and K. Harris. *Fundamentals of Queueing Theory.* John Wiley, New York, 2nd edition, 1985.

[80] A. Ha. Incentive-compatible pricing for a service facility with joint production and congestion externalities. *Management Science*, 44:1623–1636, 1998.

[81] A. Ha. Optimal pricing that coordinates queues with customer-chosen service requirements. *Management Science*, 47:915–930, 2001.

[82] J.M. Harrison. Dynamic scheduling of a multi-class queue: discount optimality. *Operations Research*, 23:270–282, 1975.

[83] J.M. Harrison. A priority queue with discounted linear costs. *Operations Research*, 23:260–269, 1975.

[84] R. Hassin. On the optimality of first-come last-served queues. *Econometrica*, 53:201–202, 1985.

[85] R. Hassin. Decentralized regulation of a queue. *Management Science*, 41:163–173, 1995.

[86] R. Hassin and M. Haviv. *To Queue or Not to Queue: Equilibrium Behavior in Queueing Systems*. Kluwer Academic Publishers, Boston, MA, 2003.

[87] R. Hassin and M. Haviv. Who should be given priority in a queue. *Operations Research Letters*, 34:191–198, 2006.

[88] R. Hassin, J. Puerto, and F. Fernandez. The use of relative priorities in optimizing the performance of a queueing system. *European J. Operational Research*, 2008. (forthcoming).

[89] M. Haviv. Stable strategies for processor sharing systems. *European J. Operational Research*, 52:103–106, 1991.

[90] M. Haviv and Y. Ritov. Externalities, tangible externalities, and queue disciplines. *Management Science*, 44:850–858, 1998.

[91] M. Haviv and J. van der Wal. Equilibrium strategies for processor sharing and queues with relative priorities. *Probability in the Engineering and the Informational Sciences*, 11:403–412, 1997.

[92] D. Heyman and M. Sobel. *Stochastic Models in Operations Research, Volumes I and II*. McGraw-Hill Book Co., New York, 1982.

[93] F. Hillier. The application of waiting-line theory to industrial problems. *J. Industrial Engineering*, 15:3–8, 1964.

[94] C. Hommes. Adaptive learning and roads to chaos: The case of the cobweb. *Economics Letters*, 36:127–132, 1991.

[95] J. Jackson. Networks of waiting lines. *Operations Research*, 5:518–521, 1957.

[96] V. Jacobson. Congestion avoidance and control. In *Proc. ACM SIGCOMM '88*, pages 314–329, 1988.

[97] J. Jaffe. Flow control power is nondecentralizable. *IEEE Trans. Communications*, COM-29(9):1301–1306, September 1981.

[98] O. Jahn, R. Mohring, R. Schulz, and N. Stier-Moses. System-optimal routing of traffic flows with user constraints in networks with congestion. *Operations Research*, 53:600–616, 2005.

[99] R. Jensen. Classical chaos. *American Scientist*, 75:168–181, 1987.

[100] S. Johansen and S. Stidham. Control of arrivals to a stochastic input-output system. *Adv. Applied Probability*, 12:972–999, 1980.

[101] R. Johari and J. Tsitsiklis. Efficiency loss in a network resource allocation game. *Math. Operat. Res.*, 29:407–435, 2004.

[102] E. Kalai, M. Kamien, and M. Rubinovitch. Optimal service speeds in a competitive environment. *Management Science*, 38:1154–1163, 1992.

[103] F. Kelly. *Reversibility and Stochastic Networks*. John Wiley, Chichester, New York, 1979.

[104] F. Kelly. Network routing. *Philos. Trans. Roy. Soc., Ser. A*, 337:343–367, 1991.

[105] F. Kelly. Charging and rate control for elastic traffic. *Euro. Trans. Telecomm.*, 8:33–37, 1997.

[106] F. Kelly. Models for a self-managed Internet. *Phil. Trans. Roy. Soc. Lond.*, 358:2335–2348, 2000.

[107] F. Kelly. Mathematical modelling of the Internet. In Engquist B. and W. Schmid, editors, *Mathematics Unlimited: 2001 and Beyond*, pages 685–702, Berlin, 2001. Springer-Verlag.

[108] F. Kelly. Fairness and stability of end-to-end congestion control. *European J. Control*, 9:159–176, 2003.

[109] F. Kelly, A. Maulloo, and D. Tan. Rate control in communication networks: shadow prices, proportional fairness, and stability. *J. Operational Research Society*, 49:237–252, 1998.

[110] Y. Kim and M. Mannino. Optimal incentive-compatible pricing for M/G/1 queues. *Operations Research Letters*, 31:459–461, 2003.

[111] M. Kitaev and V. Rykov. *Controlled Queueing Systems*. CRC Press, Boca Raton, FL, 1995.

[112] L. Kleinrock. Optimum bribing for queue position. *Operations Research*, 15:304–318, 1967.

[113] L. Kleinrock. *Queueing Systems vol. I and II*. Wiley Intersciences, New York, 1975.

[114] L. Kleinrock. Power and deterministic rules of thumb for probabilistic problems in computer communications. In *Proceedings of the International Conference on Communications*, volume 43, pages 1–10, 1979.

[115] N. Knudsen. Individual and social optimization in a multiserver queue with a general cost-benefit structure. *Econometrica*, 40:515–528, 1972.

[116] E. Koenigsberg. Stochastic models of oligopoly and buyers' co-operatives: Modelling dispatchers and brokers. Technical Report, School of Business Administration, University of California at Berkeley, submitted to Management Science, 1980.

[117] E. Koenigsberg. Uncertainty, capacity, and market share in oligopoly: A stochastic theory of product quality. *J. of Business*, 53:151–164, 1980.

[118] E. Koenigsberg. Queue systems with balking: A stochastic model of price discrimination. *RAIRO, Recherche Operationelle/Operations Research*, 19:209–219, 1985.

[119] Y. Korilis and A. Lazar. Why is flow control hard: optimality, fairness, partial and delayed information. In *Proc. 2nd ORSA Telecommunications Conference*, 1992.

[120] Y. Korilis and A. Lazar. On the existence of equilibria in noncooperative optimal flow control. *J. Assoc. Computing Machinery*, 42:584–613, 1995.

[121] Y. Korilis, A. Lazar, and A. Orda. Capacity allocation under noncooperative routing. *IEEE Trans. Automatic Control*, 42:309–325, 1997.

[122] Y. Korilis, A. Lazar, and A. Orda. Avoiding the Braess paradox in noncooperative networks. *J. Applied Probability*, 36:211–222, 1999.

[123] G. Latouche. On the trade-off between queue congestion and server's reward in an M/M/1 queue. *European J. of Operational Research*, 4:203–214, 1978.

[124] P. Lederer and L. Li. Pricing, production, scheduling, and delivery-time competition. *Operations Research*, 45:407–420, 1997.

[125] J-W. Lee, R. Mazumdar, and N. Shroff. Non convex optimization and rate control for multi-class services in the internet. *IEEE/ACM Trans. Networking*, 13:827–840, 2005.

[126] D. Levhari and I. Luski. Duopoly pricing and waiting lines. *European Economic Review*, 11:17–35, 1978.

[127] L. Li. The role of inventory in delivery time competition. *Management Science*, 38:182–197, 1992.

[128] L. Li and Y. Lee. Pricing and delivery-time performance in a competitive environment. *Management Science*, 40:633–646, 1994.

[129] S. Li and T. Başar. Distributed algorithms for the computation of noncooperative equilibria. *Automatica*, 23:523–533, 1987.

[130] T. Li and J. Yorke. Period three implies chaos. *American Mathematical Monthly*, 82:985–992, 1975.

[131] S. Lippman and S. Stidham. Individual versus social optimization in exponential congestion systems. *Operations Research*, 25:233–247, 1977.

[132] S. A. Lippman. Applying a new device in the optimization of exponential queuing systems. *Operations Research*, 23:687–710, 1975.

[133] H.-W. Lorenz. *Nonlinear Dynamical Economics and Chaotic Motion*. Springer-Verlag, Berlin, 1989.

[134] I. Luski. On partial equilibrium in a queueing system with two servers. *Rev. Economic Studies*, 43:519–525, 1976.

[135] J. MacKie-Mason and H. Varian. Pricing the Internet. In B. Kahin and J. Keller, editors, *Public Access to the Internet*, Englewood Cliffs, NJ, 1994. Prentice-Hall.

[136] J. MacKie-Mason and H. Varian. Pricing congestible network resources. *IEEE J. Selected Areas in Communications*, 13:1141–1149, 1995.

[137] A. Mandelbaum and N. Shimkin. A model for rational abandonment from invisible queues. *Queueing Systems: Theory and Applications*, 36:141–173, 2000.

[138] M. Mandjes. Pricing strategies under heterogeneous service requirements. *Computer Networks*, 42:231–249, 2003.

[139] M. Marchand. Priority pricing. *Management Science*, 20:1131–1140, 1974.

[140] P. Marcotte and L. Wynter. A new look at the multiclass network equilibrium problem. *Transportation Science*, 38:282–292, 2004.

[141] Y. Masuda and S. Whang. Dynamic pricing for network service: equilibrium and stability. *Management Science*, 45:857–869, 1999.

[142] Y. Masuda and S. Whang. Capacity management in decentralized networks. *Management Science*, 48:1628–1634, 2002.

[143] H. Mendelson. Pricing computer services: queueing effects. *Comm. Association of Computing Machinery*, 28:312–21, 1985.

[144] H. Mendelson and S. Whang. Optimal incentive-compatible priority pricing for the M/M/1 queue. *Operations Research*, 38:870–883, 1990.

[145] I. Milchtaich. Social optimality and cooperation in nonatomic congestion games. *J. Economic Theory*, 114:56–87, 2004.

[146] P. Milgrom and J. Roberts. Rationalizability, learning, and equilibrium in games with strategic complementarities. *Econometrica*, 58:1255–1277, 1990.

[147] B. Miller. A queueing reward system with several customer classes. *Management Science*, 16:234–245, 1969.

[148] B. Miller and A. Buckman. Cost allocation and opportunity costs. *Management Science*, 33:626–639, 1987.

[149] J. Murchland. Braess's paradox of traffic flow. *Transportation Research*, 4:391–394, 1970.

[150] A. Nagourney. *Network Economics: A Variational Inequality Approach*. Kluwer Academic Publisher, Dordrecht, The Netherlands, 1993.

[151] P. Naor. On the regulation of queue size by levying tolls. *Econometrica*, 37:15–24, 1969.

[152] A. Orda, N. Rom, and N. Shimkin. Competitive routing in multi-class communication networks. *IEEE/ACM Trans. Networking*, 1:614–627, 1993.

[153] V. Pareto. *Manuel d'Economie Politique*. Giard et Brière, Paris, 1909.

[154] I. Paschalidis and Y. Liu. Pricing in multiservice loss networks: static pricing, asymptotic optimality, and demand substitution effects. *IEEE/ACM Trans. Networking*, 10:425–438, 2002.

[155] G. Perakis. The price of anarchy under nonlinear and asymmetric costs. In *Integer Programming and Combinatorial Optimization*, volume 3064, pages 46–58. Springer Berlin/Heidelberg, 2004. Lecture Notes in Computer Science.

[156] L. Perko. *Differential Equations and Dynamical Systems*. Springer-Verlag, New York, 2001.

[157] M. Pinedo and X. Chao. *Operations Scheduling with Applications in Manufacturing and Services*. Irwin/McGraw-Hill, New York, 1999.

[158] I. Png and D. Reitman. Service time competition. *RAND J. Economics*, 25:619–634, 1994.

[159] S. Rao and E. Peterson. Optimal pricing of priority services. *Operations Research*, 46:46–56, 1998.

[160] A. Raviv and E. Shlifer. Utilization of waiting time in a queue for additional service. In *Developments In Operations Research*. Gordon and Breach Science Publisher, 1971.

[161] J.B. Rosen. Existence and uniqueness of equilibrium points for concave n-person games. *Econometrica*, 33:520–534, 1965.

[162] R. Rosenthal. A class of games possessing pure-strategy Nash equilibria. *International J. Game Theory*, 2:65–67, 1973.

[163] T. Roughgarden. The price of anarchy is independent of the network topology. In *Proc. ACM Symp. Theory of Computing*, volume 34, pages 428–437, 2002.

[164] T. Roughgarden. *Selfish Routing and the Price of Anarchy*. MIT Press, Cambridge, MA, 2005.

[165] T. Roughgarden. On the severity of Braesss paradox: designing networks for selfish users is hard. *J. Computer and System Science*, 72:922–953, 2006.

[166] T. Roughgarden and E. Tardos. How bad is selfish routing? *J. Association of Computing Machinery*, 49:236–259, 2002.

[167] M. Rubinovitch. The slow server problem. *J. Applied Probability*, 22:205–213, 1985.

[168] C. Rump. A Nash bargaining approach to resource allocation in a congested service system. *Operations Research Letters*, 2001. under review.

[169] C. Rump and S. Stidham. Asymptotic behavior of a relaxed sequential flow-control algorithm for multiclass networks. In *Proc. 33rd Allerton Conference on Communications, Control and Computing*, pages 939–943, University of Illinois at Urbana-Champaign, 1995.

[170] C. Rump and S. Stidham. Stability and chaos in input pricing at a service facility with adaptive customer response to congestion. *Management Science*, 44:246–261, 1998.

[171] C. Rump and S. Stidham. Relaxed asynchronous flow-control algorithms for multiclass service networks. *IIE Transactions*, 32:873–880, 2000.

[172] J. Sandefur. *Discrete Dynamical Systems*. Clarendon Press, Oxford, England, 1990.

[173] A. Schulz and N. Stier-Moses. On the performance of user equilibria in traffic networks. In *Proc ACM-SIAM Symp on Discrete Algorithms*, volume 14, pages 86–87, Baltimore, MD, 2003.

[174] R. Serfozo. Optimal control of random walks, birth and death processes, and queues. *Adv. Applied Probability*, 13:61–83, 1981.

[175] G. Shanthikumar and S. Xu. Asymptotically optimal routing and service rate allocation in a multiserver queueing system. *Operations Research*, 45:464–469, 1997.

[176] W. Sharpe. *The Economics of Compufers*. Columbia Univ. Press, New York, 1969.

[177] G. Shaw. *Rational Expectations*. St. Martin's Press, New York, 1984.

[178] Y. Sheffi. *Urban Transportation Networks*. Prentice-Hall, Englewood, NJ, 1985.

[179] S. Shenker. Fundamental design issues for the future Internet. *IEEE J. Selected Areas in Communications*, 13:1176–1188, 1995.

[180] M. Smith. The existence, uniqueness and stability of traffic equilibria. *Transportation Research*, 13B:295–304, 1979.

[181] M. Sobel. Optimal operation of queues. In A.B. Clarke, editor, *Mathematical Methods in Queueing Theory*, volume 98, pages 145–162, Berlin, 1974. Springer-Verlag. Lecture Notes in Economics and Mathematical Systems.

[182] S. Stidham. Stochastic design models for location and allocation of public service facilities: Part i. Technical Report, Department of Environmental Systems Engineering, College of Engineering, Cornell University, 1971.

[183] S. Stidham. Socially and individually optimal control of arrivals to a GI/M/1 queue. *Management Science*, 24:1598–1610, 1978.

[184] S. Stidham. Optimal control of admission, routing, and service in queues and networks of queues: a tutorial review. In *Proc. ARO Workshop: Analytic and Computational Issues in Logistics R and D*, pages 330–377, 1984. George Washington University.

[185] S. Stidham. Optimal control of admission to a queueing system. *IEEE Trans. Automatic Control*, 30:705–713, 1985.

[186] S. Stidham. Scheduling, routing, and flow control in stochastic networks. In W. Fleming and P.L. Lions, editors, *Stochastic Differential Systems, Stochastic Control Theory and Applications*, volume IMA-10, pages 529–561, New York, 1988. Springer-Verlag.

[187] S. Stidham. Pricing and capacity decisions for a service facility: Stability and multiple local optima. *Management Science*, 38:1121–1139, 1992.

[188] S. Stidham. Decentralized rate-based flow control with bidding for priorities: equilibrium conditions and stability. In *Proc. 35th IEEE Conference on Decision and Control, Kobe, Japan*, pages 2917–2920, 1996.

[189] S. Stidham. Pricing and congestion management in a network with heterogeneous users. *IEEE Trans. Automatic Control*, 49:976–981, 2004.

[190] S. Stidham. The price of anarchy for a single-class network of queues. Technical Report, Department of Statistics and Operations Research, University of North Carolina at Chapel Hill, 2009.

[191] S. Stidham and N. Prabhu. Optimal control of queueing systems. In A.B. Clarke, editor, *Mathematical Methods in Queueing Theory*, volume 98, pages 263–294, Berlin, 1974. Springer-Verlag. Lecture Notes in Economics and Mathematical Systems.

[192] D. Topkis. Minimizing a submodular function on a lattice. *Operations Research*, 26:305–321, 1978.

[193] D. Topkis. Equilibrium points in nonzero-sum n-person submodular games. *SIAM J. Control and Optimization*, 17:773–787, 1979.

[194] R. Varga. *Matrix Iterative Analysis*. Prentice-Hall, Englewood Cliffs, NJ, 2000.

[195] A. Veltman and R. Hassin. Equilibrium in queueing systems with complementary products. *Queueing Systems: Theory and Applications*, 50:325–342, 2005.

[196] J. Wardrop. Some theoretical aspects of road traffic research. *Proc. Inst. Civil Engin., Part II*, 1:325–378, 1952.

[197] D. Whitley. Discrete dynamical systems in dimensions one and two. *Bulletin of the London Mathematical Society*, 15:177–217, 1983.

[198] W. Whitt. Large fluctuations in a deterministic multiclass network of queues. *Management Science*, 39:1020–1028, 1993.

[199] D. D. Yao. *S*-modular games, with queueing applications. *Queueing Systems: Theory and applications*, 21:449–475, 1995.

[200] Z. Zhang and C. Douligeris. Convergence of synchronous and asynchronous algorithms in multiclass networks. In *Proceedings of the IEEE INFOCOM '91*, pages 939–943, Bal Harbor, FL, 1991.

Index